W9-CNC-642

Topics in Mathematics

The INTEXT EDUCATIONAL PUBLISHERS Series in MATHEMATICS under the consulting editorship of

Topics in MATHEMATICS

MODELS/LOGIC/NUMBER THEORY

GRAPH THEORY/PROBABILITY/STATISTICS

COMPUTER PROGRAMMING

Edward P. Merkes

University of Cincinnati

INTEXT EDUCATIONAL PUBLISHERS New York

Published simultaneously in Canada by Fitzhenry & Whiteside, Ltd., Toronto.

Library of Congress Cataloging in Publication Data

Merkes, Edward P
Topics in mathematics.

(The Intext Educational Publishers series in mathematics)
Includes bibliographical references and index.
1. Mathematics–1961–   I. Title.
QA39.2.M46 1975   510   75-5997
ISBN 0-7002-2474-2

INTEXT EDUCATIONAL PUBLISHERS

666 Fifth Avenue
New York, New York 10019

Typography by Christopher Simon, Simon-Erikson Associates

MANUFACTURED IN THE UNITED STATES OF AMERICA

Contents

Foreword

An important obligation of the mathematical community is to create greater public awareness of the utility of mathematics and the variety of its roles in society. One sound long-term method of dealing with this obligation is to develop course content of valid general interest for those college and university students who are not among the traditional users of mathematics. To be effective, we cannot just tell these students what mathematics is all about; we must seek to involve them in mathematical thought and stimulate them to tackle interesting problems at their level. They deserve an introduction to a wide range of mathematics from pure to applied in new and traditional areas.

In this book, Professor Merkes realistically addresses himself to such problems of content and exposition. With topics ranging from logic and number theory to statistics and a computer programming language, he involves the student in mathematical problems and activities that show the power of mathematics in many of the ways in which it touches our lives.

There never will be a definitive book for such students, but we, the editors, feel that Professor Merkes' book is a useful and provocative contribution that will benefit and stimulate those who study it.

In preliminary form, the text has been used by Professor Merkes for several semesters at his university. It is serving his students well and we hope it will serve yours equally well.

R. D. ANDERSON
Louisiana State University

Preface

Some years ago I began developing notes for a survey course offered to students in Liberal Arts and Education. We wished to provide an introduction to the true nature of mathematics by dealing in an appropriately mathematical fashion with a number of topics; that these topics should have proved useful when applied to real-world problems was a second but equally prominant goal.

My notes, and the experience of teaching what I hoped to be a judicious selection of topics, led naturally to the present book. It will be seen that I have tried to rely as little as possible upon prior knowledge. To be sure, the student is expected to have been exposed to the rudiments of arithmetic, to linear equations and their graphs, and to an introductory course in geometry. But I do not equate "exposure" to thorough familiarity with all the concepts and methods discussed in high school courses. The student need only have the capacity to become reacquainted with the concepts needed to use this book. It would not, however, be reasonable to ask those who never took the standard high school curriculum in mathematics to master these concepts and to follow their development in this textbook.

The early chapters in this text are independent of each other. However, Chapters 5 through 8 form a unit of interrelated topics in probability and statistics. Chapters 9 through 12 introduce programming along with certain applications to business problems and should be followed in sequence.

For a three quarter survey course in mathematics this text can be divided into three *independent* units; namely, Chapters 1-4; 5-7; 9-12. A student can enter the sequence in any of the three quarters without knowledge of the material in the other quarters. For a two semester course, the arrangement of Chapters 2, 4, 5, 6, 7, 8 into one course and Chapters 1, 3, 9, 10, 11, 12 into a second semester provides the student with the advantage of being able to take the two courses in any order.

In order to learn mathematics, one must DO mathematics. For this reason, much of the learning is accomplished through the exercises. There is little emphasis on drill type problems in the text; instead the exercises stress a creative use of the ideas and methods in the sections. Almost every section has some exercise that should provide a challenge for each individual in the class.

The author would like to thank the many students and faculty members at the University of Cincinnati for their helpful comments and discussions during the years spent writing this text. In particular, I owe much to James Breyer, Ronald Heibert, Arthur Kean, David Meyers, Charles Pinzka, and Tim Timmons. I acknowledge the excellent editing and proofreading of Dennis Berkey and of Sylvia Newburger, who also typed the manuscript.

List of Symbols

Defined Symbols

Defined Symbols (*continued*)

Symbol	Description	Page
$\nmid$	does not divide	59
$\sqrt{\ }$	square root	66
$n!$	n factorial	68
(n, m)	greatest common divisor of n, m	74
$l(n, m)$	number of steps in the Euclidean algorithm	75
P_r^n	permutation of n objects taken r at a time	131
C_r^n	combination of n objects taken r at a time	131
$P(A)$	probability of an event A	137
$n(A)$	number of points in A	137
$P(A \mid B)$	probability of A given the occurrence of B	149
$P(t \leqslant a)$	probability of t not exceeding a	184
$P(a \leqslant t \leqslant b)$	probability of t being between a and b	184
	FORTRAN SYMBOLS	
*	multiplication	206
+	addition	206
-	subtraction	206
/	division	206
=	equals	206
**	exponentation	206
.GT.	greater than	207
.GE.	greater than or equal to	207
.LT.	less than	207
.LE.	less than or equal to	207
.NE.	not equal	207
.EQ.	equality	207

Topics in Mathematics

Chapter 1 Mathematics—Art and Science

1.1. Why Should You Study Mathematics?

Some aspects of mathematics are important for anyone who exists in a complex society. Grade school arithmetic and simple geometry suffice for most people and they have never seen, at least since their high school education, an equation or an exponent. This group includes millionaires as well as paupers, business executives as well as secretaries, and tradesmen as well as laborers. So, if you have already mastered grade school mathematics, why study more?

Some college students certainly will never use higher forms of mathematics. Others, even though they never suspected it, discover that higher mathematics has penetrated a number of disciplines. This is particularly true in the social sciences, psychology, and business areas. Statistics, computing techniques, graph theory, matrix theory, and many seemingly esoteric mathematical topics are now found even in the early stages of in-depth studies of numerous diverse subjects.

When one is faced with the task of mastering certain abstract mathematical topics in conjunction with material in the discipline, often only confusion results. It may be wiser, indeed, to acquaint oneself with some of the modern tools of mathematics prior to the particular application. For instance, can you imagine the surprise of an English major who discovers some knowledge of group theory, a mathematical topic, is a prerequisite to understanding a discussion in a linguistics course? Unless this student has some understanding of abstract mathematics, he will be frustrated and significantly delayed in his attempt to master the material. It is not necessary that this student have had a course in group theory itself to avoid the problem. If he is familiar with the abstraction and the axiomatic approach of mathematics, the student should be able to proceed in the study of linguistics without being sidetracked for long in learning the necessary background mathematics.

Even the student who will never use higher mathematics might benefit from additional mathematics courses. Mathematics, after all, is an exercise in systematic thinking, analysis, and abstraction. Perhaps the experience can be fruitful in the inquiry and critical analysis that forms the essence of any education.

1.2 Mathematics Is Hard?

Basically, mathematics is the simplest systematic discipline ever created by man. In the real world complexity abounds and many things will never be understood by man. Mathematicians, through observation and abstraction, have invented entities and notions for their study that are far simpler than the real objects from which they were derived. A point does not have real existence, but it is much easier to comprehend and understand than an atom or an electron.

In spite of its essential simplicity, mathematics is difficult. Some of the difficulties are superficial; others are not. A student who waits until the night before an examination to study his mathematics may find his task hopeless; the student who waits to study mathematics until he must apply certain mathematical concepts to his own field of endeavor may encounter the same complication. Mathematics builds its structures upon previous ones. The ability to understand a topic today often requires mastery of previous topics since each new discussion, after the initial one, often depends upon the previous work. A prime source of difficulty many students have in learning mathematics is their failure to keep up with the discussion on a day-to-day basis. Any college student can learn much on the subject provided he can enjoy success in any other discipline. Indeed, an ability to reason is the only prerequisite and this ability is essential in all areas of higher education.

The massiveness of mathematics is a major reason that mathematics is difficult. After all, significant achievements in the subject date back to 3000 B.C. and there have been many important contributions since its early beginning and remain to be more in the future. There is no way to surmount this problem of massiveness, even should you become a professional mathematician. The only hope is that whenever you encounter a need for mathematics either you will have the required mathematics in your background or else you will have enough mastery of the techniques of mathematics to acquaint yourself with the particular aspect of the discipline needed. If you are lucky, you may even have a chance to create new mathematics!

Nobody can predict precisely which part of mathematics, if any, you will find important. Hopefully, however, the mathematics in this book will at least prove to you that there is no need to fear the sudden appearances of mathematics in your studies, even though the topic was not one included in these notes. Mathematical approaches in a discipline are attempts at simplification of the subject and are not malevolent plots intended to render the discipline incomprehensible to all except the elite. If ever you appreciate this point, you may even seek mathematical discussions in order to better understand your subject.

1.3 Man and Mathematics

Man is an animal with a greater facility to reason, to abstract, and perhaps to remember than any other member of the animal kingdom on earth. It is precisely these capabilities of mankind, in conjunction with curiosity, that generated the science of mathematics. The origin of mathematics likely dates back to the time when man abstracted from what he observed to be similar objects and introduced the concept of counting numbers. Clearly, geometric ideas such as "similar objects" may have preceded numbers but it is fruitless to debate who's on first. Both are abstractions from reality.

Mathematical entities such as numbers, points, or lines, enjoy no real existence themselves; they are creations of the mind of man. After, or perhaps during, the creation of mathematical objects, man observed relationships among these abstract objects. For instance, we speak of one number *being larger* than another or we state that a certain point *lies on* a line. Next, man applied a form of logic or reasoning to conclude certain truths about his objects and their relations, and this he called mathematics. Some of these conclusions aided him in his dealings in his society whereas others served only to satisfy his curiosity. In time, man discovered that, should he accept certain statements which he believed to be self-evident, many others could be deduced from those that were assumed. This represented the birth of the axiomatic method which has become the foundation of today's mathematics.

Around 300 B.C. Euclid collected, in the 13 books of his *Elements*, much of the enormous logical structure that his predecessors had discovered concerning the known truths (his theorems) of geometry from what man believed for centuries to be basic self-evident principles (his axioms). The works of Euclid had been studied and extended for more than a thousand years before it was detected that the original axioms were, in a sense, incomplete. Certain unstated assumptions were used in the geometry which, by the standards of modern mathematics, rendered the axiom system logically defective. It was not until the turn of this century that a complete set of axioms for Euclidean geometry was introduced by the great German mathematician, David Hilbert (1962–1943).

Until recent times the only branch of mathematics that was considered suitable for the axiomatic method was geometry. In the past two centuries, however, satisfactory systems of axioms for other branches of the subject were discovered. Even the counting numbers of early arithmetic were axiomatized by G. Peano (1858–1932), an Italian mathematican.

Hope that all mathematical truth could be derived from axioms was ended by Kurt Gödel (1906–) when he proved in 1931 that no finite set of axioms strong enough to imply the basic facts of arithmetic can answer all questions; certain formulas cannot be proved or disproved by deduction from the axioms. In other words, there is always some "theorem" that cannot be verified from the axioms with logic; something is always, so to speak, out of reach. These "theorems" that cannot be proved are known to exist but are hard to identify. In only a few instances have they related to questions that mathematicians previously attempted to resolve. Even with some inherent limitation, however, the axiomatic method remains today the basis of all branches of mathematics.

From the Age of Reason (late seventeenth century) to the present, an axiomatic approach has been attempted, sometimes with little accomplished, in many areas of human endeavor. Interest in the method was stimulated by the phenomenal success of Isaac Newton (1642–1727) in postulating certain laws of physics. Whereas the methods of Euclid were known to scholars for centuries, the axiomatic method previously appeared, even for mathematicians of the period, to be confined to the area of geometry, an area which is somewhat remote from reality as perceived by mankind. Newton, however, displayed that the method can, indeed, aid us in better understanding our universe and the laws of nature. A number of intellectual leaders of the time followed the success of Newton and attempted to apply the axiomatic method to such diverse areas as psychology, economics, and religion. These subjects, however, are not as suitable for the approach as physics and it is likely that there is no simple set of axioms that covers an entire social science or religion. The axiomatic method can, nevertheless, enjoy success when applied to fragments of a social science.

In Sections 1.8 and 1.9 we illustrate this use of deductive reasoning for some special problems in economics.

Exercise

1. Thomas Jefferson idolized men like Newton and was influenced by his axiomatic approach. Read the American Declaration of Independence, authored by Jefferson, and list the axioms that you find in this document.

1.4. Mathematics as an Art Form

Bertrand Russell once stated "Mathematics, rightly viewed, possesses not only truth–but supreme beauty–a beauty cold and austere, like that of sculpture, without appeal to any part of our weaker nature, without the gorgeous trappings of painting or music, yet sublimely pure, and capable of a stern perfection such as only the greatest art can show." Mathematicians, philosophers, and scientists generally agree with Russell's flowery passage but, unlike most art forms, the beauty of mathematics is not readily observed by other people.

The problem is that appreciation is generally not immediate or even cultivated after a short period. One most often must study the topic for years before even a glimpse of the beauty is obtained. The fact is, however, that it is not mathematics as an art form that society knows or accepts; rather, it is the practical side of the subject that motivates most individuals to study it. Mathematics is beautiful, it is intellectually challenging, it is enjoyable, but, most important for society, it is useful.

There is a story that claims a gentleman asked Euclid if he could study mathematics under the master. After reading the first theorem in Euclid's *Elements*, the man asked what value additional study of the *Elements* would provide for him. Euclid reportedly turned to his slave and said, "Pay this man so that he can gain from his study." The art form of mathematics apparently was a major factor in the early development of mathematics, as this story emphasizes.

1.5. Axioms and Theorems

The starting point of any logical treatment of a subject must be a set of undefined elements and relations and a set of unproved propositions, called axioms, involving them. The axioms are simply assumed to be true and formal methods of logic are used to derive other propositions, called theorems.

At first the statement about undefined elements and relations seems to imply that the axioms are empty of content. The problem is overcome once you realize that we can have intuitive meaning for these concepts, although we cannot logically define them. Indeed, a vicious circle of reasoning is obtained if you attempt to define all terms because each term requires other terms to define it. For example, try to define a word like "affection" and then attempt to define each word used in the definition. Repetition of this process will eventually bring you to a point in which some word is defined in terms of words that have already appeared in the sequence of steps. In other words, you eventually must use words that are simply undefined although you have an intuitive "feeling" as to their meaning. Affection might be defined as a mild form of love, and love might be defined as a strong affection. The circular nature of the definition is then completed; affection is a form of love and love is a form of affection. If you accept the word "love" as undefined, then affection can be defined in terms of love. But, you cannot logically *define* each in terms of the other.

The same vicious circle is obtained if one attempts to prove all the axioms of a system. Indeed, a proof of an axiom requires other propositions and, hence, the original axiom is then a theorem derived from a new set of axioms. Whereas we may desire to extend backward the particular system by replacing its axioms by those that we consider more basic, there is no way to verify, logically, all axioms utilized without completing a circle in reasoning. Mathematics is, thus, that subject in which one never "knows" what he is talking about, although what he says is true.

Axioms usually are, but need not be, stated with abstract undefined terms. Consider, by way of illustration, the following axioms concerning the activities of a certain organization:

Axiom 1. No proposal is favored by all members of the organization.

Axiom 2. Every proposal is favored by at least two distinct members of the organization.

Axiom 3. There is at least one proposal.

The undefined terms are the organization and the proposals. The undefined relation is favoring a proposal. The other words in the axioms are considered as defined, of course, in terms of other undefined terms.

Let us deduce a theorem from these axioms. Axiom 3 is an existence statement so there is a proposal and by Axiom 2 this proposal is favored by two distinct members of the organization. Thus, "there are at least two members in the organization" is a theorem; this statement being a logical consequence of the axioms.

Let A and B represent labels (in lieu of names) for the two members that exist by the theorem. Now A and B both favor the proposal that exists by Axiom 3. Hence, by Axiom 1 there must be at least one additional member of the organization; otherwise, all would favor a proposal. Call this additional member C. We have, therefore, a new theorem that states there are at least three members of the organization. This new result includes the former one and it is, therefore, redundant to state our first result.

We can build a model of an organization that obeys these axioms. Let A, B, and C be the three members who exist by the theorem and represent these individuals as distinct points in the plane. The proposals are taken to be lines in the plane and for a member "to favor a proposal" is described by having the line (proposal) pass through the point representing the member. In this manner a picture of the situation is obtained which can help guide our intuition for additional inquiry into the axiom system.

For the case at hand we now have three points in the plane, one for each member of the organization that we have identified. There is a proposal that is favored by A and B and, hence, this is represented by a line through the points A and B. This line should not pass through the point C since C is a member who does not favor the proposal. The model now satisfies all three axioms in the system.

Although this is not the only model of our axiom system, Figure 1.1 is a minimal model in the sense that it contains exactly one proposal and exactly three members of the organization. We have already proved that each model must have *at least* one proposal and *at least* three members. Hence, there is no model with fewer proposals or fewer members than the one represented in Figure 1.1.

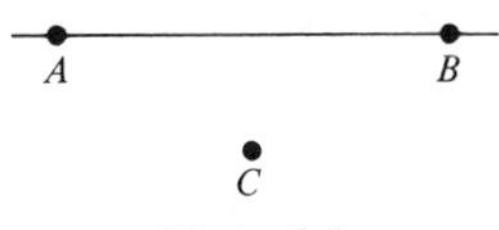

Figure 1.1

The geometric model of the organization assists us in obtaining conclusions about the axiom system itself. Indeed, it displays a case in which the axioms are satisfied and, thereby, shows that the axioms cannot be self-contradictory. When an axiom system is long and complex, it is possible that contradictory results might be derived from the system. For example, let us add to the previous axioms on our organization a fourth axiom that states, "Each organization has at most two members." There can no longer be a model that satisfies all four axioms since the theorem that there are at least three members contradicts the fourth axiom. This theorem, moreover, is a logical conclusion of the first three axioms.

Whenever an axiom system is free of built-in contradictions it is called a *consistent* system. A model that satisfies all the axioms of a system, such as the model in Figure 1.1 of the three axioms of an organization, proves that the system is consistent. You cannot build a model of a system of axioms that are inconsistent.

Another fact about our axiom system on an organization can be deduced by considering *only* the points A and B along with the line joining these points in Figure 1.1. This is a model that satisfies Axiom 2 and Axiom 3 but not Axiom 1. This model shows that there is no way in which the first axiom can be logically deduced from the others. Indeed, if it *was* a consequence of the other axioms, every model that obeyed the commands of Axiom 2 and Axiom 3 would automatically satisfy Axiom 1. We describe this fact about the axiom system by stating that Axiom 1 is *independent* of the other axioms. It is not a disaster to have dependent axioms, although it somewhat defeats the purpose of having axioms. Should some "axioms" be derived from the others, it is better to rename these "axioms" as "theorems" in the system.

The establishment of consequences of a given axiom system can be an adventure into the world of pure reason for an individual. The excitement, however, is usually not noticed when the axiom system is spelled out in advance and the step-by-step reasoning is displayed for each theorem, free of the motivating models. Unfortunately, this is the case in most textbooks on mathematics and is often the case in a mathematics lecture. The enjoyment of the discovery of a set of axioms for a problem that interests you and the attempts to verify a conjecture, which originates from our intuition or a model, is sometimes lost when too much of the effort in establishing theorems is written down. The real enjoyment is your personal involvement with the system, either by creation of the axioms or through your own attempts at proving your conjectures. Even exercises, which provide some guide to the reasoning process, fail to provide the full pleasure of discovery.

Exercises

1. Adjunct the axiom, "Each member of the organization favors at least one proposal" to the three axioms in this section. Build a model and refine your model, if necessary, to determine the smallest number of members of the organization and the smallest number of proposals.
2. Construct a model that satisfies Axiom 1 and Axiom 3 but does not satisfy Axiom 2. This model verifies that Axiom 2 is independent of the other axioms.
3. Show that ordinary Euclidean geometry is a model of the axioms of this section if a member (point A) favoring a proposal (line L) is represented by the point A being on the line L. This model and reality suggest an additional axiom, namely, "Each organization has a finite number of members." Does this imply there are then only a finite number of proposals?
4. Euclid attempted to define all terms in his *Elements*. Some examples are the following:
   a. A *point* is that which has no part.
   b. A *line* is breadthless length.
   c. A *straight line* is a line which lies evenly with the points on itself.

   List the concepts that are undefined if these definitions are accepted.
5. Consider the following axioms:
   A1. Given any two members of a team, there is exactly one game that both enjoy playing.
   A2. Given two games, there is a member of the team who enjoys playing both of them.
   A3. There exists at least one game.

A4. Each game has exactly three members of the team who enjoy playing it.
A5. All members of the team do not enjoy playing the same game.
Show that there are at least seven members of the team. What is the smallest number of games the members of the team enjoy playing?

6. Try to create a simple small set of axioms for some problem of personal interest to you.

1.6. Mathematical Models

Real-life situations are often too complex for a complete mathematical description. Problems of economic inflation, politics, human behavior, poverty, or warfare stimulate volumes of discussion that hopefully aid in the resolution of difficulties. General principles have been, and are, discovered from experiences and judgments that are reformulated in mathematical terms. This step is the initial one in the creation of a *mathematical model* to approximate the real situation.

Once a problem has been at least approximately described quantitatively in mathematical terms, the next step in its resolution is to use available mathematical methods, or invent new ones, to solve the mathematical problem. The third step is to explore the results in the context of the original real-world problem. Often the result suggests that the model is inadequate, and then a refinement of the model is necessary. This process is repeated until some realistic solutions of the original problem are obtained.

If one is to enjoy success with model building of real-world problems, experience, originality, and a thorough understanding of the original nonmathematical problems are prerequisites. Some experience can be achieved with problems of very limited scope which are in this text. Meaningful models for physical, economic, political, and psychological problems, however, require a good knowledge of the particular discipline as well as certain aspects of mathematics.

Some models are *deterministic* in that certain axioms or laws are assumed and these, along with some special conditions pertaining to the problem at hand, exactly determine the response. Other models are *probabilistic* and these apply to instances in which a degree of uncertainty prevails in the real-world problem as, for example, in the case of politics. At times the model could lead to mathematical problems that have yet to be resolved, in which case computer simulation often proves useful.

It is for these reasons that this book is divided into essentially three parts. First, some deterministic models are discussed with the aid of logic and graph theory. This beginning provides us also with an opportunity to introduce what may appear to be irrelevant mathematics, namely number theory, which, perhaps, will have the side effect of granting us a better understanding of mathematical approaches in general. (Number theory also has meaningful applications although they are often deeper than the applications cited in this text.) The computer and the statistical considerations each comprise an additional third of this text.

1.7 The Parking Lot Problem

In order to illustrate model building, consider the following "real-life" problem for a certain parking lot owner. The owner has a square plot of land that is 64 ft on each side in the downtown area of a city. Each car parked on the lot requires a rectangular space of 8 by 16 ft and, hence, the owner can park 32 cars per day in his lot. The city council, which controls the city's public transportation (as is the case in any progressive city), decides to provide a shelter for bus passengers on the corner of the parking lot. The shelter will require an 8 by 8 ft plot on a corner of the parking lot. The owner notices that this reduces the number of cars he can park each day to 31 but it leaves him an 8×8 ft piece of land that is not used in parking cars. If he decides to build a collection booth on this plot, where should he place his booth? Because of the morning traffic patterns, it appears to be more convenient to place his booth on the corner diagonally opposite the bus shelter. Is this a wise choice?

The first step toward the solution of this problem is the creation of a mathematical model. Note that if we call 8 ft a unit, then the parking lot is an 8-unit square, the space for each car is 2 units by 1 unit, and the bus shelter is 1 unit by 1 unit. (Such a convenient unit is unlikely to be available in a less artificial problem.) Alas, this unit suggests using an 8×8 standard checkerboard to serve as a model of the lot. The cars are two by one, just the size of a domino.

The problem of parking cars is now transcribed into one of filling the checkerboard with nonoverlapping dominoes. The city wants one square in the corner of the checkerboard and, in our model, this square is either a black or a red one. Furthermore, each domino covers two squares, one of these is red and the other is black, if the cars are parked in one of two perpendicular directions (East-West or North-South, say) according to the location of the squares on the model. Now should the owner build his booth on the corner diagonally opposite that for the bus passengers, his booth would be on the same color square of the model as the bus shelter. Indeed, opposite corners of the checkerboard are of the same color; either both are black or both are red. Since there is an equal number of red and of black squares on the checkerboard, there are two less squares of one color after the diagonally opposite corners have been removed. In the illustration (Fig. 1.2) there are 32 red squares and only 30 black squares after the city and the owner build shelters as indicated. Now each domino (car) covers two squares and these, being adjacent, are of opposite colors. Hence, there must be an equal number of squares of each color when the remaining squares are to be covered

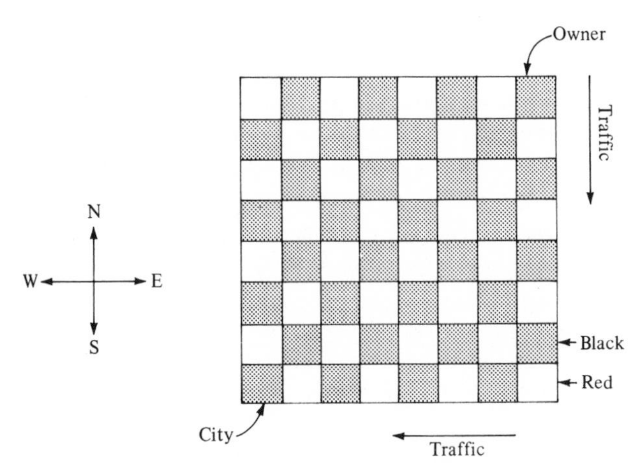

Figure 1.2

with dominoes. This is not the case when diagonally opposite corners have been removed.

Thus, the owner should be warned that building his booth on the corner opposite the bus shelter will leave him two 8 by 8 squares that are not suitable for parking a car and will reduce the number of cars he can park to 30. On the other hand, should he build his booth on what corresponds to the square of the checkerboard that is a different color from the square reserved for the bus shelter, he will then have an equal number of squares of each color and he is still able to park 31 cars on the lot each day. (Try it!)

Suppose the traffic flow indicates the owner should build his booth on a square as close as possible to the corner square. (Perhaps he wishes to encourage the potential customers to park in his lot before they become involved in a usual moving traffic jam at his South–East corner.) He must, however, pick a square on the board of opposite color to the corner square or suffer the loss of a parking space. There are two such choices, one to the left and the other down from the upper corner, so additional information is needed to make the solution unique. If the streets are one-way streets as indicated in Figure 1.2 and if morning traffic passes on the street East of the lot, the owner can attract the customers best by building the booth on the first square South of the corner square.

Some refinement of the model might be desired if this problem were actually encountered. For example, cars are not all the same size and a statistical knowledge of the expected number of small cars each day should help the owner plan to maximize the number of cars he can park. Another problem is to arrange the cars in a pattern that most efficiently enables him to remove a car from the lot. It is apparent that some cars have to be moved to permit others to exit from the lot.

Exercises

1. Consider the problem of this section if we assume the parking lot is a 2-unit square. Do the same for a 3-unit square and then a 4-unit square. In particular, consider how many 2×1 cars can be parked if opposite 1×1 corners are removed in each case.
2. Suppose each car covers a 3×1 square unit rectangle. Can an 8×8 parking lot be completely filled with cars? What is the result if the lot is 9×9?
3. Suppose the parking lot is a 9-unit square. How does the problem of this section change for this case? In particular, how many cars can be parked on the initial lot and on the lot with one corner removed? Assume two opposite corners are removed. How many cars can now be parked on the lot?
4. Does parking the cars (2×1) in one of two perpendicular directions parallel to the sides of the lot on an 8×8 parking lot maximize the number of cars that can be parked? Why?
5. Suppose the parking lot is an 8×9 rectangular plot. How does this lot change the problem in this section?
6. If three of the four corner squares of a 9×9 checkerboard are removed, can the remainder of the board be covered by 3×1 rectangles such that each rectangle covers three squares of the board and each square is covered exactly once?
7. What is the maximum number of cars (2×1) that can be parked on the lot in the text (8×8) if (a) each car is parked in a North–South or East–West direction, (b) each car can be removed from the lot by driving it along a straight line path, and (c) each car can be removed without moving any other car?
8. Change (c) in problem 7 to the statement (c′) each car can be removed by moving no more than one other car. What is the number of cars that can be parked in this case if we also assume that a pair of diagonally opposite corners of the lot are occupied by shelters?
9. (a) Can one slide a checker from one corner of a checkerboard to the diagonally opposite corner in such a way that each square of the board has been entered once and only once? (b) Can this be accomplished if diagonal moves are not admissible?
10. Can you fill an 8×8 checkerboard with T-shaped objects (airplanes) where the horizontal bar in the T is three squares (3×1) and its center square rests on a 1×1 square?

1.8. The Three Grocery Store Problem

As a further illustration of deterministic mathematical models, we consider a simplified version for the problem of locating a warehouse to supply a number of stores. Such problems are encountered when a chainstore, such as Sears or Kroger, decides to build a central warehouse for its stores in an area. The locations of the stores themselves are influenced by such factors as population densities and proximity to other similar stores. Once the location of the stores is established, the cost of shipping from a warehouse to each store is not negligible and must be considered in respect to the location of the warehouse. The company wants to minimize the cost of transporting supplies from the warehouse to the stores so the selection of the site for the warehouse would take this factor into consideration during the planning stages.

Before attempting a mathematical approach for this problem, it is necessary that the meaning of the term "proportional" be recalled. Two measures are proportional if there is a fixed constant such that, when a certain one of these measures is multiplied by the constant, the other measure is obtained. For example, if x is the amount of money you have in dollars and y is the amount measured in pennies, then $y = 100x$. Thus y and x are proportional since y can be obtained from x by multiplication by the constant 100. The constant is independent of any particular choice of x, the amount measured in dollars. If we state that a cost is proportional to a distance, we mean there is a constant such that when it is multiplied by the distance, given in miles, say, a cost in dollars is obtained. This constant does not change when the mileage changes.

We are now prepared to state the simplified warehouse location problem. Suppose a grocery chain has three stores, A, B, and C, on the main East-West street of a certain town and assume store B is between A and C. These stores are to be supplied from a single warehouse and the shipping cost from a warehouse to any store is proportional to the product of quantity shipped by the distance to the warehouse. (The constant of proportionality is the same for all three stores.) If the stores A, B, and C, receive supplies in lots of 100, 75, and 20 units, respectively, where should the warehouse be located so that the sum of the shipping costs is minimized?

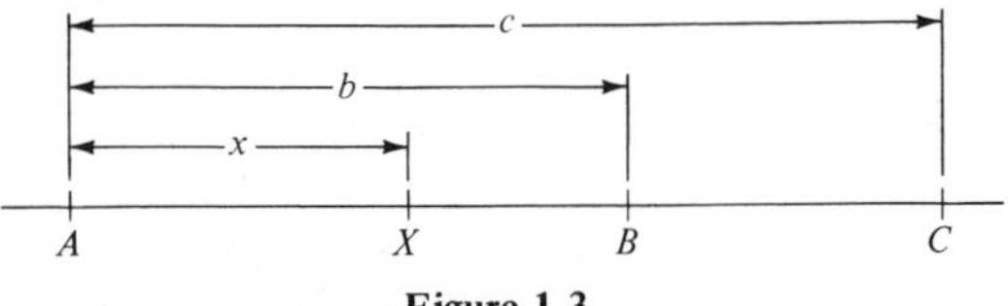

Figure 1.3

Once again we appeal to a geometric model. The main street is represented by a horizontal line (since it is an East-West street it is assumed to be straight) and the stores are denoted by points on this line. Let b denote the distance, which was not given in the problem, from store A to store B and let c be the distance from A to C. The problem is the location of the warehouse X of distance x from the store A.

We assume the warehouse is to be located on the main street. (In the exercises, it is proved that this must be the case if we are to minimize the cost.) Could the warehouse be on the main street to the right of C in the model? If this were the case, we could move it slightly to the left and thereby decrease the distance to each of the stores and, hence, decrease the costs of shipping. It follows that the warehouse cannot be located to the right of C if costs are to be minimized. Similarly, it cannot be located to the left of store A. Hence, it should be between A and B or between B and C. Suppose X is between A and B. Then the costs of shipping to stores A, B, and C, are respectively proportional to $100x, 75(b - x)$, and $20(c - x)$ since x, $b - x$, and $c - x$ are the distances from X to each of the stores. (See Figure 1.3.) The total shipping costs are, therefore, proportional to

$$100x + 75(b - x) + 20(c - x) = 75b + 20c + 5x$$

by the assumptions in the statement of the problem and by algebra. What choice of x makes this as small as possible? Clearly, $x = 0$ is the choice since any positive value of x will make the cost greater than $75b + 20c$. (A negative choice of x is meaningless in this problem.) The conclusion that $x = 0$, translated back into the original problem means that the warehouse optimally should be part of the store A itself. If this is impossible, the costs formula indicates it should be as close as possible to store A and between A and B, perhaps in the first available lot to the right of store A.

In the analysis, so far, we assumed X was between A and B. Suppose X is between B and C (Figure 1.4). Then the shipping costs are now proportional to

$$100x + 75(x - b) + 20(c - x) = 20c - 75b + 155x$$

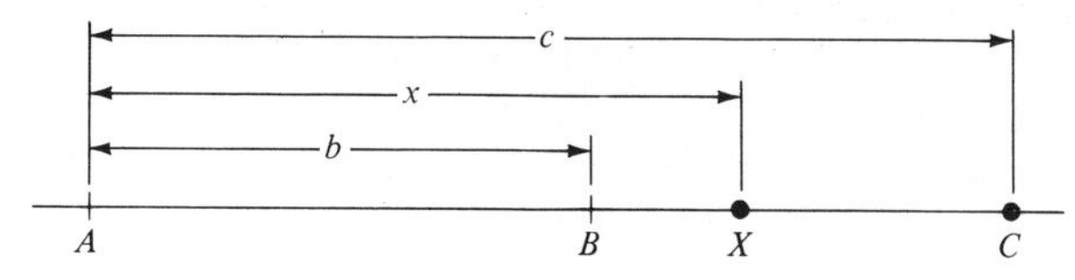

Figure 1.4

Again $x = 0$ would make this as small as possible, but the assumption that X is between B and C and the definition of x (the distance to store A) does not permit us to use the value $x = 0$. Indeed, x is $> b$ when the warehouse is to the right or at the store B. The cost of transportation is, however, smallest when x is as small as is admissible. Hence, we must take $x = b$, that is, locate the warehouse at the store B if it is to be between B and C and the cost is minimized. This places the shipping cost at a figure proportional to

$$20c - 75b + 155b = 20c + 80b$$

But this, since $b > 0$, is larger than the $20c + 75b$ cost figure obtained with the assumption that X is between A and B. The conclusion is that it is best to locate the warehouse at store A, but should you be required to locate between B and C, it would minimize the cost to locate the warehouse at store B. Furthermore, you can show from this analysis that the cost increases the farther to the right (East) of A that the warehouse is located.

As proposed, this problem is not practical. It can, however, be generalized to include more than three stores and to the case in which the stores are not all on the same street. In a city in which all streets run in one of two perpendicular directions, unlike Boston or Cincinnati, a solution can be obtained by methods that are not any deeper than the ones in the previous argument. The results can be adjusted to cases in which the city has only a few diagonal streets like Chicago.

For certain other situations a mathematical solution has been obtained, although the analysis is much more difficult than that of our three grocery store problem. In such cases, a computer simulation can be programmed in which the costs are calculated and compared for all essentially different paths from warehouse to stores. Even here a prior mathematical analysis is often needed. Computer time, itself, is costly, and the capacity of the computer is not unlimited. Rather than let the computer try all paths, therefore, a computer program should be written that considers as few paths as necessary to solve the problem.

Exercises

1. Assume the lots to be shipped to stores A, B, and C in the problem of this section are respectively 100, 90, and 40. What then is the optimum location for the warehouse on the main street?

2. Assume the lots to be shipped are each 100 units for the three stores. What is the solution to the warehouse problem in this case?

3. If the lots to be shipped to A, B, and C, are respectively 100, 50, and 50 units, then prove that the warehouse can be located anywhere between A and B on the main street and the shipping costs will be minimized.

4. Prove that the warehouse must be located on the main street in the grocery store problem of this section.

5. Suppose there are four stores and they are located such that one store is on each corner of a square. In addition, assume that the city streets form a grid such that this square is an enlarged version of a checkerboard (Figure 1.2). Where would you locate the warehouse for these stores if each store is to receive lots of 100 units and if cost is to be minimized?

6. Assume there are three stores A, B, and C, located in a city which has only East–West and North–South two-way streets. Solve the warehouse location problem in this case under the assumption that stores A, B, and C, receive, respectively, lots of 100, 75, and 20 units. (Place each store on a street in

the city. Locate the store with a pair of coordinates and solve the problem using these coordinates.)

7. Add the two diagonals of the square in exercise 5 to the city streets. Where would you now locate the warehouse if the stores receive equally sized lots?
8. In how many different ways can a truck travel from the store in the North-West corner of a 3 block by 3 block grid to the store in the South-East corner if the truck only travels in either of two directions, South or East, along the city streets?

1.9. Simplified Nim

The playing of games is an ageless form of relaxation for man. Many games are susceptible to mathematical discussions, although as the number of possible actions and reactions of the players increases, a complete analysis may prove impractical, if not impossible. Complex games like bridge and chess, however, do have strategies that, to some degree, can be mathematically formulated.

In this section we introduce an extremely simple version of the classical game of Nim. The game is played between two players with a pile of sticks. The rules are

1. On each turn, a player must pick up at least one stick from the pile and he cannot pick up more than a certain number that is fixed in advance of the game.
2. The turns alternate between the two players.
3. The person picking up the last stick loses.

Suppose, for example, there are initially 21 sticks in the pile. On each turn the player whose turn it is must pick up at least one stick, but cannot remove more than four from the pile. If you are the second to take a turn, you can completely control the outcome and, unless you are the benevolent type, win. Why?

The strategy is simply to be sure that exactly five sticks have been removed after each pair of turns, your opponent's and yours. This is possible since you can remove four sticks if he removes one, three if he removes two, etc. After a total of eight turns your partner will be facing a one stick pile since five sticks have been removed in each pair of turns, and, therefore, 20 sticks have been removed in eight turns.

Now assume we have N sticks in the initial pile and the game permits a player to remove no more than n sticks during a turn. Since each player must remove at least one per turn, the second player can make certain that exactly $n + 1$ sticks have been removed in each pair of turns. Divide N by $(n + 1)$ and you obtain a quotient which is the number of pairs of turns until there are n or fewer sticks in the remaining pile. The remainder in this division is equal to the number of sticks after these turns. If this remainder is one, then the second player, using the strategy to insure the removal of $n + 1$ sticks in each pair of turns, wins the game. For example, with 91 sticks and a rule whereby no more than eight can be removed on a turn, we have $91 = 9 \cdot 10 + 1$ so the second player will win since the remainder is one. Moreover, there will be 10 pairs of turns or a total of 20 turns in all before the first player is facing a one stick pile. If, however, there were 93 sticks in the original pile, the strategy would have left the first player facing a pile with three (the remainder of $93 \div 9$) sticks. Not a very enviable position for the second player to be in!

Exercises

1. If you have 29 sticks, what is the maximum number(s) of sticks you should permit to be removed on each turn in order that you can always win, as the second to play, using the strategy of this section?
2. Suppose there are 42 sticks in the initial pile. Is there any number that you can permit to be removed on each turn that assures you will win using the strategy of this section when you are second to play?
3. Devise a strategy that will enable you to win when you are the first player of a game of Nim with 42 sticks and a limit of four in the removal of sticks in any one turn.
4. Suppose the remainder upon division of N by $n + 1$ is greater than one. Show you can always reduce the corresponding game of Nim to the one in the discussion if you are the first mover in the game.
5. Suppose there are 50 sticks in the initial pile and the rule is that not more than four, but at least one, is to be removed on each turn. Can you pick your playing position and devise a strategy to insure that you will win the game?
6. The following is an elementary two-person game. Find a strategy whereby you can always win if you start the game. The first person writes down a counting number from 1 to 9 and then adds it to another counting number from 1 to 9. The second person adds to this total a counting number from 1 to 9, of his choice. Thereafter, alternate turns, each time adding a counting number between 1 and 9. The first person to obtain a total that is 100 or greater loses the game.
7. A teacher ends disputes between two children in his grade school class by permitting each of the two students to select a number from 1 through 9 with the restriction that each cannot select the same number. The teacher than randomly draws a numbered card from a bag containing nine cards, each contain a number from 1 to 9. The child whose selected number is

closest to the drawn number wins the debate. The leading troublemaker in the class seemed to win an unusually number of times although he was always forced to select his number after his opponent. Why?

8. You play a game with a friend. Each writes a number from 1 through 10 on a piece of paper and then displays the results. If they are the same number, you win $10. Otherwise, he wins $1. Is there a strategy that you can use to help you if you plan to play this game 100 times?

References

1. R. Courant. Mathematics in the modern world. *Scientific American* September. 1964.
2. R. Courant and H. Robbins. *What Is Mathematics? An Elementary Approach to Ideas and Methods*. New York, N.Y.: Oxford Univ. Press. 1941.
3. B. Ghiselin, ed. *The Creative Process*. New York, N.Y.: New American Library. 1955.
4. I. Greenberg and R. A. Robertello. The three factory problem. *Mathematics Magazine*, 38 (1965): 67–72.
5. T. Saaty and F. J. Weyl, ed. *The Spirit and Uses of the Mathematical Sciences*. New York, N.Y.: McGraw-Hill. 1969.
6. Edward Spitznagel. *Selected Topics in Mathematics*. New York, N.Y.: Holt, Rinehart and Winston. 1971.

Chapter 2 Logic for Rational Animals

2.1. Inductive and Deductive Arguments

In Chapter 1 the dependence of mathematics on logic was stressed. There is, however, more to mathematics than applied logic, and there is more to logic than that which is exemplified by mathematics.

The logic of mathematics is entirely deductive. Deductive reasoning begins with premises and draws conclusions directly from these premises. An example of a deductive argument is the syllogism:

Premises: All doctors are in favor of socialized medicine. The president of the American Medical Society is a doctor.

Conclusion: The president of the American Medical Society is in favor of socialized medicine.

The logic is impeccable and this argument is valid. Nonetheless, the conclusion is not necessarily true. The problem, of course, is that the first premise is false. A valid argument is one with premises which, *if true*, force the acceptance of the conclusion. By not having each premise true, false results can be deduced by valid arguments.

In mathematics, the premises are either axioms, which are assumed to be true, or else theorems which have been deduced from the axioms by valid arguments. Unless the mathematician has made a logical error or has started with contradictory axioms, his conclusions are mathematical truths. For other sciences one does not have this freedom because, in order to have sound reasoning, the premised must agree with our knowledge of reality.

Every day we are exposed to argumentation from the media, as well as individuals, which is intended to persuade us to accept certain ideas or products. Many of these arguments are examples of inductive reasoning rather than the deductive methods emphasized in mathematics.

An inductive argument is one that draws conclusions from an analogy, historical information, or statistical considerations. Validity in such argumentation is, more often than not, subjective as it is not always simple to determine. Bias and emotions, as well as knowledge, play a role in our evaluation of the validity of an inductive argument. "The Governor did not keep many of his campaign promises in 1972. Therefore, if reelected, he will not keep many of his campaign promises in 1976." This could be a valid inductive argument in some state. Before the argument is accepted, however, one should expect some justification of the word "many" in the statement. Perhaps a list of kept and unkept campaign promises is in order. Furthermore, one should ask what conditions produced any failure to keep a promise and if those conditions will prevail during a second term in the office of the Governor. The validity is a complicated matter.

It may come as a surprise but most practitioners of mathematics use inductive reasoning to develop new results. Indeed, only through examination of particular situations can general theories be proclaimed in any area of endeavor. Whereas the mathematicians have results in their field suggested to them by inductive reasoning, the demands of mathematics require that each assertion, other than an axiom or a definition, be verified by deductive argumentation. More often than not, the deductions of a creative mathematician are based on previously established facts that are often quite far removed from the actual axioms of the subject. Except for the initiators of new mathematical areas, the working mathematician is attempting to extend, to generalize, to improve, to collect, or to simplify the results of many other mathematicians. This continuous building process rapidly produces a superstructure that is high above its foundation.

There are numerous examples in our everyday life of invalid inductive reasoning. Distortions of history, suppression of evidence, false or misleading statistics or cited information, and doubtful comparisons commonly appear in inductive argumentation. For example, the role of the American black in the development of this country has been, until quite recently, suppressed in many history and civics textbooks used in secondary schools. Other examples abound. Consider the statement "There is a 90 percent chance that our country shall have a woman president within the next twelve years." Perhaps it is true but it is currently an invalid argument. How could any statistic, such as the 90% quoted, ever be proved? What evidence is being used to support this claim? If, in the future, a party nominates a strong woman candidate for the presidency, then such a statistic could be tested. But that is in the future and the argument is invalid today.

Guilt by association, name-calling, and ridicule are other forms of invalid argumentation. No echelon of society is free of such attacks. One famous example is the statement of Spiro Agnew in a speech delivered in New Orleans, October 19, 1969: "A spirit of national masochism prevails, encouraged by an effete corps of impudent snobs who characterize themselves as intellectuals."

For now, additional examples of fallacious inductive arguments are left for your discovery from your exposure to the media, politics, and education. Mathematics, after all, does not concern itself with inductive reasoning although one branch of mathematics, namely statistics, can appear in these arguments. We shall resurrect the concept of inductive argument in our later discussions of statistics.

Exercises

1. Attempt to find an invalid deductive argument from the material you read this week.
2. Find three examples of what you consider to be invalid inductive reasoning. (Letters to the editor in newspapers are particularly good sources.)
3. Discuss the validity of the arguments in each of the following quotations:
   a. "Remember: Once you organize people around something as commonly agreed upon as pollution, then an organized people is on the move. From there it is a short and natural step to political pollution." Saul Alinsky. *Rules for Radicals*. New York, N.Y.: Random House. 1971. p. 23.
   b. A group of citizens who were attempting to repeal the Ohio Income Tax used the following argument. "Do you really believe that repealing a tax could ever cost you more than keeping one?" (*Cincinnati Post*. November 6, 1972. p. B9)
   c. Norman Mailer in his book *Marilyn* compares playwrite Arthur Miller with Tennessee Williams and then goes on to add: "He would be inhuman if he did not have sufficient ambition to recognize that a marriage to Monroe would be theatrically equal to five new works by Williams."
   d. "Many Americans have incomes in excess of $1,000,000 per year."
   e. Assume the president of a small prestigeous college states, "The average income of last year's graduates of our college was $25,000 per year."

4. Record invalid arguments used during some situation comedy on TV.
5. Give an example of illustrate invalid argumentation from the media ads of each of the following products: gasoline, cigarettes, toothpaste, detergents.
6. Find an example of an argument by ridicule such as the one by Agnew.
7. Three men paid $10 each for a joint room in a hotel. The desk clerk noticed that the room rental price should have been $25 instead of the $30 that was charged the three men. A bellboy was dispatched to return the $5 overcharge to the men. Being a little dishonest the bellboy returned only $3 and kept $2 for himself. Now, in the end, each man paid $9 for the room, which, with the $2 taken by the bellboy accounts for $29. Which crook stole the extra dollar?
8. Consider the following inductive argument: A scientist had collected a number of grasshoppers. He carefully placed each on a table, and, in a loud voice, said "Jump." Each time he conducted the experiment the grasshoppers jumped as commanded. Then the scientist removed the hind legs of each of the grasshoppers and repeated his experiment. In no case did any grasshopper jump on his second turn. The scientist concluded, "A grasshopper with his hind legs removed cannot hear." (Adapted from H. Eves. *The Other Side of The Equation*. Prindle, Weber, and Schmidt. 1969.)

2.2. Truth, What Is It?

Throughout the ages tolerance for others was often lost when a man, or group of men, would convince a population that they knew a social truth. There may well be no immutable truths in a society. The truth of one generation may not be acceptable to the next generation.

In mathematics we do not have such difficulties since we *assign* a value, true or false, to each simple statement and proceed to discover, by the accepted logic of the day, the implications of statements. For example, in Euclidean geometry we assign the value "true" to the parallel axiom: "Given a line and a point not on the given line, there is exactly one line through the given point parallel to the given line." Suppose you substitute for the parallel axiom some axiom that contradicts it. For example, assume there are at least two lines through the given point parallel to the given line or assume each pair of lines intersect. Suppose furthermore, you accept without change the other axioms of Euclidean geometry. The results would be another type of geometry, called a non-Euclidean geometry, which has every bit as logical a structure as the more classical one.

In the history of the development of mathematics, precisely this was done by Nicolai Ivanoritch Lobachevski (1793–1856) and Bernard Riemann (1826–1866). The fact that new consistent geometries could result from this substitution terminated the centuries old search by the mathematicians for a proof of the parallel axiom based on the other axioms of Euclidean geometry. (The parallel axiom never seemed as "obvious from nature" as the other axioms.) In this century, Einstein used the non-Euclidean geometry of Riemann in his relativity theories which among other things indicates that Riemannian geometry is a better approximation to reality in large scale problems than the more classical geometry of Euclid.

The assignment of one of two values, true or false, to each statement used by mathematicians in their axioms builds the principle of contradiction into the logic of mathematics. More precisely, titled as the *law against contradiction*, it simply requires that "No acceptable statement is both true and false." We have already taken advantage of this principle with some of the arguments in Chapter 1 and intend to utilize the law throughout this text. It is a natural law and when one violates it in any argument in life, one is faced with charges of inconsistency.

There are, however, some difficulties if we attempt to assign a truth value to every statement. By way of illustration, consider the simple sounding sentence "This statement is false." If it is true, then, as the sentence requires, it is false. Since we insist on one truth value for the statement, we cannot assign a "true" to the sentence. On the other hand, if we assign a "false" to the sentence, then it is false that this statement is false, that is, this statement is true. Again we have a difficulty in tying down a single truth value for the statement. It is clear that no assignment of true or false can be made to the sentence if only one of these values is permitted. In logic this phenomenon is called a paradox and, as we have illustrated, paradoxes do exist with the unrestricted use of language. A simplistic solution to the dilemma is to avoid, if possible, paradoxes in our mathematical statements.

Whereas this simple solution to the paradox problem suffices for our study of mathematics, work on the foundations of the subject has been stimulated by the discovery of paradoxes that appear to exist within the system. More precisely, the axioms of logic, used at the turn of the last century, led to a rather simple mathematical paradox uncovered by Bertrand Russell in 1901. After suitable adjustments in the axiom system, another, and much more difficult, problem was discovered by Kurt Gödel in 1931. Gödel's work put an end to all hope that all branches of mathematics can be rigorously developed by means of the axiomatic method. His proofs brought about the astounding revelation that it is impossible to establish absolute freedom from contradiction of complex deductive systems unless you assume principles of logic with internal freedom of contradiction. Like revolutionary scientific discoveries, the impact of Gödel's result was to promote deeper studies into the foundations of mathematics and logic. The subject now occupies one of the more important and esoteric frontiers of mathematical research. A proof that a system of axioms is inconsistent forces the researchers to restructure the system, but an argument that consistency questions in some sense cannot be answered does not curtail activity in the discipline.

Exercises

1. Assign a truth value to each of the following sentences in this set of six sentences.
   a. Exactly one of these sentences is false.
   b. Exactly two of these sentences are false.
   c. Exactly three of these sentences are false.
   d. Exactly four of these sentences are false.
   e. Exactly five of these sentences are false.
   f. All these sentences are false.

2. Which of the following are paradoxes? Assign a truth value if the statement is not a paradox.
   a. What is true is false.
   b. Frederic, the protagonist in *The Pirates of Penzance*, has reached the age of 21 after passing only five birthdays.
   c. All authors are liars.
   d. This statement is not true.
   e. No sentence is true.
   f. Nonnegotiable demands are negotiable.
   g. It is a rule that every rule has an exception.
   h. It rains tomorrow.

3. Joe states, "What John says is false." John says, "What Joe states is true." Is what Joe states true or false?

4. Is there a paradox in the following three sentences? This paper is blue. This note was written by Plato. All three of these statements are false.

5. Achilles attempts to overtake a tortoise with a head start. Every time Achilles reaches a point where the tortoise was, it has moved to a new point. Can Achilles ever catch the tortoise?

6. In a certain village there is a man who is a barber. Each day this barber shaves all, and only those men in the village, who do not shave themselves. Who shaves the barber? Can anyone in the village have a beard? (The first question is one of Bertrand Russell.)

7. In a letter to the editor of a local newspaper, a group of nuns wrote, "The remarks made about the Archbishop were very insulting and certainly not true. Even if they were true, we are very sorry that things like this would get into print." Is their argument consistent?

8. A market research organization reports that based on a sample of 1000 executives interviewed, 62% read the *Wall Street Journal* regularly, 53% read *Time* regularly, and 16% read both. If they also report that 4% read neither on a regular basis, is there a contradiction in their figures? Explain your answer.

2.3. Sets and Logic

When mathematicians attempt to formalize their logic, they are driven into a discussion of an area called "set theory" and then beyond into a realm of metamathematics. The importance of set theory in logic was first noted by the nineteenth century mathematician George Boole of Queen's College, Dublin, and announced with his publication, *The Laws of Thought* (1854). Boole showed how the logic of Aristotle, taught for centuries in universities throughout the world, could be subjected to formal methods of calculations among sets, much like the manipulations used in a high school algebra course. This algebra of logic is now called Boolean algebra in honor of its discoverer. Although we have no intention of initiating a formal study of Boolean algebra, we will attempt to indicate the analogy between logic and set theory that this algebra suggests.

Since the concept of "set" or "collection" is so basic, we accept this notion without a formal definition. (Indeed, you cannot define every term since each term requires other terms to define it.) The items in the collection or set are called *elements* or, sometimes, *points*, and other times just *members*. The terminology "point" is particularly appropriate when a geometric configuration is utilized to guide our formal reasoning. Whenever x is an element of the set A we write $x \in A$, and use the symbols $x \notin A$ to indicate that x is not an element of A. A *subset* B of a set A is, itself, a set such that whenever $x \in B$ then $x \in A$ also, that is, each element of B is also an element of A. The notation for this is $A \supset B$ or $B \subset A$. In order to avoid paradoxes, Bertrand Russell showed, in 1901, it is important to begin with a set, called the "universal" set, denoted by I, and to discuss only subsets of this universal set. It is convenient, moreover, to include the "null" or "empty" set, written as $\emptyset$, in discussions of set theory and to require that the null set be a subset of every set. As the name implies to any German, the null set is a set with no elements.

In order to illustrate definitions, a graphic device has been invented and is generally referred to as a Venn diagram in honor of John Venn (1834–1923), although many of his mathematical predecessors had also used these diagrams. Let us agree, although this is not standard in mathematics, to represent the universal set I by the interior of a rectangle in the plane. The members of I are represented by the points inside this rectangle. There can already be a flaw in this construction. Suppose, for example, I is the set of all live humans on earth. The collection is clearly finite; in fact, it contains fewer than 100,000,000,000 members. Yet, the rectangle has infinitely many points in its interior. Remember, however, that the Venn diagram is a visual aid and is not intended to be an accurate model of the real situation. The Venn diagram can, even with its flaws, assist us in the deductive reasoning process but it does not displace the argument itself. Those that doubt this are challenged to solve the problems in the supplementary exercises.

A subset A of I is any collection of points in the rectangle I. It is convenient, again for visual purposes, to separate the points in A from those in I that are not in A by using some geometric curve at least when A is not empty and when A is not identical with I. Shaded portions of Figure 2.1 show some subsets of I.

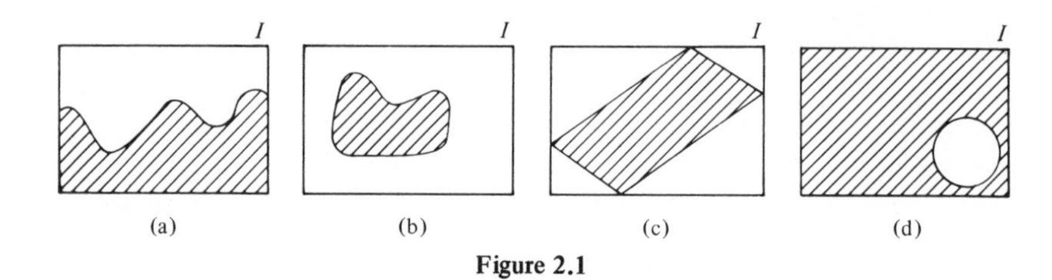

Figure 2.1

The boundary points of these curves can be thought of as either part of the subset or not. You decide in each instance, but remember the decision whenever relations of the subsets with other sets are to be established.

If A and B are subsets of I we say that A equals B, written $A = B$, if A is a subset of B and B is a subset of A, that is, A and B are sets such that each contains the same elements. The set of all live humans and the set of all live humans with heads are equal.

There are some operations with sets that, with the advent of "new mathematics," you may have been forced to learn in early grade school. One is the union of two sets, written $A \cup B$ and read "A cup B." It is defined as the subset of I containing all elements x such that x is a member of at least one of the sets, A or B. The union of all human male and female students in this mathematics class is the set of all human students in this class. A Venn diagram illustrating the union of two sets A and B (interior of circles) is given in Figure 2.2(a). The intersection of two sets A and B, written $A \cap B$ and read "A cap B," is the subset of I containing all elements x such that $x \in A$ and $x \in B$. Figure 2.2(b) illustrates this definition. Hopefully, the intersection of male and female students in this class is the empty set.

Finally, the complement of a set A, written A' and read "A prime," is the set of all $x \in I$ that are not in A. Figure 2.2(c) illustrates this concept. The universal set being considered determines the extent of this complement. For example, if I is all students in this class, the complement of the set of male students is the set of female students in the class. However, if I is the set of all students at this university, the complement of the set of male students in the class includes not only the females in the class but also all other students at the university who are not taking this class.

This discussion has apparently carried us away from logic. That this is really not the case is perhaps illustrated by the following example. A survey of 100

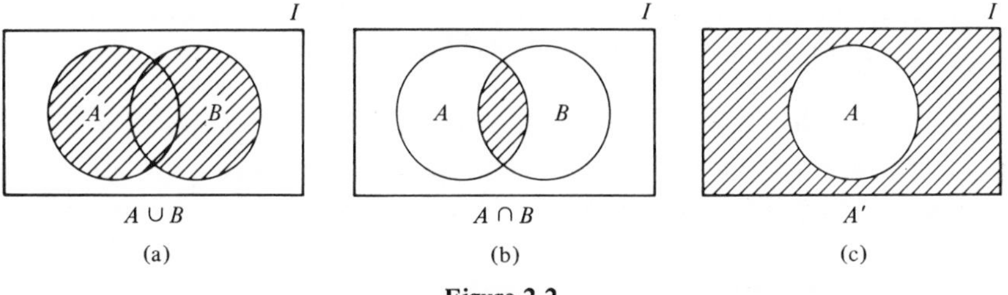

Figure 2.2

college freshmen finds that 80 take English, 60 take mathematics, and 5 take neither mathematics nor English. Should we challenge the results of this survey since $80 + 60 + 5 = 145$ which exceeds the number of students supposedly contacted? Let I be the set of 100 freshmen in the survey, E the set of those taking English, and M the set of those taking mathematics. The survey does not claim that the sets E and M have nothing in common, that is, $E \cap M$ could contain elements. In fact, if the survey is accurately reporting the information, there must be freshmen who are taking both English and mathematics and, hence, were counted twice in the total 145. Since there were only 100 students surveyed, $145 - 100 = 45$ students must have been counted twice, once because they are taking English and once because they are taking mathematics. We conclude that apparently there is nothing wrong with the survey, and, furthermore, that 45 students in the survey are taking both English and mathematics. Also, $60 - 45 = 15$ are taking mathematics and not English while $35 = 80 - 45$ are taking English and not mathematics. The Venn diagram in Figure 2.3 illustrates the analysis of this problem. Expect more illustrations.

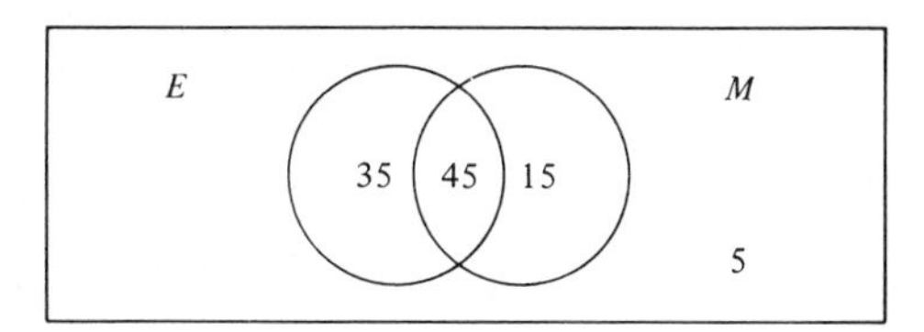

Figure 2.3

Exercises

1. Let A be the set of letters $\{a, b, c\}$ and let B be the set $\{c, x, z\}$. If I is the set of all letters in the alphabet, determine each of the following sets:

   a. $A \cap B$ b. $A \cup B$ c. A' d. $A' \cap B$ e. $A' \cup B'$ f. $A \cup A'$

   Draw Venn diagrams for each of the sets a. through e.

2. Let I be the set of counting numbers not exceeding 30. Let A be the odd

members in I and let B be the multiples of 3 in I, that is, B contains 3, 6, 9, 12, etc. List the elements in each of the following sets.

a. A' b. $A \cup A'$ c. $A \cap A'$ d. $A \cap B$ e. $A \cup B$
f. $A' \cap B$ g. $A' \cup B$ h. $A' \cup B'$

3. A Gallup Poll investigator reported speaking to 100 persons, of whom 20 were students, 65 were women, and 30 were men. Why was this investigator fired?

4. Another investigator reported speaking to 100 persons, of whom 20 were students, 60 were males, and 30 were females who were not students. To how many male students did he speak?

5. After a wage freeze, a company with 420 employees decided to raise the salary of 240 employees. Also, 115 people were promoted while only 60 of these employees got a raise. How many employees received neither a raise nor a promotion?

6. A realtor has 55 new houses, each of which has at least three bedrooms and two baths. There are eight that have more than three bedrooms, three baths, and a fireplace. Five houses have more than three bedrooms, three baths, and no fireplace. Three houses have more than three bedrooms, a fireplace, and two baths. Eight homes have more than three bedrooms and exactly two baths but no fireplace. Twenty-four houses have a fireplace but only three bedrooms and two baths. Finally, seven houses have only three bedrooms but three baths and a fireplace.
   a. How many houses have more than three bedrooms?
   b. How many houses have more than two baths that have only three bedrooms and no fireplace?
   c. How many houses have a fireplace?
   d. How many houses have more than two baths?

7. A set is called *normal* if it is not an element of itself and it is *abnormal* otherwise. An example of an abnormal set is the set of all sets that can be defined in fewer than 50 words. Which of the following are abnormal sets?
   a. The set of all counting numbers.
   b. The set of all sets that contain fewer than 10 elements.
   c. The set of all sets that contain more than 10 elements.

8. Is the set of all normal sets a normal or an abnormal set? (Russell's paradox.)

2.4. To Be or Not To Be

If we are to succeed in our study of mathematics, care must be exercised concerning the meaning of our terms. We cannot tolerate subjective interpretations of terms in mathematics; otherwise the exactness of the science is lost.

A universal word like "all" or "every" has a meaning (no exceptions) that hardly requires discussion. It is convenient, however, to assign no existence to the use of "all," that is, we permit phrases such as "All purple men" even though, to the extent of our current knowledge, there are no creatures in existence that could be classified as naturally purple men. The set of all purple men, therefore, is empty.

With the word "some," which is to mean "at least one," there is an implied existence. To state that "Some purple men are fat" is equivalent to "At least one purple man exists who is fat." If there are no purple men, then a false value must be assigned to the statement which asserts the existence of at least one purple man. Often "some" is interpreted by individuals to mean "at least one but not all." It is unlikely that a politician who asserts that "Some politicians are honest" intends to include the possibility that all politicians are honest. If this politician believed the last statement he would have asserted that all, rather than some, politicians are honest. In our use of "some," however, we mean "at least one" and possibly "all." The statement "Some even integer is divisible by two" is true even though it is not as informative as "All even integers are divisible by two."

Language does not appear to be sympathetic to our interpretation of the word "some." There is a problem with the singular and the plural that we frequently have to face. For example, if "some men are honest" is to mean "at least one man exists who is honest," then it is better to formulate the original statement as "some man is honest." Rather than get tangled in a tense-singular-plural dispute, it is perhaps advisable, and it is often done, to define a phrase such as "some men" to mean "at least one member of the set of all men." For someone who is

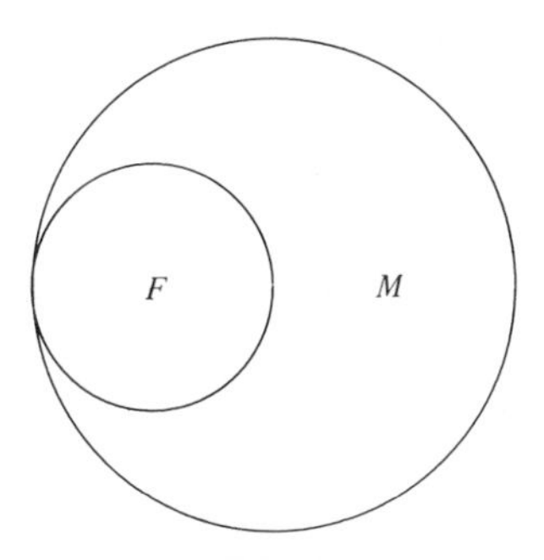

Figure 2.4

not aware of this convention, however, "some men" admittedly may mean "two or more men."

The discussion of terms can be simplified by considering the statements in terms of sets and by representing these sets by Venn diagrams. For example, in Figure 2.4 the set of all men might be represented by the set of points M in and on a circle in the plane. The set of all fat men comprises some subset, that is, a particular collection drawn from the set of all men, and it is represented by the points F within and on a second circle. There are points in M that are not also in F if we believe that there are men who are not fat.

Recall that the empty set $\emptyset$ is taken to be a subset of every set. Since the set of all purple men has been asserted to be empty, it must be a subset of M and also of F. The fact that $\emptyset \subset M$ translates in this instance into the statement "All purple men are men" and $\emptyset \subset F$ can be interpreted as "All purple men are fat men." Now both these quoted statements are taken to be true in order to complete the parallel between logic and set theory. In fact, you can make any statement about the set of all purple men, except, of course, that there is such a creature, and the statement will be true. Indeed, all purple men are old, fat, vain, hardheaded, male chauvinist pigs.

Words like "none" or "no" do not imply existence in mathematics. For instance, consider the statement: "No purple man is fat." Indeed, there is no purple man among the fat men because there is no purple man anywhere! It seems plausible to accept the truth of the given statement about the empty set of purple men.

Exercises

1. Assign truth values (true or false) to each of the following statements:
   a. All men have two legs.
   b. No man has twenty legs.
   c. No women's liberationist visits a Playboy Club.
   d. All purple men are blue.
   e. Some triangles have angles that total 180°.
   f. Every live man uses his head.
   g. No purple man is green.
   h. Some positive numbers are smaller than one.
2. Draw a Venn diagram and indicate each of the following sets in the diagram:
   a. The set I of all humans.
   b. The set of all females.
   c. The set of all fat humans.
   d. The set of all Germans.
   e. The set of all Americans.
3. Frequently words like "all" and "some" are omitted from statements, possibly for impact. There is an implied "all" in most of these statements whenever this practice is followed. Insert the universal word (all, none) or the existential word (some) to make, in *your opinion*, each of the following statements true.
   a. Politicians are dishonest.
   b. Republicans favor sales taxes.
   c. Hippies are dirty.

d. College students are interested in ecology.
e. Poor people are democrats.
f. Policemen are pigs.
g. Animals consume food.
h. Green men are from Ireland.

2.5. Deduction with the Aid of Pictures

A Venn diagram can serve as an excellent guide for valid deductive reasoning. Consider the premises: "All intelligent students can pass mathematics," and "You can pass mathematics." Can we conclude, on the basis of these premises, that "You are intelligent?" Although this simple example has a transparent fallacy, it can serve as an illustration of the use of Venn diagrams in reasoning. We let I be the set of all students and T the subset of intelligent students. Let M be the subset of students that can pass math. A Venn diagram which illustrates the first premise is given in Figure 2.5 since the statement is simply that T is

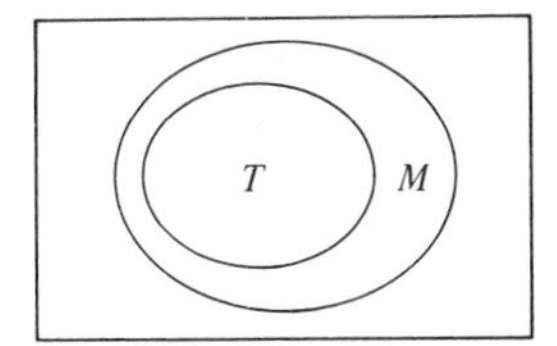

Figure 2.5

a subset of M. Now the second premise states that you are some point in the set M and nothing more. Since it does not state that you must be a point in T, you could be in M yet not in T. Thus, the statement "You are intelligent" is not a valid conclusion from the premises and the premises neither establish nor contradict the proposed conclusion.

If the second premise in the previous argument is replaced by "You are intelligent," then you become a member of the set T of intelligent students. Since all points in T are also in M, it follows that you are also in M, that is, "You can pass mathematics." This is a valid conclusion from the new premises.

Suppose we consider the premise "All intelligent students can pass mathematics" in conjunction with the premise that "Some student who passed history cannot pass mathematics." If H denotes the set of students who passed history, the second premise means that there is at least one element in H that is not in M. Any one of the diagrams in Figure 2.6 are among the possible candidates for the Venn diagram of these two premises. The information is insufficient for us to make a logical choice.

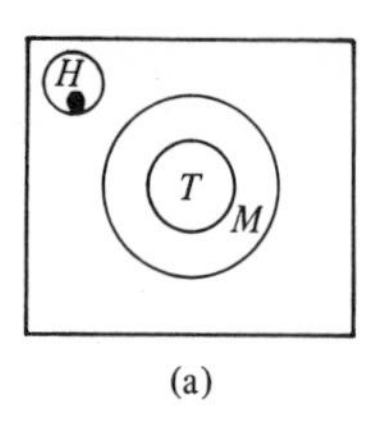

(a)

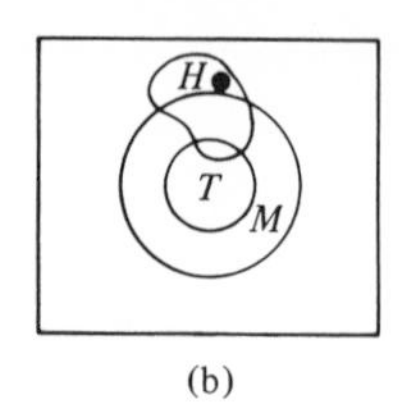

(b)

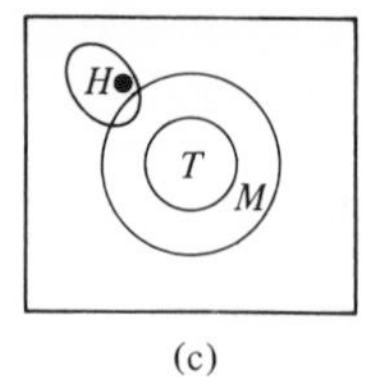

(c)

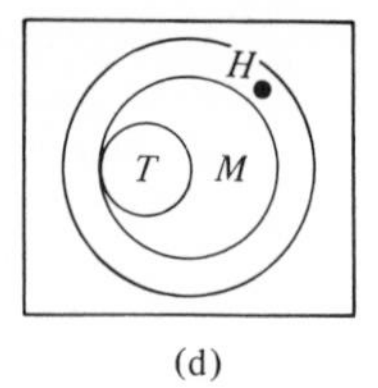

(d)

Figure 2.6

(The dots in Figure 2.6 indicate a student who can pass history and not pass mathematics. There is at least one such student by the premise using the definition of "some.")

For example, add the third premise "All students who can pass mathematics passed history." In our set notation this premise is $M \subset H$. Now only diagram (d) in Figure 2.6 accurately describes all three premises. One conclusion is that $T \subset H$, that is, all intelligent students have passed history.

Lewis Carroll of *Alice in Wonderland* fame was actually a nineteenth century English mathematician named Charles Dodgson. In a book on symbolic logic that he authored, many of the illustrations follow the delightful nonsense humor of his children's stories. One example has the following premises: (a) Babies are illogical, (b) Nobody is despised who can manage a crocodile, and (c) Illogical persons are despised. These premises permit us to conclude that "Babies cannot manage a crocodile" and also that "Babies are despised." Indeed, let B be the set of all babies, C be the set of all crocodile managers, D be the set of all despised persons, and L be the set of all illogical people. They are all subsets of the universal set I of all humans. We next analyze each premise and place it in our set theory notation. (a) states that $B \subset L$. (b) states that if $x \in C$, then $x \notin D$. This means $C \subset D'$, that is, crocodile managers are not despised. (c) states that $L \subset D$. Draw the Venn diagram for statements (a), (c), and (b) in that order

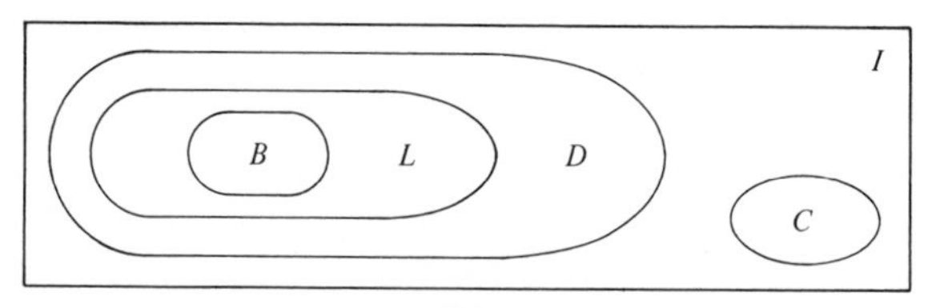

Figure 2.7

(to avoid an error). The result is given in Figure 2.7. The conclusions are $B \subset D$, babies are despised; $L \cap C = \phi$, an illogical person cannot manage a crocodile; and $B \cap C = \phi$, babies cannot manage a crocodile.

One should not expect the Venn diagram to suggest answers for all logical puzzles, as the last three exercises in this section suggest. Sometimes other methods are more convenient. The supplementary problems at the end of this chapter contain a number of additional puzzles, most of which do not depend on Venn diagrams for solutions.

Exercises

1. Which conclusions follow from the premises:
   a. Some college students are clever. All freshmen are college students. *Conclusion:* Some freshmen are clever.
   b. All hippies hate war. Some students are hippies. *Conclusion:* Some students hate war.
   c. Richard is the son of Smith. Smith is a five letter word. *Conclusion:* Richard is the son of a five letter word.
   d. All murders are immoral acts. Some murders are justified. *Conclusion:* Some immoral acts are justified.
   e. All men are animals. Some animals have legs. *Conclusion:* Some men have legs.
   f. A U.S. president must violate some of his principles. Anyone who violates some of his principles is not trustworthy. Only trustworthy people should be elected president of the U.S. *Conclusions:* Nobody should be elected president of the U.S. President Johnson was not trustworthy. (Remark: Substitute Nixon for Johnson, if you wish, or George Washington, for that matter.)
   g. Doctors are in favor of legalized abortions. Pope Paul is a Doctor of Theology. *Conclusion:* Pope Paul favors legalized abortions.
   h. If someone is here, then he is not in England. Someone is here. Therefore, it is not the case that someone is in England.
2. Another Lewis Carroll puzzle is the following:
   a. The only types of food my doctor permits me to eat are not very rich.
   b. Nothing that agrees with me is unsuitable for supper.
   c. Wedding cake is always very rich.
   d. My doctor permits me to eat all types of food that are suitable for supper.
   What conclusions can you draw from these premises?
3. A final appropriate Lewis Carroll puzzle states the following premises:
   a. When I work a logic example without grumbling, you can be sure it is one that I understand.
   b. These examples are not arranged in a regular order like the examples I am used to.
   c. No easy example ever makes my head ache.
   d. I cannot understand examples that are not arranged in regular order like those I am used to.
   e. I never grumble at an example unless it gives me a headache.
   Conclude that these examples are not easy for you.

4. There are three men, Mr. Rep, Mr. Dem, and Mr. Ind. One is a Republican, one is a Democrat, and one is an Independent. The Democrat often beats Mr. Ind at cards. Mr. Ind is the brother-in-law of the Republican. Mr. Dem has more children than the Democrat. Who is the Republican?
5. Sue, Bill, Mary, and John attend a local University at which each course carries three credit hours. Exactly one of these students takes a course entitled "Topics in Mathematics." Identify this student if the following facts are known.
   a. All four students are taking 15 credit hours of course work.
   b. Sue and Bill are freshmen.
   c. Only juniors and seniors can take Italian.
   d. Mary is in John's tennis course.
   e. Calculus is a co-requisite to physics.
   f. All four students went directly from high school to college and get good grades.
   g. John takes chemistry and physics.
   h. All four students take a language.
   i. All freshmen must take English and history.
   j. Mary hates chemistry.
   k. Sue takes art.
   l. Mary graduated from high school in June of this year whereas John graduated in June of last year.
   m. Bill is in Sue's French class.
   n. Sue and Mary are in Bill's biology class.
   o. Sue is in Mary's history class.
6. Two couples are on a camping trip in Canada and they have one two-person canoe to portage. After a few days, each male believes his wife is becoming too friendly with the other male in the party. How must they cross each lake they encounter thereafter if both men insist that no women is in the company of a man unless her husband is also present? The canoe cannot carry more than two people on a single trip.

2.6. Compound Statements

A statement is a meaningful collection of words in sentence form. A statement is compound if it can be subdivided into two or more separate statements. A statement that cannot be subdivided in this manner is called a simple statement. For example, "Today is Monday and I hate it" is compound since it is two simple statements connected by the word "and." "If it is Monday, then I will hate this day" is also compound since it is two simple statements placed in the form of an implication.

The acceptable simple statements that will be considered in this discussion of logic are those for which there is a definite truth value assignment. Paradoxes, even though they are simple statements such as "This sentence is false," are to be avoided. Acceptable compound statements are those composed of acceptable simple statements. The assignment of a "definite truth value" to simple statements is often subjective. For instance, some individuals would assign a false value to the statement "Johnson was a good president" whereas others claim it to be true. Perhaps the truth of the statement can only be judged in a historical sense at some future date. As far as we are concerned, the statement can be assigned either a true or a false value, but not both simultaneously. Which truth value is assigned is unimportant since the logic can work with either choice. What must be avoided are simple statements that are paradoxes, but we need not avoid simple statements that are controversial.

One basic connector of simple statements that produces compound statements is the word "and." "Jack and Jill are over the hill" is really the two simple statements "Jack is over the hill" and "Jill is over the hill" connected by "and." The truth value of the compound statement is dependent upon the assigned truth values for each of these two simple statements. Since there are two possibilities for each, there are four possible cases, listed in Table 2.1, called a truth table. The first row means that each simple statement, now abbreviated as A for "Jack is over the hill" and B for "Jill is over the hill," has been assigned truth value T, for true. The third entry in this row asserts that the compound statement is also true in this case. The second row assigns a true to A but a false, F, to B. The compound statement is, in this case, false. The other rows have similar interpretations.

All four possible permutations of the truth value assignments of the two simple statements have been listed along with the appropriate truth value of the compound statement as recorded in the third column. The table provides us with a definition of the connector "and" in that it assigns truth values to compound statements built from simple statements connected by this word. More generally, the statements A and B in and of themselves can be compound statements and the table gives a truth value assignment for A and B.

TABLE 2.1

A	B	A and B
T	T	T
T	F	F
F	T	F
F	F	F

Most English-speaking people agree with the definition of "and" in Table 2.1.

A second basic connector is "or" and there is less agreement as to its definition. We define the connector "or" in Table 2.2. The controversy lies in the first row of the table where the assignment is made of "true" to "A or B" when both A and B are true. The word "or" is sometimes used in language to mean "either one or the other but not both." For instance, consider the statement "We must conserve our resources or we are doomed." Our definition of "or" assigns a true value to this statement when "We must conserve our resources" and "We are doomed" are simultaneously true. The impact of the statement is perhaps lost since the author undoubtedly intended something other than to have his statement true in all cases except when we do not conserve our resources and are not doomed!

TABLE 2.2

A	B	A or B
T	T	T
T	F	T
F	T	T
F	F	F

In common usage "Either . . . or . . ." statements have the same meaning as an "or" statement; the "either" is superfluous. It would be convenient and reasonable to insist on this to represent the exclusive "or," that is, to have the statement "Either A or B" false when both A and B are true. But, alas, others in the world are not so inclined and, hence, "either . . . or . . ." remains equivalent to an "or" connector.

Connectors are *binary operations* in that only two statements can be joined together at a time. In arithmetic the operations of addition and multiplication are binary in this sense. Indeed, should you wish to add three numbers, 4, 6, and 8, say, you first add 4 and 6 to obtain 10, and then add 8 to 10. Each step of the way you are combining just two integers. Because of these pairwise connections it is necessary for clairty to use parentheses to indicate the order in which pairs are joined. For example, $(4 + 6) + 8$ means to first calculate $4 + 6 = 10$ and then take this answer and add it to 8. On the other hand, $4 + (6 + 8)$ means 6 is first added to 8 to obtain 14 and then $4 + 14$ is computed.

In some cases the result obtained by the various ordering in the performance of the operations is always the same. These are called the associative cases. You may recall that addition of integers was associative, that is, $(a + b) + c = a + (b + c)$ for any three integers a, b, and c. Multiplication is also associative but subtraction is not. For example, $7 - (3 - 2) = 7 - 1 = 6$, whereas $(7 - 3) - 2 = 4 - 2 = 2$. These answers are quite different. Sometimes we even encounter nonassociative cases with language. What is the meaning of "fat dog lovers?" Is it lovers of fat dogs or is it fat people who are dog lovers? In order to convey your intention a dash could be inserted, fat-dog lovers" or "fat dog-lovers," for removal of the ambiguity in this nonassociative situation.

The connectors "and" and "or" are associative operations for statements. This is really quite apparent. For example, A and (B and C) has the same meaning as (A and B) and C since each is true only when all three statements A, B, and C are simultaneously true. A formal verification of the associativity of the connectors, however, is obtained by constructing truth tables for the compound statements.

There is an interesting and important analogy between the connectors "and," and "or" of statements and the "intersection, union," respectively, of sets. By way of illustration consider Table 2.3:

TABLE 2.3

$x \in A$	$x \in B$	$x \in (A \cap B)$	$x \in (A \cup B)$
T	T	T	T
T	F	F	T
F	T	F	T
F	F	F	F

We assume in Table 2.3 that A and B are sets and x is or is not an element of them. The statement $x \in A$ is true or false according to the values in the table. The third column defines $A \cap B$ in the sense that $x \in (A \cap B)$ only when x is in both A and B. Comparing the third column in Table 2.3 with Table 2.1 shows that the truth values are the same. In other words, the intersection of sets behaves exactly like the connection of propositions by the word "and." Similarly, the last column in Table 2.3 is identical to the last column in Table 2.2 except, of course, for the column headings. Thus the union of sets behaves exactly like the connection of propositions by the word "or."

The analogy between set theory and compound statements enables us to use the cup and cap notation instead of "or" and "and," respectively, when speaking of propositions. Thus $A \cup B$ means "A or B," and $A \cap B$ means "A and B," providing A and B are propositions rather than sets. There should be no confusion with this dual interpretation of the operations, and some merit can be derived later. In formal logic, however, there are special symbols for "or" and "and"; the most common of which are $\vee$ and $\wedge$. Set theory symbols are not commonly used in mathematical logic.

Exercises

1. Assign truth values to each of the following compound statements:
   a. One plus one is two and two plus two is four.
   b. One plus one is two or two plus two is six.
   c. Either Ford or Nixon is our president.
   d. Some Americans have incomes in excess of $1,000,000 and pay no taxes.
   e. The moon is round or the world is flat.
2. Define the connector "neither . . . nor . . ." by a truth table.

3. Show that division "÷" is not an associative binary operation on nonzero numbers.
4. The operations of cup and cap in set theory are associative. Illustrate the associativity of $\cup$ and $\cap$ for the sets A, B, and C in the Venn diagram in Fig. 2.8.

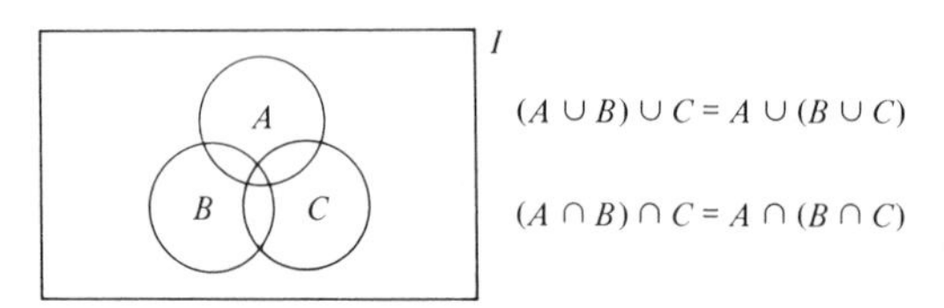

Figure 2.8

5. Construct a truth table which proves "(A or B) or C" and "A or (B or C)" and have the same set of truth values. (Remark: To consider all possible cases there must be eight rows in the table.)
6. Use Venn diagrams to convince yourself that the following are identities for any three sets A, B, and C: $A \cup (B \cap C) = (A \cup B) \cap (A \cup C)$; $A \cap (B \cup C) = (A \cap B) \cup (A \cap C)$. What are logical analogs of these set theory results?

2.7. Logical Equivalence and Negation

There are many different ways to say precisely the same thing. For example, we have already agreed to equate "some men are honest" with "some man is honest," and "Jack and Jill are over the hill" with "Jack is over the hill and Jill is over the hill." We define a concept in logic to cover this situation, called logical equivalence, by a truth table as in Table 2.4. Two statements A and B are logically equivalent, written $A \equiv B$, when they can only be simultaneously true or simultaneously false.

TABLE 2.4

A	B	$A \equiv B$
T	T	T
T	F	F
F	T	F
F	F	T

The negation of a statement A, written $\sim A$, is a statement derived from the original by inserting the phrase "It is false that" before the statement A. The negation $\sim A$ is true whenever A is false and $\sim A$ is false whenever A is true. Consider the statement "All men are happy." The negation "It is false that all men are happy" is clumsy and, therefore, should be replaced by a logically equivalent statement. One might be tempted to assume "All men are not happy" is this equivalent. Remember, however, that the negation must be true *whenever* the original statement is false. Now "All men are happy" is false when there is at least one poor fellow that is not happy. Since other men might well be happy, this one gentleman is not enough for us to accept as true the statement "All men are not happy." This discussion, and our agreement about the word "some," indicate that the proper negation is "Some men are not happy" or, equivalently, "Some man is not happy." Indeed, the minimum that need occur for "All men are happy" to be false is for one man not to be happy, as in Figure 2.9.

Figure 2.9

More generally, we can conclude from this argument that the negation of an "All" statement is equivalent to a "Some . . . not . . ." statement. The negation of "All women hate housework" is "*Some* women do *not* hate housework."

In terms of sets an "all" statement merely means one set is a subset of another. For example, if A is the set of all women and B is the set of human housework haters, then "All women hate housework" is equivalent to $A \subset B$. The latter is not true if there is an $x \in A$ while $x \notin B$, that is, there is some woman who does not hate housework, poor fool that she is.

In English certain "all" statements are ambiguous. For example, "All men are not happy" could, on one hand, mean that the set of all men is included in the set of individuals that are not happy. On the other hand, it could mean "there is some man who is not happy." These two interpretations are far from being logically equivalent. One solution to this problem is to replace an ambiguous "all" statement by a logical equivalent that has no second interpretation. For instance, if we choose the first meaning of "All men are not happy," a less awkward and equivalent substitute is the sentence "No man is happy." The sentence "Some man is not happy" conveys, without ambiguity, the second interpretation of our "all" statement.

The negation of a "some" statement is obtained by replacing the "some" with "no." The negation of "*Some* professors are idiots" is "*No* professors are idiots" or, equivalently, "No professor is an idiot."

A double negation $\sim(\sim A)$ of a statement A is logically equivalent to the original statement A as verified by Table 2.5. It has the verbal form "It is false that it is false . . ." and it is clear that the first "It is false" annihilates the second such phrase. Common usage sometimes violates this rule. For example, "I ain't got no education" may be logically equivalent to "I have an education" but there is considerable doubt that the person quoted in the first statement intended to convey anything other than the fact that he was not educated. Even should you believe that language is a form of communication and, as such, proper usage is anything that conveys an idea, there is sometimes a problem in attempting to determine the intended meaning of statements containing a number of negative terms such as "not" and "no."

TABLE 2.5

A	$\sim A$	$\sim(\sim A)$
T	F	T
F	T	F

The negation of the compound statement "A and B" is logically equivalent to "($\sim A$) or ($\sim B$)." The rather long truth table (Table 2.6) will convince you of this equivalence.

TABLE 2.6

A	B	$\sim A$	$\sim B$	A and B	$(\sim A)$ or $(\sim B)$
T	T	F	F	T	F
T	F	F	T	F	T
F	T	T	F	F	T
F	F	T	T	F	T

The fifth column is obtained from the definition of "and." Entries in the third and fourth columns are consequences of the meaning of negation, whereas the last column follows from these and the definition of "or." From this table we discover

$$\sim(A \text{ and } B) \equiv (\sim A) \text{ or } (\sim B)$$

since $\sim(A$ and $B)$ reverses the truth values in the fifth column.

In worded examples the negation of "Jack and Jill are over the hill" is "Jack is not over the hill or Jill is not over the hill," or, equivalently, "Jack or Jill is not over the hill." The result should be reasonable to you since one or the other (or both) not being over the hill constitutes a contradiction of the statement that they are both over the hill.

Similarly $\sim(A$ or $B) \equiv (\sim A)$ and $(\sim B)$ as is verified by a truth table. This illustration, and others, are left for the exercises.

There are set theory analogs of equivalence and negation. Equivalence and negation are, in fact, similar to equality of sets and to complementation, respectively. A comparison of Table 2.7 with the truth table for $A \equiv B$ and for $\sim A$ should convince you of this analogy.

TABLE 2.7

$x \in A$	$x \in B$	$x \in A$ and $x \in B$	$x \in A'$
T	T	T	F
T	F	F	F
F	T	F	T
F	F	T	T

Consequently, set theory results can be transcribed into logical results. For example, we spoke of the negation of the compound statement "A or B." The set theory analog of this negation is $(A \cup B)'$ since "or" is replaced by cup, $\cup$, and negation is replaced by complement, $'$. It is easy to see by Venn diagrams that $(A \cup B)' = A' \cap B'$.

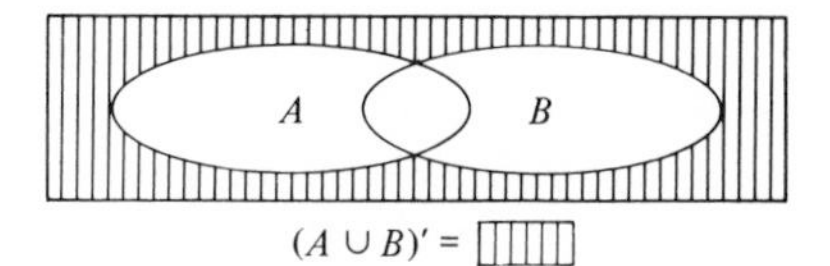

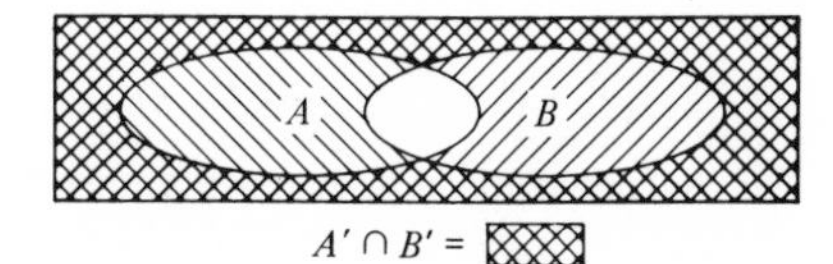

Figure 2.10

In Figure 2.10 $(A \cup B)'$ is the shaded area in the first part whereas $A' \cap B'$ is the doubly shaded area in the second. They are the same.

Exercises

1. Identify all the logically equivalent statements in the set of statements below:
   a. All people are not wise.
   b. No person is wise.
   c. Some people are wise.
   d. Some people are unwise.
   e. All people are dumb.
   f. Some person is not wise.
   g. Some person is wise.
   h. Not all people are wise.
2. Negate each of the following simple statements:
   a. All men are against the women's liberation movement.
   b. Playboy clubs are decadent.
   c. Some politicians are smart.

d. Not all doctors are idiots.
e. All men are equal.
f. Nixon was an excellent president.
g. The professors on this campus are not human.
h. All administrators on this campus are deaf.
i. No M.D. favors medicare.

3. Construct a truth table to establish the equivalence:

$$\sim(A \text{ or } B) \equiv (\sim A) \text{ and } (\sim B)$$

4. Negate the following compound statements:
   a. There will be a Black or a woman vice-presidential candidate for the Democratic party for the next presidential election.
   b. Nixon and Agnew are not ideal candidates for any office.
   c. Either we stop warfare or this world will come to an abrupt end.
   d. Greed and government spending cause inflation.

5. Use a truth table to establish the equivalence (A and B) and $C \equiv A$ and (B and C). (Remark: There must be eight entries in each column in order to obtain all possible cases for three statements, A, B and C.)

6. Negate the statements (A and B) or C, A and (B or C).

7. A fallacious yet common argument is to assume that two wrongs make a right. For example, the killing of civilians was practiced by the Viet Cong before the involvement of the U.S. in Vietnam. Can this be used to justify U.S. forces killing civilians? Perhaps this type of argument is a distortion of the double negative $\sim(\sim A) \equiv A$. In any case, find a current example of this type of fallacious reasoning.

8. Opposites, or antonyms, are not necessarily the same as negations when dealing with words. The negation of a word W is "not W." We accept as an opposite something else, as in the case of "good" which has the antonym "bad," whereas in logic "not good" is really not equivalent to "bad." List two words or word phrases which have antonyms that are equivalent to the negation of the word, and two words of word phrases which are not so equivalent. (An antonym of a word is based on common usage and in reality cannot be uniquely defined.)

9. Are "Some men are not happy" and "Some men are sad" logically equivalent?
   Are "Some men are not alive" and "Some men are dead" equivalent?
   Are "Not all men are good" and "Some men are bad" equivalent?

10. Draw Venn diagrams for each of the following:
    a. $(A')'$ and A
    b. $(A \cup B)'$ and $A' \cap B'$
    c. $(A \cap B)'$, $A' \cup B'$ and $A' \cup (A \cap B)'$
    d. $A \cup A'$ and I
    e. $A \cup (A' \cap B)$ and $A \cup B$

    The sets in each of the above problems are equal. Write the analog of each of the above in the propositional calculus.

2.8. Implications

"If you work, then you shall be paid" is an implication. In common usage the statement is often judged for validity only when you in fact work. When you do not work the statement may go unchallenged and no truth value is assigned (Table 2.8). This is unsatisfactory for our purposes since each acceptable statement must have exactly one truth value and implication should be included to make our discussion complete.

TABLE 2.8

A	B	If A, then B
T	T	T
T	F	F
F	T	
F	F	

TABLE 2.9

A	B	If A, then B
T	T	T
T	F	F
F	T	T
F	F	T

The actual definition of an implication, or an "If . . . , then . . ." compound statement, is given by Table 2.9. Once again there are individuals who do not always wish to accept this definition. Indeed, a statement such as "If Communism is world-wide, then peace will finally come" sounds anti-American when accepted as true. Yet it presently is a true implication because the premise, "Communism is world-wide," is false. According to Table 2.9 the implication is therefore true, regardless of the truth or falsity of the second statement, "peace will finally come."

If an implication is to serve any useful role in logic, it is necessary to define it as in Table 2.9. On one hand, if no truth value is assigned when the statement A is false, as in Table 2.8, the compound statement "If A, then B" is not an acceptable statement for logic. On the other hand, there are only three other possible ways to assign the truth values in the open positions in the last two rows of Table 2.8 and they are listed in Table 2.10 under the columns headed by C, D, and E.

TABLE 2.10

A	B	C	D	E	$A \equiv B$	A and B	If A, then B
T	T	T	T	T	T	T	T
T	F	F	F	F	F	F	F
F	T	T	F	F	F	F	T
F	F	F	T	F	T	F	T

Notice that the first of these assignments, C, is just equivalent to B; hence, as a definition of implication this would be a redundancy. The next column, D, is identical with the definition of $A \equiv B$; hence, this selection would make implications the same as logical equivalence. (Sometimes, yet not always, common usages of an implication conveys this meaning. For example, "if you flip the switch, the light goes on.") The column E is equivalent to our definition of "A and B." In all three cases, we already have introduced logical forms that cover the situation. By selecting one of these as a definition we are not adding any useful tools to our methods of constructing compound statements. Only the last column in Table 2.10 is a reasonable definition of implication.

In working with implications it is better not to attempt to memorize a table for a definition. Indeed, Table 2.9 states that an implication is true in all cases *except* when the antecedent, A, is true while the consequence, B, is false. This is all that one need remember. You can fill in a table involving implication if you search for the case in which the implication is false and place a true value into the other positions in the column.

It is perhaps interesting to notice that only the last column in Table 2.10, which is the definition of "If A, then B," is logically equivalent to $(\sim A)$ or B. Indeed, Table 2.11 established this equivalence. This means that there is already an equivalent form to an implication and, hence, the introduction of implications is unnecessary. Whereas this has validity, it does not consider the convenience of having a statement to replace the clumsy "$(\sim A)$ or B" statement. It is much easier to state and understand "If she is human, then she will react to my approach" than the logically equivalent "She is not human or she will react to my approach."

TABLE 2.11

A	$\sim A$	B	$(\sim A)$ or B	If A, then B
T	F	T	T	T
T	F	F	F	F
F	T	T	T	T
F	T	F	T	T

Much of mathematics is expressed in implication form. Hence, symbols have been introduced for it. We write $A \Rightarrow B$, where A and B are statements, for the implication "If A, then B." Instead of "If A, then B" the symbol $A \Rightarrow B$ is often read as "A implies B."

The converse of an implication $A \Rightarrow B$ is the implication $B \Rightarrow A$. As you may well have discovered in your previous study of mathematics, converses are not always true, even though the original implication is true. "If an integer is four times another integer, then it is even" is true, but the converse "If an integer is even, then it is four times another integer" is false since, for example, the integer six is even. Sometimes, however, an implication and its converse are both true. A brief analysis shows that the statements A and B in the implication must be logically equivalent in this case. For this purpose consider Table 2.12.

TABLE 2.12

A	B	$A \Rightarrow B$	$B \Rightarrow A$	$(A \Rightarrow B)$ and $(B \Rightarrow A)$
T	T	T	T	T
T	F	F	T	F
F	T	T	F	F
F	F	T	T	T

Now $A \Rightarrow B$ and $B \Rightarrow A$ are simultaneously true only for the first and last rows in this table, that is, both implications are only true in case both A and B are true or both are false. Being true or false together, the two statements A and B are logically equivalent.

When $A \Rightarrow B$ and $B \Rightarrow A$ are both true, mathematicians write $A \Leftrightarrow B$ and read this as "A if and only if B." It is simply another symbol for logical equivalence. Another common form for logical equivalence is "A iff B," invented by some lazy mathematician who found it shorter than "A if and only if B," and easier to write than "$A \Leftrightarrow B$."†

Although the converse of an implication need not be true when the implication is true, there is a pseudo-converse, called the contrapositive, which is logically equivalent to the implication. The contrapositive of $A \Rightarrow B$ is defined as $\sim B \Rightarrow \sim A$. "If he is a hippie, then he is against war" written in contrapositive form as "If he is not against war, then he is not a hippie" and this is a second way to convey the same statement. The equivalence of an implication and its contrapositive is verified in truth Table 2.13.

TABLE 2.13

A	B	$A \Rightarrow B$	$\sim B$	$\sim A$	$\sim B \Rightarrow \sim A$
T	T	T	F	F	T
T	F	F	T	F	F
F	T	T	F	T	T
F	F	T	T	T	T

↑——EQUIVALENT——↑

In order to negate an implication $A \Rightarrow B$, it is perhaps best to first write the implication in the equivalent form $(\sim A)$ or B. From our rules for the negation of an "or" statement and from our result on double negation we have

$$\sim[(\sim A) \text{ or } B] \equiv \sim(\sim A) \textit{ and } (\sim B) \equiv A \text{ and } (\sim B)$$

†There is a tale, perhaps a tall one, that ascribes the beginning of our minus sign " – " also to lazy mathematicians. Before the symbol was invented, "2 minus 1" was used for 2 – 1. As time progressed, it became common to abbreviate this as 2 m 1. But mathematicians, with contempt for trivial operation, accelerated the speed in which the m was written with the evident result: m ⌒ ⌒ ⌒ ~ –. Plus and times symbols could also have such a strange origin.

A truth table could help convince you that "A and $(\sim B)$" is the proper negation of "$A \Rightarrow B$." It may also be obvious if you recall that the only case for which an implication $A \Rightarrow B$ is false is when A is true while B is false. The negation of "If Nixon is a communist, then Ford is a socialist" is, therefore, "Nixon is a communist and Ford is not a socialist."

Exercises

1. Let A be false and let B be a true proposition. Which of the following compound statements are true?
   a. A or B
   b. A and B
   c. If A, then B
   d. If $(\sim A)$, then B
   e. $(\sim A)$ or $(\sim B)$
   f. $(\sim A)$ and B
   g. If B, then A
   h. $A \equiv B$
   i. If $(\sim A)$ then $(\sim B)$
2. Assign a truth value for each of the following implications:
   a. If the student–faculty ratio is 2 to 1, then all classes are small.
   b. If an individual is a professor, then that person is a democrat.
   c. If the student government is run by students, then it cannot accomplish anything.
   d. If all our buildings are domes, then this campus is beautiful.
   e. If a man is not a conservative, then he is a liberal.
   f. If the world is flat, then the world is round.
   g. If you vote, then you are a good citizen.
   h. If life is a dream, then everyone is asleep.
   i. If all people are alike, then there are no problems.
3. Write the converse and the contrapositive of each of the implications in Exercise 2.
4. Negate each implication in Exercise 2.
5. Negate each of the following implications:
   a. If all men are equal, then some men are more equal than others.
   b. If you obtain an A in a course, then you know as much as the teacher.
   c. If all students become students, then grades should be abolished.
   d. If grades are the motivation for study, then nothing positive comes from an education.
   e. If you smoke Salems, then you will live and love in a forest by a stream.
   f. If you try it, then you will like it.
   g. If you want to know a person, share an inheritance with him. (Benjamin Franklin)

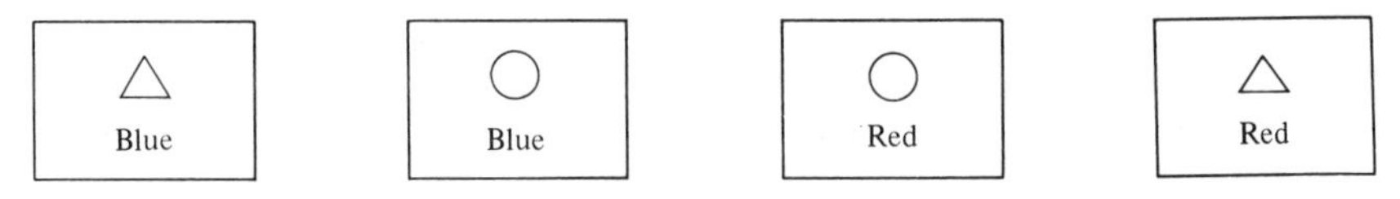

Figure 2.11

6. There are four cards in Figure 2.11. Each side of every card contains a triangle or a circle colored red or blue. What is the smallest number of cards that you must turn over in order to *prove* that there is a red triangle on the opposite side of each card containing a blue circle.

7. A tautology is a statement that is always true. For example, "$A \equiv \sim (\sim A)$" is true when A is true as well as when A is false. Which of the following statements are tautologies?
   a. $(A \text{ or } B) \Rightarrow A$
   b. $(A \Rightarrow \sim A) \equiv A$
   c. $A \Rightarrow (A \text{ and } B)$
   d. $(A \Rightarrow B) \equiv \sim (\sim A) \text{ or } B$
   e. $A \text{ or } (\sim A)$

8. a. Suppose $A \Rightarrow B$ and $\sim A \Rightarrow B$ are true. What can be stated about B?
   b. Suppose $A \Rightarrow B$, $B \Rightarrow C$ and $C \Rightarrow A$. What can be stated about A and C?

9. Find a set theory analog of an implication.

10. What is the set theory analog of a tautology?

11. $A \Rightarrow (A \text{ or } B)$ is a tautology. By means of set theory, use the results in problems 9 and 10 to indicate that this is the case.

12. Two tribes inhabit an island. The members of one tribe always tell the truth whereas the members of the other tribe always lie. A traveler on the island comes to a fork in the road and has to ask a nearby native which branch he should take to reach a certain village. Although the traveler does not know from which tribe this native comes, he manages to formulate a single question from which the reply assures him of the correct road to take. What was his question?

13. Senator Sam Ervin, Chairman of the Senate Select Committee to investigate the Watergate caper, was quoted as saying, "If these tapes establish that Dean is a liar, the committee ought to have them. If they establish the converse, that he was not a liar, we ought to have them so we can make a speedy determination on this point." (*UPI*, July 30, 1973) Has Senator Ervin correctly stated the converse? Do the two implications, connected by "and" form a tautology?

2.9 Set Theory Proofs

We have drawn an analogy between logic and set theory. In set theory, however, appeal was made to Venn diagrams which, as previously mentioned, are useful to envision results but do not really prove these results. It is not exciting logic to establish these formulas and, consequently, little stress will be placed on the basic arguments in this section. What is hoped is that you realize that set theory results can be proved.

EXAMPLE 1. Let A and B be sets. $A \subset B \Rightarrow B' \subset A'$.

Proof: The hypothesis $A \subset B$ means each element $x \in A$ is also in B. Now let x be *any* element in B'. Then x is not in B and, hence, x is not in A since $A \subset B$. x not being in A, however, means $x \in A'$. Thus, all elements $x \in B'$ are also in A' or $B' \subset A'$.

EXAMPLE 2. $(A \cap B)' = A' \cup B'$.

Proof: Let $x \in (A \cap B)'$. Then $x \notin (A \cap B)$ so x is not in *both* A and B. Hence, $x \notin A$ or $x \notin B$. If $x \notin A$, then $x \in A'$ and, hence, $x \in (A' \cup B')$. On the other hand, if $x \notin B$, then $x \in B'$ and, hence, $x \in (A' \cup B')$. In either case, therefore, $x \in (A' \cup B')$ and we have shown that all elements of $(A \cap B)'$ are also in $A' \cup B'$, that is, $(A \cap B)' \subset (A' \cup B')$.

We must next show there is nothing in $A' \cup B'$ that is not already in $(A \cap B)'$ in order to prove the sets equal. Now $x \in (A' \cup B')$ implies $x \in A'$ or $x \in B'$. If $x \in A'$, then $x \notin A$ and if $x \in B'$, then $x \notin B$. Thus either $x \notin A$ or $x \notin B$ so $x \notin (A \cap B)$ since x would have to be in both A and B to be a number of the intersection. It follows that $x \in (A \cap B)'$. Each element in $A' \cup B'$ is also in $(A \cap B)'$. In conjunction with the fact that $(A \cap B)' \subset (A' \cap B')$, this proves $(A \cap B)' = A' \cup B'$, the conclusion desired.

These proofs are long and tedious, but they flow in a rather natural way. Each step is a result derived from the previous step, usually by means of a definition. By knowing the precise meaning of each operation, you should be able to create your own argument. Perhaps an accompanying Venn diagram can assist as a guide for the formal reasoning.

Exercises

1. Suppose $A \subset B$ and $A' \subset B'$. Prove that $A = B$.
2. Prove $A' \subset B \Rightarrow B' \subset A$.
3. Establish each of the following identities:
   a. $(A \cup B)' = A' \cap B'$
   b. $[A \cap B) \cap A]' = A' \cup B'$
   c. $(A \cup B') \cup A' = I$
   d. $[(A \cap B) \cap C]' = (A' \cup B') \cup C'$
   e. $(A')' = A$
4. Prove $(A \subset B \text{ and } B \subset C) \Rightarrow A \subset C$

2.10 Applications of Boolean Algebra

An important and somewhat surprising application of Boolean algebra is in the analysis and design of electrical switching circuits which, among other things, plays an important role in the development of computers. An applied mathematician, Claude Shannon, showed that by studying circuits algebraically the circuit designers could remove unnecessary and costly relays with a minimum of effort. Removal of one switch may not seem at first to be important, but remember that the cost accumulates when tens of thousands of circuits of the same type are produced by a company. Of course, few, if any, of you will ever be involved with circuit or computer design in your lifetime. Nonetheless, there is some general knowledge in all of this which, if extended, would persuade individuals to change their perspective concerning computers. The computer would no longer be envisioned as a "magic box" in an air-conditioned wing of a building.

Suppose we have a source of electrical energy connecting two terminals (ends of a wire), T_1 and T_2. It is also assumed T_1 and T_2 are connected by a conducting wire through which electrical current flows. If we cut the wire between T_1 and T_2, current will no longer flow. Let us insert a switch A, with an "on" and an "off" position, at the point where the wire was spliced as in Figure 2.12. When the switch is on, current flows. Otherwise, it does not flow.

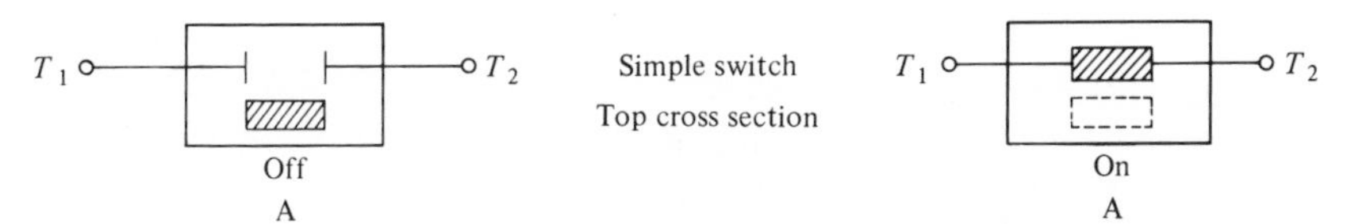

Figure 2.12

We denote a switch by a letter such as A, B, or C and designate it by a dot on the line (wire) in a diagram. Since we are concentrating on circuits with switches, we will label the circuit as well with the letter we have associated with the switch in the circuit. Two circuits A and B (switches A and B) can be used to form a new circuit $A \cap B$ by connecting the end of one wire in the circuit A to the end of a wire in the circuit B. When this is accomplished the switches are said to be in *series* (Figure 2.13(a)).

Another way to form a new circuit $A \cup B$ is to connect the beginning terminals of A and B together as well as to connect the ending terminals together (see Figure 2.13(b)). This produces a circuit where the switches are said to be *parallel.* Notice that electrical current flows in a series circuit $A \cap B$ iff A and B are *on* whereas it flows in the parallel circuit $A \cup B$ iff A or B is *on.* We could build, therefore, an "on-off" table for these circuits and they would be analogous to the truth tables for "and" and for "or." (See Table 2.14.)

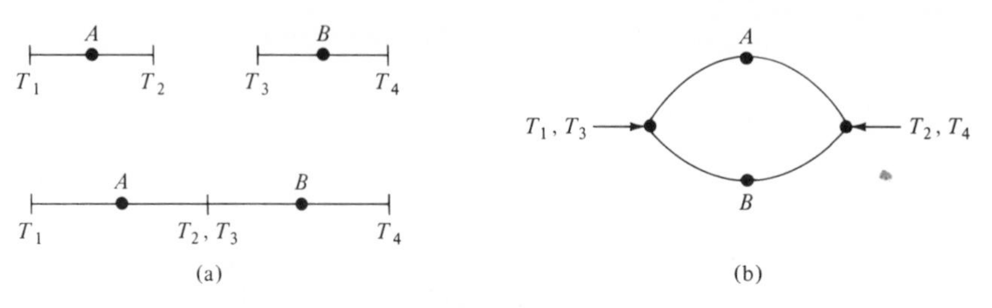

Figure 2.13

The last two columns in Table 2.14(a) correspond to current flowing (on) or not flowing (off) in the compound circuits. The entries in these columns are the same as those in the last two columns of Table 2.14(b) provided "on" is replaced by "T" and "off" by "F."

TABLE 2.14

A	B	$A \cap B$	$A \cup B$	A	B	A and B	A or B
on	on	on	on	T	T	T	T
on	off	off	on	T	F	F	T
off	on	off	on	F	T	F	T
off	off	off	off	F	F	F	F
		(a)				(b)	

Two circuits P and Q are *equivalent*, written $P = Q$, whenever they are either both on or both off for each of the possible positions of the switches in their circuits. This is the same as saying they must have the same truth table values as in the equivalence of statements in logic. For example, the two circuits in Figure 2.14 are equivalent:

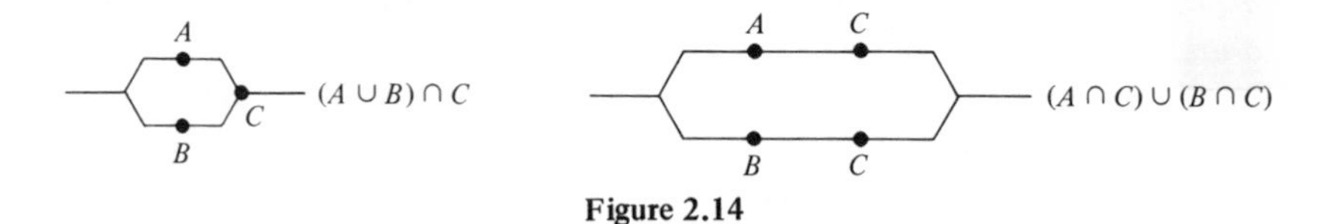

Figure 2.14

A truth table or the Boolean algebra identity $(A \cup B) \cap C = (A \cap C) \cup (B \cap C)$ can be used to verify this. It is, however, apparent if one considers the flow of electricity directly from the figures.

In Figure 2.14 there were two switches labeled C in one of the circuits. Switches need not always perform independently of one another. We can couple two or more switches together so that they open and close simultaneously. Coupled switches enable us to have essentially the same switch in various locations since coupled switches are either both on or both off. (See Figure 2.15.)

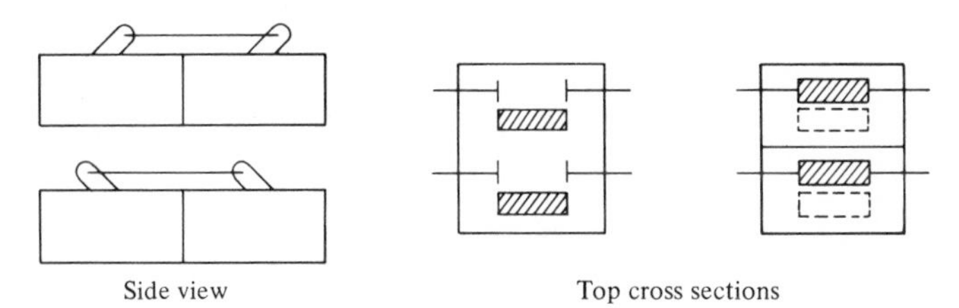

Figure 2.15

We can symbolize any circuit using set theory. For example, consider the circuit in Figure 2.16 along with a set theory description of it.

Additional switches can easily be adjuncted in series or in parallel with, of course, a corresponding change in the set theory description.

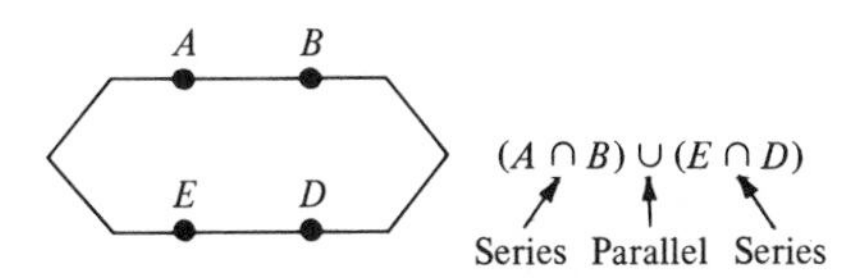

Figure 2.16

In logic we have negation and in set theory we have complementation. To introduce an analogy in circuit theory, we mention the dual or three-way switch. Each position in a dual switch permits the flow of electrical current along a particular circuit. For example, when you flip the switch from position A in Figure 2.17, you break the connection of T_1 and T_2 and replace it with a connec-

Figure 2.17

tion between T_3 and T_4. Two flips of the switch A returns you to the original state, that is, $(A')' = A$.

Dual switches can be introduced into circuits. When A is on, A' is off and vice versa.

If you have a stairway in your house, it is likely there is a dual switch on the top and on the bottom of the stairs. The circuit for this stairway is represented symbolically in Figure 2.18. The various positions for the dual switches, A and B are included in Table 2.15.

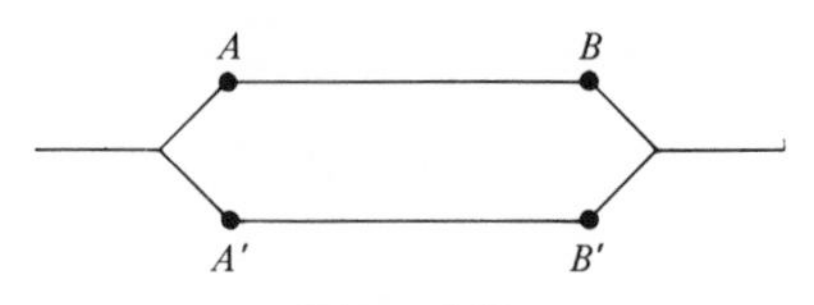

Figure 2.18

When the stairway light is off A and B have one of the two middle positions in the rows of Table 2.15. Change either A or B and we will produce either the top row or bottom row in the table with the end result that the light goes on. Another change of either A or B produces the middle rows again and off goes the light.

TABLE 2.15

A	B	A'	B'	$A \cap B$	$A' \cap B'$	$(A \cap B) \cup (A' \cap B')$
T	T	F	F	T	F	T
T	F	F	T	F	F	F
F	T	T	F	F	F	F
F	F	T	T	F	T	T

What is done in circuit design is to attempt to use as few switches and wires as is possible to accomplish a certain circuit. For example, consider the two circuits in Figure 2.19.

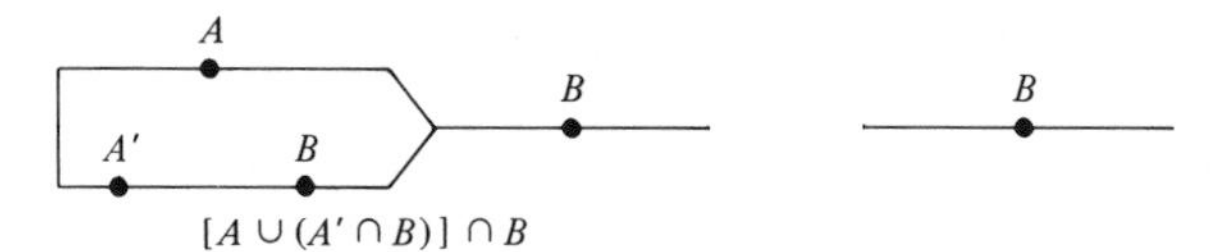

Figure 2.19

Using set theory, we can prove the equivalence of these two circuits. For,

$$[A \cup (A' \cap B)] \cap B = [(A \cup A') \cap (A \cup B)] \cap B = [I \cap (A \cup B] \cap B$$
$$= (A \cup B) \cap B = B$$

(A Venn diagram might be helpful here.) Of course, it is not difficult to detect the equivalence of these simple circuits directly from the diagrams in Figure 2.19.

EXAMPLE. A three man committee wishes to employ an electric circuit to record a secret simple majority vote. Each member pushes a button for a "yea" vote and otherwise does not push it. A signal light is to go on if a majority of the committee members vote yes. Design the circuit.

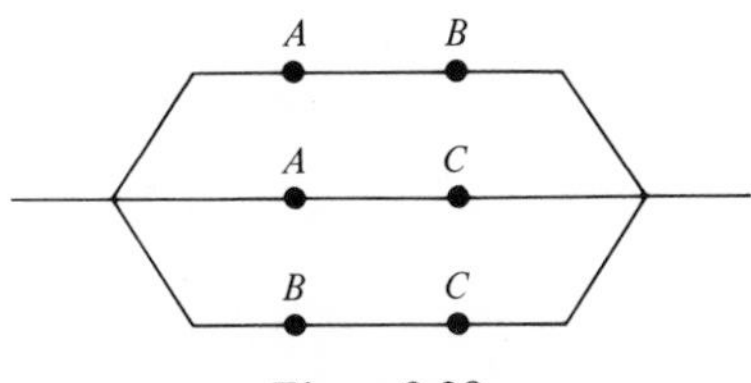

Figure 2.20

Solution: Let the switches for the members be A, B, and C, respectively. We want the circuit for $(A \cap B) \cup (A \cap C) \cup (B \cap C)$, that is, for ($A$ and B) or (A and C) or (B and C) since these are the possibilities for a majority of "yea" votes. A circuit that accomplishes the task is diagrammed in Figure 2.20. However, this circuit uses more switches than necessary. Can you redesign the circuit to use fewer switches? (Exercise 4.)

Exercises

1. Show that the circuits in Figure 2.21 are equivalent.

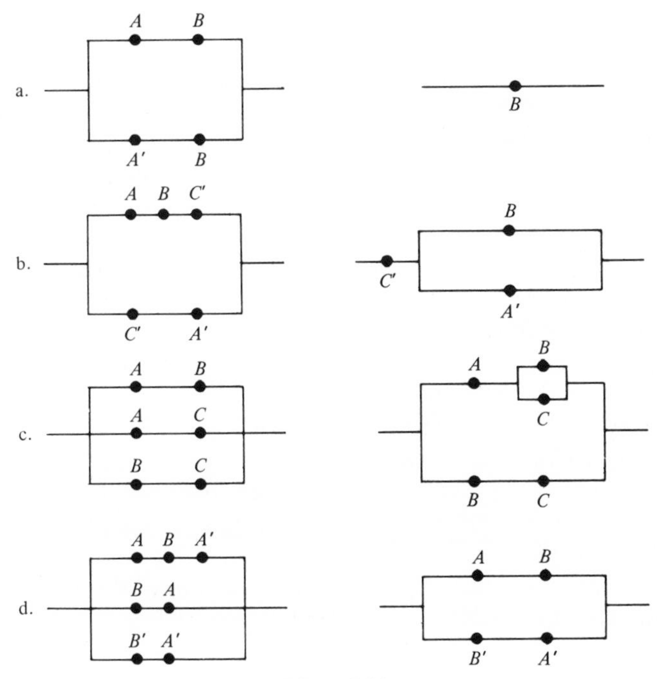

Figure 2.21

2. Write the set theory analogs of each of the circuits in exercise 1 and use Venn Diagrams to justify the stated equivalence.

3. Simplify each of the circuits in Figure 2.22.

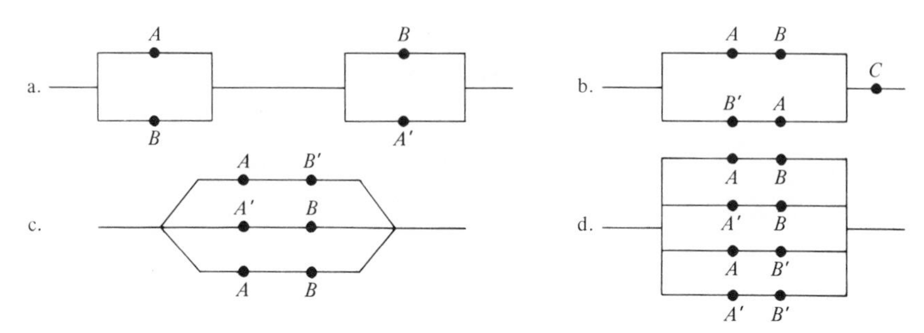

Figure 2.22

4. Can you find a circuit equivalent to the one in Figure 2.20 that has fewer than five switches, where coupled or dual switches count as two?

Supplementary Problems

1. The English nobleman, Sir Thomas Shout, was found shot and killed on his estate. After considerable efforts, Inspector Shylock Houses of Scotland Yard discovers that Lady Shout was having an affair with cousin Harold while Sir Thomas Shout had been deeply involved with Lady Chatterly. However, Sir Thomas had refused to divorce his wife. Furthermore, the maid and butler, being loyal servants of Sir Thomas for many years, were favorably included in his will. Inspector Houses interviewed all these suspects together one afternoon. If each person made the following three statements, two of which were true and the other false, whodunit?

Cousin Harold: I certainly did not kill Sir Thomas. The Butler did it. Lady Chatterly and I are former lovers.

Butler: I did not kill Sir Thomas. I never owned a pistol. I did not approve of Cousin Harold and Lady Shout's affair.

Maid: I am innocent. I hate Lady Shout. Lady Shout is guilty.

Lady Shout: I did not kill my husband. Lady Chatterly did it. The maid lied when she said I am guilty.

Lady Chatterly: I am innocent. Lady Shout did it when she found out about Tommie and me. I never met Cousin Harold before and was not a former lover of his.

2. Three persons, one of whom is blind, are accused of a crime. The judge in

their trial offers to free them all if anyone could succeed in the following identification. Five hats, two of which are red and the others white, are taken into a dark room with the accused and one of these hats was placed on each of these three persons. Then the accused were together taken into the light. If any of the accused could correctly state what color hat he himself was wearing, without removing it from his head, they would all be freed. On the other hand, if anyone made an incorrect statement concerning his hat color, all would be convicted. When the judge successively asked each of the two accused who could see the others if he knew the color of the hat on his head, each answered no. The blind person, however, correctly stated the color of his own hat when it was his turn to answer the judge's question. How did he know? (Adapted from a problem in I.M. Copi. *Introduction to Logic*. New York, N.Y.: Macmillan. 1961.)

3. Prove that there were at least two healthy ten-year old or older oak trees in the United States that had exactly the same number of leaves on July 15 last summer.

4. Three college professors, a mathematician, an economist, and a physicist, each have sons who are also teachers. Their surnames are Euclid, Keynes, and Newton. The physicist and Newton, Jr. each have 70 students. The economist has 55 students. Euclid, Jr. has 6 fewer students than the physicist whereas the mathematician has 66 students. The mathematician has exactly half as many publications as the man who has the nearest number of students as he has. The economist's son has 13 publications. Keynes, Jr.'s father has more hair than the physicist. Who is the mathematician?

5. Suppose an examination in this course is to be held next week on one of the regularly scheduled meetings of this class. In order to avoid last-minute cramming, the exact day of the examination is to be a secret. However, if all the students agree before class that the examination is to be given on that day, and they are correct, then the examination will be cancelled. Can the examination be given? (If the teacher waits until Friday, all would know and, hence, it could be cancelled. Since it cannot be given on Friday, then it must be given on Wednesday or Monday. But if it is not given on Monday, all would know it is to be given on Wednesday for Friday is already out of consideration. Thus, it is not possible to give the test on Wednesday since all would know the test is scheduled on this day. Monday leads to the same complication.)

6. Suppose that there are three towns, A, B, and C, on a road and that the distance between A and B equals the distance between B and C, each being 42 miles. If you drive at an average speed of 60 mph between A and B and at an average speed of 70 mph between B and C, what exactly is your average speed for the entire trip from A to C?

7. Another Lewis Carroll example is the following: No kitten that loves fish is unteachable. No kitten without a tail will play with a gorilla. Kittens with whiskers always love fish. No teachable kitten has green eyes. No kittens have

tails unless they have whiskers. Conclude that green-eyed kittens will not play with a gorilla.

8. Where in the world, besides at the North Pole, can you travel due south for 100 miles, due east for 100 miles, followed by due north for 100 miles and end up at your starting point?

9. I sold some stock for $4000 and then bought it back for $3500. Later I sold it again for $4500. How much did I gain on these transactions?

10. A vagabond is walking across a railroad bridge. When he is 40 percent of the way along the bridge, he hears a train behind him. If he can just make it to either end of the bridge and if the train is traveling at 50 mph, how fast can the vagabond run?

References

For interesting and contemporary examples of fallacies in inductive arguments, the following are excellent references:

1. H. Kahana. *Logic and Contemporary Rhetoric.* New York, N.Y.: Wadsworth. 1971.
2. G. Allport. *The Nature of Prejudice.* New York, N.Y.: Doubleday Anchor Book. 1948.
3. J. Corry. *The Los Angeles Times.* New York, N.Y.: Harper & Row. December, 1969.

The second book also contains a number of examples of fallacies in deductive reasoning. Prejudice often stems from the application of deductive logic to false general principles about certain loose groupings of people. References for mathematical logic, in addition to those already cited in the text, include:

4. H. Gross and F. Miller. *Mathematics, A Chronicle of Human Endeavor.* New York, N.Y.: Holt, Rinehart and Winston. 1971.
5. A. Douglis. *Ideas in Mathematics.* Philadelphia, Pa.: Saunders. 1970.
6. E. Nagel and J. Newmann. Godel's Proof. *Scientific American*, June, 1956.
7. W. Quine. *From a Logical Point of View.* Cambridge, Mass.: Harvard University Press. 1953.
8. K. Kemeny, J. Snell, and G. Thompson. *Introduction to Finite Mathematics.* Englewood Cliffs, N.J.: Prentice-Hall. 1957.

Chapter 3 Introducing the Queen

3.1. Number Theory

Mathematics has often been admiringly called the Queen of Mathematics, an idea that dates at least back to Gauss (1777–1855). If there is a queen within mathematics, it is certainly the subject called number theory. Number theory, which is a study of properties of the counting numbers, 1, 2, 3, and so on, has been a constant source of mathematical inquiry throughout the ages. In no other area of mathematics can one ask such simple questions that are so astoundingly difficult to resolve. There is no lack of challenge for the amateur or the professional mathematician who turns his/her attention toward number theory.

Counting numbers, or natural numbers, as they are often called, have always fascinated mankind. Many people have even assigned mystic significance to certain of these numbers. We still consider numbers such as 13 unlucky (who knows why?) and select others to be lucky numbers.

About 500 B.C. the Pythagoreans, a group of Greek philosophers and mathematicians, not only attached mystic significances to numbers but built a philosophy based upon them. They, for example, considered "one" as the essence of reason and "two" as the initial state of opinion. Complex conclusions about man and the universe were drawn from such identifications of numbers with attributes of man. Their labors, however, were not all futile; it was the Pythagoreans who first called attention to the real beauty of numbers themselves. Some 200 years later, even though no simple systematic way was available for the expression of larger numbers, many general facts were known to the Greeks and were recorded in Books seven to nine of Euclid.

Significant contributions have appeared concerning counting numbers throughout the centuries. An acceptable axiomatic approach to counting numbers, however, awaited discovery by Giuseppe Peano, an Italian mathematician, in the late nineteenth century. Mathematicians often laid their foundations after their structures were built. Indeed, the axiomatic approach to any area of mathematics except geometry is relatively new for a subject that dates back to early man.

There are, of course, number systems in mathematics in addition to the counting numbers. For example, the set of *integers*, is the collection of numbers 0, +1, −1, +2, −2, +3, −3, +4, −4, ... that you encountered in grade school. The positive integers +1, +2, +3, ... behave just like the counting or natural numbers 1, 2, 3, ... We shall make no distinction between these collections.

A logical development of numbers forces us to define each new number system in terms of a known one. Beginning with Peano's axioms for the natural numbers, we define the integers in terms of natural numbers. The next set of numbers, the rational numbers, is defined in terms of integers. Descriptively rational numbers are numbers of the form a/b where a and b are integers with $b \neq 0$. After the rational numbers comes the real numbers, which intuitively are all decimal expressions, and they have been logically developed from the rational numbers. In fact, each of the classes of numbers discussed, except the natural numbers, includes a subset that behaves like the class from which it was derived. For example, the real number 0.5 behaves like the rational number $\frac{1}{2}$ and, except for axiomatic discussions, is usually identified with $\frac{1}{2}$. The rational number 3/1 or 6/2 behaves like the integer +3. The integer +100 behaves like the natural number 100. Although we plan to study the natural numbers at this point, it is helpful, and sometimes necessary, to have these other classes of numbers on hand.

3.2. To Divide or Not To Divide

Following a somewhat historical development rather than an axiomatic approach, we initially assume you are familiar with basic operations and comparisons of counting numbers such as addition, multiplication, and ordering by size. This starting point will enable us to develop some divisibility tests for the natural numbers.

Let a and b be counting numbers. We say that *a divides b*, or b is a *multiple* of a, or a is a *factor* of b, or b is *divisible* by a, if there is a counting number x such that $b = ax$. For example, 6 divides 24 since there is a number, 4, such that $24 = 6 \cdot 4$. If a divides b, we write $a \mid b$ and if a does not divide b, we write $a \nmid b$. There are some basic properties of this concept, among which are the following: (In each case, a, b, and c are counting numbers.)

Law 1. $a \mid b \Rightarrow a \mid (bc)$

Law 2. $a \mid b$ and $a \mid c \Rightarrow a \mid (b + c)$

Law 3. $a \mid b$ and $a \mid (b + c) \Rightarrow a \mid c$.

Each of these properties can be verified from the definition. Consider, for example, property 2. The hypothesis $a \mid b$ means, according to the definition of the concept, that there is a counting number x such that $b = ax$. Similarly $a \mid c$ means there is a counting number y such that $c = ay$. (We use a different letter since the x and the y can be, but need not be, different numbers.) Thus $b + c = ax + ay = a(x + y)$, which proves $a \mid (b + c)$ since there is a counting number $(x + y)$ such that $b + c = a(x + y)$.

We have used the fact that the sum of two counting numbers is itself a counting number in the preceding argument. We have also used a property of numbers known as the distributive law, namely, $a(x + y) = ax + ay$. The Pythagoreans, who tied numbers to geometry as well as providing them with mystical significance, justified properties like the distributive law by geometric methods as indicated by the diagrams in Figure 3.1.

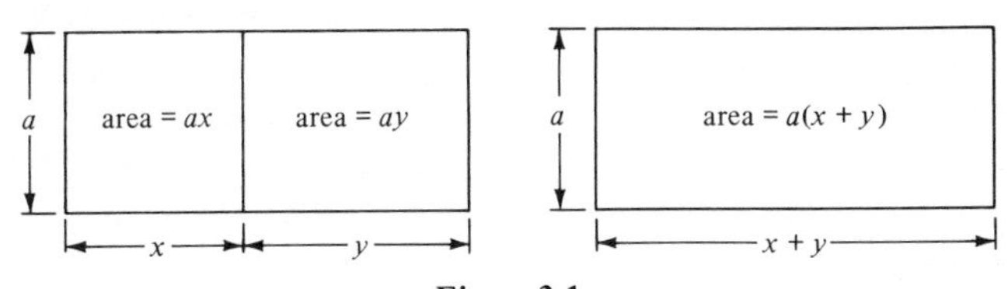

Figure 3.1

(Properties such as

$$ab = ba, \qquad (a + b) \cdot (a - b) = a^2 - b^2, \qquad (a + b)^2 = a^2 + 2ab + b^2$$

can similarly be justified geometrically to the satisfaction of at least a Pythagorean.)

A counting number a is even if $2 \mid a$ and it is otherwise odd. One of the most familiar of all divisibility tests is the determination of when a number is even.

TEST 1

A counting number is even iff the digit in the units place is even.

Let us first prove that if the digit in the last place (the units place) is even, then the number is even. We begin by noting that $10 = 2 \cdot 5$ is even and, hence, by Law 1 every multiple of 10 is even. (Indeed, if $2|10$, then $2|10c$ for all choices of c among the counting numbers.) Now a number with zero in the units position is a multiple of ten, a fact which we will assume. Suppose the number a has the even digit f in the units position. Then $2|f$ and $a - f$ has zero in the units position. Thus $a - f$ is a multiple of 10 and it follows that $2|(a - f)$. Since $2|f$ also, we conclude by Law 2 that the number 2 divides $(a - f) + f = a$.

You might object to the assumptions in the above proof. After all, the assumptions are perhaps more advanced than what is being proved. Since we do not wish to force an axiomatic approach upon you, some assumptions like these must be made. As long as you keep a record of these assumptions, so that you will not later attempt to prove them from results which are, in fact, a consequence of these hypotheses, there is no harm in accepting them without a proof. On the other hand, you may prefer to accept certain other premises. It is up to you; it is all mathematics.

It remains to prove that if the number a is even, then its last digit is even. For this we write $a = x + f$ where f is the last digit in a and, hence, x is a multiple of ten. Since $2|x$ and since $2|a$, where $a = x + f$, it follows by Law 3 that $2|f$, that is, that 2 divides the last digit of a.

TEST 2

A counting number a is divisible by 4 iff the number obtained from its last two digits, in order, is divisible by 4.

For example, the number 1,793,462,981,716 is divisible by 4 since $4|16$. The last two digits were taken in the order of their appearance in the number and were not permuted. This is the meaning of the "in order" phrase of Test 2.

We now turn to a verification of Test 2. It is assumed in this proof that a counting number with the digit 0 in the last two places is a multiple of 100.

Let f be the number obtained from the last two digits, in order, of a. Then $a = x + f$ where x is a multiple of 100. Now $4|100$, so by Law 1, 4 divides all multiples of 100; in particular, 4 divides x. Hence, if $4|f$ we conclude by Law 2 that $4|a$ since $a = x + f$. Conversely, if $4|a$ and, as we have found, $4|x$, then by Law 3, $4|f$ since $a = x + f$.

TEST 3

A counting number a is divisible by 8 iff the number obtained from its last three digits, in order, is divisible by 8.

The proof is left as an exercise.

TEST 4

A counting number a is divisible by 3 iff the sum of all its digits is divisible by 3.

For example, 21, 342, 981, 712, 014 is divisible by 3 since the sum of the digits is 45, which is 15 times 3.

You will eventually notice in using this test that if you discard multiples of three as you add the digits, the final number obtained will be a multiple of 3 iff the original number was such. In the example, for instance, we need not include the first three digits in your sum of the digits since they add up to 6, which is a multiple of 3. The next two add up to 6; the next digit is 9, itself a multiple of 3; the next two digits are 8 and 1 which total 9, and so on. You can drop any multiple of three as you form the sum of the digits for Test 4. The outcome is not affected by this conservation of effort.

For the proof of this test, we assume that if you subtract 1 from a multiple of 10, the resulting number is a multiple of nine. By way of illustration, $10 - 1 = 9$, $100 - 1 = 99 = 9 \cdot 11$, $1000 - 1 = 999 = 9 \cdot 111$. By the distributive law, it follows that if a digit t is subtracted from $10t$, $100t$, $1000t$, etc., a multiple of 9 results. Indeed, $10t - t = (10 - 1)t = 9t$, $100t - t = (100 - 1)t = 99t$, $1000t - t = (1000 - 1)t = 999t$, etc.

THEOREM. Let a be a natural number and let s be the sum of the digits in a. Then the difference $x = a - s$ is a multiple of nine.

Proof: We can write a as a sum, where each term other than the last is a digit times a multiple of ten. For example, if $a = 5843$, then we write $a = 5(1000) + 8(100) + 4(10) + 3$. Subtract from a the sum of the digits s in a in the following manner. Do not accumulate the sum of the digit; instead write s formally as a sum of digits. For instance, with $a = 5843$ we write $s = 5 + 8 + 4 + 3$ rather than as the actual sum $s = 20$. Then subtract each of the digits in the formal sum s from the corresponding digit times a multiple of 10 in a, for example,

$$\begin{aligned} a - s &= 5(1000) + 8(100) + 4(10) + 3 - (5 + 8 + 4 + 3) \\ &= 5(1000) - 5 + 8(100) - 8 + 4(10) - 4 \\ &= 5(1000 - 1) + 8(100 - 1) + 4(10 - 1) \\ &= 5 \cdot 999 + 8 \cdot 99 + 4 \cdot 9 \end{aligned}$$

The result is that each digit in a, except for the last one, is subtracted from a multiple of ten times that digit. This always produces a multiple of nine and, by Law 2, the sum of multiples of nine is also a multiple of nine. The last digit in a has been canceled by this subtraction process. Since all other terms in $a - s$ are multiples of nine, we conclude $a - s$ is itself a multiple of nine.

The theorem is quite general; if you subtract from *any* counting number the sum of its digits, a multiple of nine remains. This is precisely what is needed to justify Test 4. Indeed, let $x = a - s$ where a is a given natural number and s is the sum of its digits. We know $9 | x$ no matter what choice was made for a. It follows that $3 | x$. Now if $3 | a$ $(= x + s)$, we have $3 | s$ by Law 3. Conversely if $3 | s$, we have $3 | a$ $(= x + s)$ by Law 2. That is it—the elusive proof of Test 4 has been found!

In view of the theorems in this section it is relatively easy to justify the rule for nines.

TEST 5

A counting number is divisible by nine iff the sum of its digits is divisible by nine.

Some additional divisibility tests appear in the exercises. There exist relatively simple criteria for divisibility tests of the integers two through 12, except for seven. Perhaps seven is not a lucky number.

Exercises

1. Which of the following numbers are divisible by four and which are divisible by six?

   a. 3,969,597,891,345
   b. 51,333,351,324
   c. 81,912,331,714,572
   d. 1,268,127,991,314

2. Determine all the digits x in each of the following numbers such that the number is divisible by three.

   a. 9,37x,503,821
   b. 3,125,179,x11

   Find a choice of x that makes the numbers divisible by nine.

3. Is it true that if a counting number is divisible by three, then any number resulting from a rearrangement of the digits of the original number is also divisible by three? Why?

4. State simple divisibility tests for the integer five and the integer six. Prove your results. What is a divisibility test for the integer 12?

5. A counting number is divisible by 11 iff the sum of its digits in the odd numbered positions (count from either end) *subtracted* from the sum of the digits in the even numbered positions is itself an integer that is divisible by 11. Use this test to determine which one of the following numbers are divisible by 11. (An integer c is divisible by 11 if there is an integer x such that $11x = c$.)

   a. 12,317,913,806
   b. 123,456,789
   c. 942,865,731
   d. 687,942,531
   e. 642,738,195
   f. 715,349,286

   (An exceptional student might attempt a proof of this criteria.)

6. Pick any counting number a. Add twice a to the sum of the digits in a. The result is always divisible by three. Why?

7. Show that $a|c$ and $b|c$ does not imply $(ab)|c$ or $(a + b)|c$ in general.

8. For each of the following numbers, use divisibility tests to determine the set of divisors from the set $\{2, 3, 4, 5, 6, 8, 9, 10, 11, 12\}$.

   a. 123,456,789 c. 99,999,990 e. 13,547,531
   b. 1,001,001,001 d. 1,245,678,324 f. 1,234,512,345

9. Take any counting number and multiply it by 9. Circle one of the digits in the product and add the remaining uncircled digits. If the sum is larger than 9, add the digits in the sum and continue the latter process until the sum does not exceed 9. Subtract the final sum from 9 and the answer is the circled digit. Does this always happen? Why?

10. There are 37 ears of corn in a hollow stump. How long will it take a squirrel to remove them if he carries out *exactly* three ears per day?

11. Take any four digit number in which not all the digits are the same. Write the number obtained from these digits in descending order (largest digit first) and then construct the number obtained from these digits written in ascending order. (For example, if 3912 is your number, you write 9321 and 1239, respectively.) Now subtract the smaller from the larger of the constructed two numbers and then repeat the process on the answer. *No matter what number you initially choose*, the repetition of this process will eventually produce the number 6174. Try it. [D. R. Kaprekar. An interesting property of the number 6174. *Scripta Math.* 21 (1955) 304. See also C. W. Trigg. Kaprekar's constant. *Amer. Math. Monthly.* 78(1971) 197–198.]

12. In addition to the laws of this section, there is another useful rule: Let a, b, and c be natural numbers and let $b|c$. If $c|a$, then $b|a$. (a) Prove this law. (b) Show the converse is false. (c) The contrapositive states something about multiples of b. Give a verbal form of the contrapositive of the above implication.

3.3. What You Always Wanted To Know about Squares and Were Afraid To Ask

Every counting number has a square although not every counting number is the square of one of its kind. In this section we observe a few properties of the squares of natural numbers for no good reason. Perhaps you will find them interesting anyway.

Which of the digits 0, 1, 2, 3, 4, 5, 6, 7, 8, or 9 cannot be the right-most digit in the square of a counting number? First of all, when two counting numbers are multiplied, the right-most digit of the product is determined solely from the right-most digits in the original numbers. Hence, the last digit in the square of a counting number a must be determined from the last digit in a itself. Chart 3.1 shows what the last digit in the square can be for various last digits in the number a itself. The last column in the chart, therefore, shows the last digit in a square of a counting number must be 0, 1, 4, 5, 6, or 9. There are no other possibilities. We can also conclude from this chart that the square of any even integer is even, whereas the square of an odd counting number is odd. I venture that this does not surprise you.

CHART 3.1

Product	Last digit
$0 \cdot 0 = 0$	0
$1 \cdot 1 = 1$	1
$2 \cdot 2 = 4$	4
$3 \cdot 3 = 9$	9
$4 \cdot 4 = 16$	6
$5 \cdot 5 = 25$	5
$6 \cdot 6 = 36$	6
$7 \cdot 7 = 49$	9
$8 \cdot 8 = 64$	4
$9 \cdot 9 = 81$	1

Not too many years ago advertisements appeared for booklets which, presumably, taught the art of quick mental computation. Although the availability of inexpensive mini-calculators has probably supressed interest in such mental activities, there are a few "tricks" that are easy and helpful when your electronic machine is in the other room.

One is the method of computing squares via the algebraic identity $x^2 - a^2 = (x - a)(x + a)$. Consider, by way of illustration, the square of 75. Since $75 + 5 = 80$ and $75 - 5 = 70$, we need only multiply 70 by 80 to obtain 5600 and add $5^2 = 25$. Indeed, $x^2 = (x - a)(x + a) + a^2$ from the identity, and $x = 75$, $a = 5$ in this case. The choice of $a = 5$ was based on the fact that it gave us simple numbers to multiply. To square 31, select $a = 1$. We have from the identity that

$$31^2 = 30 \cdot 32 + 1 = 960 + 1 = 961$$

Selecting the a appropriately, you can square a number in your head after a little practice. Just think of your saving of time, paper, energy (if you don't use a mini-calculator) once you master this trick.

If you square an odd counting number a, the answer is always one more than a multiple of 8. To prove this fact, first note $a^2 - 1 = (a - 1)(a + 1)$ by the previous algebraic identity. Now a is odd so both $a - 1$ and $a + 1$ are even; hence their product is divisible by 4. More can be said, however, by observing that $a - 1$ and $a + 1$ are successive even integers. Now every second even integer is divisible by 4 $(2, \underline{4}, 6, \underline{8}, 10, \underline{12}, 14, \underline{16}, 18, \underline{20} \ldots\ldots)$ and, hence, one of these two successive even integers is divisible by 4. The other, being even, is divisible by 2. Thus $4 \cdot 2 = 8$ divides their product.

This observation can be used as a check for the squaring operations on integers.

$$87^2 = (87 - 3) \cdot (87 + 3) + 9 = 84 \cdot 90 + 9 = 7569$$

If we had obtained an incorrect answer, say 7567 instead, then $7567 - 1 = 7566$ and $8 | 7566$ so the error would be noticed if the calculation were checked in this manner. Of course, the method of checking is not foolproof since the incorrect answer 7649 would pass unnoticed.

It is curious, and true, that the sum of the first n odd counting numbers is always n^2. For example, the sum of the first 5 odd numbers is $1 + 3 + 5 + 7 + 9 = 25$, and the sum of the first 6 odd numbers is $1 + 3 + 5 + 7 + 9 + 11 = 36$. To the Pythagoreans this might have seemed geometrically evident from Figure 3.2.

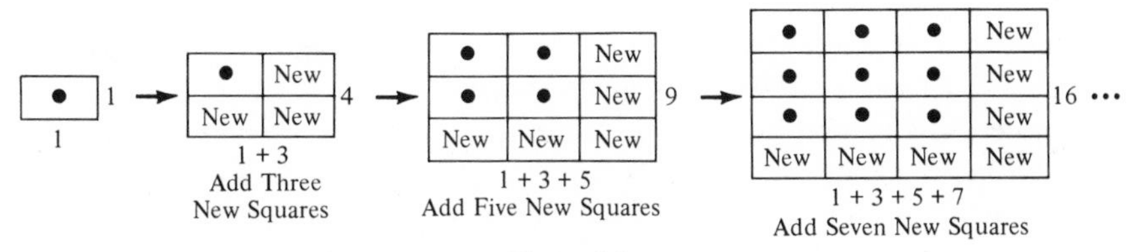

Figure 3.2

Eventually, an argument that is acceptable by modern standards will be given for this fact. One can be obtained that is related to the figure from the algebraic identity $(n + 1)^2 - n^2 = 2n + 1$. We leave this proof as a challenge to the reader with sufficient algebraic background.

Exercises

1. Show that the last digit in the square of a multiple of 4 must be 0, 4, or 6.
2. If n is an odd counting number, what possible last digit can n^2 have?
3. Compute the square of each of the following numbers using the formula $x^2 = (x - a) \cdot (x + a) + a^2$.

   a. 45 b. 125 c. 99 d. 37 e. 73 f. 81 g. 147

   Check you answer by showing each square less one is divisible by eight.
4. Show that the product of any three consecutive counting numbers bigger than one is divisible by six.
5. What is the sum of all the odd integers from 1 to 999?

3.4 Of Prime Concern

Aristotle (384–322 B.C.) as well as Euclid distinguished between numbers that have no divisors except one and the number itself and those that do not enjoy this property. A counting number larger than one is called a *prime* if it has no factors except one and itself. Thus, 2, 3, 5, 7, 11, 13, and 17 are prime numbers whereas 4, 6, 8, 9, 10, 12, 14, 15, and 16 are not. The counting numbers that are larger than one and are not prime are called *composite* numbers. Composite numbers, therefore, have a divisor other than one and the number itself.

It is frequently a difficult task to determine if a large number is prime or composite, although it is known that there are infinitely many of each type. Presently the largest known prime number, verified with the aid of a computer by a group of mathematicians at the University of Illinois, is $2^{11213} - 1$ which, when expressed in the standard notation of integers, contains more than 1000 digits. For composite numbers there is no analogous problem. Indeed, if a is the largest known composite number, then $2a$ is a larger number and it also is composite. With primes, however, there is no known general way to use a prime to construct a larger number that is assuredly prime.

Suppose c is a composite number. Then we can write $c = a \cdot b$ where a and b are counting numbers, neither of which is one. Now either $a = b$, and c is a perfect square, or one of these two numbers is smaller than the other. Let the smaller be b when they are unequal. Then $c = a \cdot b > b \cdot b = b^2$, since $a > b$, and we have $b < \sqrt{c}$.† If $a = b$ then $c = ab = b^2$, so $b = \sqrt{c}$. In either case, therefore, $b \leqslant \sqrt{c}$. This observation provides us with a test for composite numbers.

TEST 1

If c is a composite number, then it is divisible by some counting number $b \leqslant \sqrt{c}$.

The test limits the number of cases that need be considered to determine if a number is composite. From the definition alone, we would be required to check each counting number $b < c$ to determine if it is a divisor of c. The test confines this checking to those numbers $\leqslant \sqrt{c}$, a smaller collection.

When written in its contrapositive form, Test 1 becomes a test for prime numbers, namely.

TEST 1′

If c is a counting number that is not divisible by any counting number $b \leqslant \sqrt{c}$, then c is a prime.

†The square root of the natural number c is defined as the *real number* x such that $x > 0$ and $x^2 = c$. That any positive *real* number has a real square root is a fact that can be proved in mathematics.

In checking to determine if a number is prime one encounters other considerations that further decrease the number of cases which need be tested. For example, if two does not divide a number c, then no multiple of 2 (4, 6, 8, 10, ...) can divide the number. (See Exercise 12 of Section 3.2.) Once three is shown not to be a divisor of a number, it is unnecessary to try 6, 9, 12, 15, 18, 21, ... By repeated application of this observation, it is apparent that only prime numbers $\leqslant \sqrt{c}$ should be tested to determine if c is a prime. This method of filtering out only possible prime divisors of a number is called the sieve of Eratosthenes and dates back at least to 200 B.C.

Consider, for example, the number 101 and prove it is prime. Since $10 < \sqrt{101} < 11$, we are required by Test 1′ and the sieve of Eratosthenes to check only those prime numbers not exceeding 10 as possible divisors of 101. By trial, 2, 3, 5, and 7 are not divisors of 101 and, hence, 101 is a prime.

This method seems powerful, but the number of cases requiring checking becomes quite large as the size of the number, suspected to be a prime, increases. There are 25 primes less than 100, 168 primes less than 1000, and 1229 primes less than 10,000. Checking to determine if a number near 100,000,000 is prime would be quite a task by this method.

One immediate, yet important, observation about composite numbers that is derived from the sieve of Eratosthenes is that *each is divisible by a prime.* Indeed, if no prime $\leqslant \sqrt{c}$ divides the counting number c, then c is itself a prime.

A general formula for the number of primes less than an integer N has never been discovered. For particular N, of course, one could count the number of primes less than N if a list of primes is first determined. Owing to the magnitude of the task of listing the primes less than N for N large, one cannot expect much from this counting method. The German mathematician Gauss and the French mathematician Legendre (1752–1833) independently discovered around 1800 an approximation for the number of primes less than N. Near the turn of the last century a French mathematician, Hadamard, and a Belgian mathematician, de la Vallee Poussin, independently proved that the Gauss-Legendre formula is in a sense the best approximation that can be obtained.

This fact is known as the *prime number theorem* and it has been classified as one of the highlights of nineteenth century mathematics.

Exercises

1. Which of the following are prime numbers?

   a. 127 c. 137 e. 139 g. 159
   b. 313 d. 437 f. 499 h. 3021

2. Use the sieve of Eratosthenes to find all primes less than 100. What is the longest string of consecutive composite numbers between one and 101?

3. The last digit of a prime larger than five must be 1, 3, 7, or 9. (Why?) Mathew Madix, amateur mathematician, conjectures from the list of primes in Exercise 2 that the last digit of two consecutive primes is never the same digit. Prove his conjecture false.
4. A famous unsolved problem known as Goldbach's conjecture, states that every even integer greater than 2 can be expressed as the sum of two primes. Show the conjecture is correct for all even counting numbers greater than 2 and less than 20. (Actually it has been shown to be true for all even counting numbers greater than 2 and less than 33 million.)
5. Assume Goldbach's conjecture in Exercise 4 is true. Show then that every even counting number greater than four can be expressed as the sum of three primes.
6. The sum of two primes greater than two is always an even number. (Why?) Prove that the sum of two consecutive primes, however, is not two times a prime.
7. The Russian mathematician Chebyshev (1821–1894) proved that there is at least one prime between each counting number greater than one and two times that number. Check this result for counting numbers n where $5 < n \leqslant 20$.
8. Become famous. Find two consecutive counting numbers which have no prime number between their squares. (It is unknown if this is possible.)

3.5. How Many Primes Are There?

We spoke previously about the largest known prime and the number of primes less than a given integer. Could there be a largest prime N? By a simple argument, Euclid was the first to prove there is no largest prime and, hence, that there are infinitely many prime numbers.

Before formally attempting to justify Euclid's result, it might be well to observe another fact about primes. We introduce at this point the notation $n!$, read "n factoral," for the product of all the counting numbers that do not exceed the positive integer n. Thus, $3! = 1 \cdot 2 \cdot 3 = 6$ and $6! = 1 \cdot 2 \cdot 3 \cdot 4 \cdot 5 \cdot 6 = 720$. Now the factorials for $n > 2$ are certainly composite numbers, but we cannot be sure that $n! + 1$ is composite. Indeed, $3! + 1 = 6 + 1 = 7$, which is a prime.

Consider $n! + 1$ for a relatively large n; for example, take $n = 22$. Then $22! + 1$ is a very large number x. Since $22! = 1 \cdot 2 \cdot 3 \cdot 4 \cdot \ldots \cdot 20 \cdot 21 \cdot 22$, the product of the integers from 1 through 22, it is clear that $2|22!$, $3|22!$, $4|22!$, and so on up to 22 itself. Thus, $22!$ is divisible by each counting number not exceeding 22. Now can $2|x\ (= 22! + 1)$? If we consider Law 3 of Section 3.2 with

$a = 22!$, $b = 1$, we conclude that *if* $2|22!$ *and* $2|(22! + 1)$, *then* $2|1$. But the conclusion of this implication is absurd; 2 is not a divisor of 1. Hence, something is false in the antecedent. Since we already have proved $2|22!$, it must be true that $2 \nmid (22! + 1)$. In the very same manner we can show $3 \nmid (22! + 1)$, $4 \nmid (22! + 1)$, and so on up to $22 \nmid (22! + 1)$. Thus, the counting number $x = 22! + 1$ is not divisible by any natural number larger than one and not exceeding 22.

There was nothing special about taking $n = 22$ in this problem; we could indeed take n to be any integer $\geqslant 2$. The conclusion would then be summarized by the following:

REMARK 1. For a counting number $n \geqslant 2$, the number $x = n! + 1$ is *not* divisible by any counting number bigger than one and not exceeding n.

This seemingly useless remark enables us to supply a simple proof of Euclid's theorem.

THEOREM. (Euclid) There are infinitely many primes.

Proof: Suppose that this is not true and, hence, that there is a largest prime N. Consider the natural number $N! + 1$. It is larger than N and, by Remark 1, it is *not* divisible by any counting number that is larger than one and that does not exceed N. But either $N! + 1$ is a prime or $N! + 1$ is a composite. The first case is impossible since N was the largest prime and $N! + 1$ is larger than N. The second case is also impossible since N is the largest prime and, in fact, all counting numbers including the primes $\leqslant N$ do *not* divide $N! + 1$. Yet each composite is divisible by a prime. Something is wrong. It must be the hypothesis that N is the largest prime. Hence, there is no largest prime.

Consecutive primes starting with three can be as close as two units apart such as 3 and 5, 11 and 13, and 17 and 19. Such primes are called *twin primes* when they differ by two. It is suspected, but it has never been proved, that there are infinitely many twin primes. On the other hand, consecutive primes can be widely separated. For example, $100! + 2$, $100! + 3$, $100! + 4 \ldots$, $100! + 99$, $100! + 100$ is a set of 99 *consecutive counting numbers and none are prime*. Indeed, $2|(100! + 2)$ since $2|100!$ and $2|2$; $3|(100! + 3)$ since $3|100!$ and $3|3$, etc. In this way, we show that 2 divides the first member, 3 divides the second, 4 divides the third, and so on until we obtain that 100 divides the last number in the chain. What this means is that the 99 numbers in our list are consecutive and composites. There must be at least 99 composite numbers between the largest prime $< 100! + 2$ and the smallest prime $> 100! + 100$. The 100 in this argument could be replaced by any counting number so we find that there can be very large differences between consecutive primes. In fact, you can show by this method that there are consecutive primes with differences larger than any preassigned number.

Exercises

1. Show there are infinitely many composite numbers.
2. Without doing any calculations, tell why the numbers $9! + 1$, $10! + 2$, $7! + 4$, $11! + 3$, $12! + 4$ are *not* divisible by 5, 6, 7, 8, or 9.
3. Prove each of the following is a composite number.
   a. $9! + 3$ b. $10! + 11!$ c. 73,168,143
   b. $9! + 91$ e. $10^{12} - 1$ f. $5^{25} + 1$
4. List all twin primes less than 100.
5. Is there any primes $p > 3$ such that p, $p + 2$, and $p + 4$ are all prime numbers? If the answer is yes, give one such prime. If the answer is no, show why this is the case. (Such might be called prime triplets!)
6. The following result was published in 1770 by Edward Waring and attributed to an amateur mathematician, John Wilson: If n is a prime, then n divides $(n - 1)! + 1$. Show that this is the case for $n = 5, 7, 9$ and 11.
7. Prove the converse to Wilson's theorem in Exercise 6, namely, "If $(n - 1)! + 1$ is divisible by n, then n is a prime." (Hint: Assume n is a composite and reach a contradiction.)
8. Show that there exist consecutive prime numbers p and q such that $p - q > 1000$.

3.6 Prime Generating Formulas

Since the discovery of prime numbers, mathematicians certainly have been intrigued with the possibility of creating a formula which produces primes. Euler (1707–1783) pointed out that the formula $n^2 - n + 41$ is always a prime for n a counting number not exceeding 41. When $n = 41$, however, the formula yields the square of 41, a composite. Later, it was proved that there is no expression obtainable from sums of powers of n multiplied by integers which, for each counting number n, yields only primes.

Fermat (1601–1665), knowing that the numbers $2^n + 1$ are not prime unless n is a power of 2, conjectured that

$$2^{2^k} + 1 \tag{3.1}$$

for each counting number k is a prime. Euler showed, however, that when $k = 5$ we obtain

$$2^{32} + 1 = 4{,}294{,}967{,}297 = (641) \cdot (6700417)$$

and, hence, that the conjecture is false. It remains to this date unknown whether formula (3.1) is prime or not for any choice of k larger than 5.

Numbers of the form $2^n - 1$ are sometimes prime but it is unknown if there are infinitely many choices of n for which these numbers are prime. As previously noted, $2^{11213} - 1$ is currently the largest identified prime number.

In 1947, an American mathematician, W. Mills, showed that there exists a real number $r > 1$ such that r^{3n} always has a prime counting number in the part of this number that preceeds the decimal point for each choice of $n (n = 1, 2, 3, 4, \ldots)$. Recall that the collection of all decimals, at least intuitively, comprises the real number system. A real number like 4.714 has 4 as the part before the decimal point; 9314.312 . . . has 9314 before the decimal point. Mills' theorem does provide a prime generator, but it is not constructive in the sense that we cannot write down the number r (among other complications it has infinitely many decimal places) and, hence, we cannot actually use it for finding many primes.

A young Russian mathematician, Yuri Matyasevich, devised in 1971 a much more explicit, yet quite complicated, formula for generating prime numbers. The formula is obtained by certain multiplications and additions among 21 variables (letters), $a, b, c, \ldots, s, t, u$ and certain integers. The result is a particular polynomial in $a, b, c, \ldots, t, u$ which is a prime number whenever positive integers are substituted by the variables. Furthermore, Matyasevich proves that *every* prime number is obtainable from his polynomial. Thus, if we allow $a, b, c, \ldots, t, u$ to run through all the possible combinations of positive integers in the Matyásevich polynomial, all prime numbers would be generated.

Exercises

1. Show that $n^2 + n + 17$ is a prime for $n = 1, 2, 3, 4$, and 5 but is not prime for $n = 17$. (Actually it is prime for $n = 0, 1, 2, \ldots, 15$). Euler thought this was also true of $n = 16$ but $289 = 17^2$.
2. For what choices of n, $0 \leqslant n \leqslant 7$ is $n^2 - n + 5$ a prime number?
3. The numbers

$$2 + 1, \quad 2 \cdot 3 + 1, \quad 2 \cdot 3 \cdot 5 + 1, \quad 2 \cdot 3 \cdot 5 \cdot 7 + 1, \ldots$$

formed by taking the product of all the primes less than a number and adding one are sometimes prime. Are they always prime? (Euclid used these numbers, rather than the ones derived from factorials, in his original proof of the fact that there is no largest prime.) (Hint: $30031 = 59 \cdot (509)$)

3.7. What Is the Most Fundamental Theorem in Arithmetic?

The answer to the question which titles this section is, of course, subjective. There is, however, a result which traditionally carries the name "the fundamental theorem of arithmetic" although it might be more informative to replace "arithmetic" by "number theory" in this title. In any case, the theorem simply states that each counting number can be represented in one and only one way, except for the order of the factors, as a product of prime numbers. For example, $35 = 5 \cdot 7$, $75 = 3 \cdot 5^2$, $100 = 2^2 \cdot 5^2$, and $5615610 = 2 \cdot 3 \cdot 5 \cdot 7 \cdot 11^2 \cdot 13 \cdot 17$; in each case the counting number is written as a product of prime numbers in a certain order. The order is not important since multiplication is commutative ($ab = ba$) as well as associative. Thus, $35 = 5 \cdot 7 = 7 \cdot 5$ and $100 = 2^2 \cdot 5^2 = 2 \cdot 5 \cdot 5 \cdot 2$. What is essential is that there is no way to factor a number into different sets of primes. If 3, 5, and 7 are found to be factors of a number then, no matter how hard we try, they will be factors in any representation of the number as a product of primes.

The argument used to establish at least part of the fundamental theorem of arithmetic is not difficult to understand. Each prime number, on one hand, is considered as already factored into primes. On the other hand, each composite c can be written as a product $c = a \cdot b$ where a and b are not one or c. If both factors are primes, then c has been factored into primes. Otherwise, at least one of these factors is a composite number and, hence, it can itself be expressed as a product of smaller numbers. Continue this process of factoring each composite number obtained into smaller numbers larger than one. Because each factor is always smaller than the number from which it is derived, we cannot continue this process forever. Therefore, in a finite number of steps c can be written as a product of primes.

The fact that the factorization into primes is unique, except for the order of the factors, is slightly more ticklish to prove than the other part of the fundamental theorem. Although we do not intend to give a proof, the basic tool for the uniqueness of the factorization is the property that a prime number which divides a product must divide one of the factors in that product. In symbols, if p is prime and if $p | (a \cdot b)$, then $p | a$ or $p | b$.

A major difficulty with providing a uniqueness argument for the fundamental theorem is that the "tool" is itself elusive to prove; so much so, in fact, that it is assumed to be true in the sequel of this chapter.

One application of the fundamental theorem of arithmetic is to find common divisors for two or more natural numbers. Suppose we have two counting numbers, c and d. Is there, in fact, a largest number g such that $g | c$ and $g | d$? When such a number g exists, and one always does, it is called the greatest common divisor of c and d. Without attempting to prove all of these facts at this point, we can invent a method of obtaining the greatest common divisor of two numbers. Assume, for illustration, that $c = 140$ and $d = 168$. Factor each of these numbers into primes to obtain $140 = 2 \cdot 2 \cdot 5 \cdot 7$ and $168 = 2 \cdot 2 \cdot 2 \cdot 3 \cdot 7$. Now it is clear that $7 | 140$ and $7 | 168$ but 5 and 3 do not divide *both* numbers c and d. On the other hand $2^2 = 4$ does divide both, while $2^3 = 8$ does not. Hence, $2^2 \cdot 7 = 28$ divides both numbers and clearly is the largest number with this property. This means 28 is the greatest common divisor of the original numbers, 140 and 168.

Exercises

1. Express each of the following numbers as the product of powers of primes:
   a. 28 b. 64 c. 105 d. 320 e. 426 f. 1800 g. 55440
2. Find the greatest common divisor (gcd) of the following pairs of numbers:
   a. 28, 64 b. 329, 426 c. 9999, 11 d. 1800, 55440
   e. 147, 98 f. 216, 144 g. 352, 484.
3. The least common multiple (lcm) of two counting numbers c and d is the smallest number l such that $c|l$ and $d|l$. By factoring c and d into primes, devise a method of obtaining the lcm which is assumed to exist.
4. Apply the method of problem 3 to find the lcm of the pairs in exercise 2.
5. If g is the gcd of c and d and l the lcm, then $g \cdot l = c \cdot d$. Show this is the case for the problems in Exercise 2.
6. Show that the gcd of two natural numbers c and d always exists. (Hint: Any divisor of c and of d is not larger than the smaller of these two numbers. Furthermore, one is a common divisor of the pair c and d.)

3.8. How Long Is Division?

Let a and b be counting numbers with $b > a$. Consider the set of all multiples of a, namely, $a, 2a, 3a, 4a, \ldots, na, \ldots$. Either the number b is itself a multiple of a or else it falls between two successive multiples of a as indicated in Figure 3.3.

Figure 3.3

Thus, there is a counting number q such that $qa < b$ and $(q + 1)a > b$ when b is not a multiple of a. It follows that

$$b = qa + r, \qquad 0 \leqslant r < a, \tag{3.2}$$

where r is the difference between b and qa, the distance from qa to b in Figure 3.3. We permit $r = 0$ so that the case $b = qa$, when b is a multiple of a, can be included in the discussion.

The processes of obtaining Eq. (3.2) is called the *division algorithm* and the numbers q and r are known as the quotient and remainder, respectively, upon division by b by a. Formula (3.2) is valid even when $b < a$ for, in this case, we can take $q = 0$ and $r = b$. It is often cited in grade school mathematics as a check for a long division problem. For example, 125 when divided by 9 has a quotient of 13 and a remainder of 8. Check: $125 = 9 \cdot 13 + 8$.

The greatest common divisor (gcd) of two counting numbers a and b, if you recall, is the largest counting number g with the property that $g|a$ and $g|b$. Let us introduce the notation (a, b) for the gcd g of a and b. For example, the gcd of 12 and 20 is $(12, 20) = 4$. (The existence of the greatest common divisor of two natural numbers is proved by completing Exercise 6 of Section 3.7.)

Euclid, in his *Elements*, displayed a rather simple method for finding (a, b). In order to illustrate his process, let $b > a$ and apply the division algorithm (Eq. 3.2) to obtain

$$b = qa + r, \qquad 0 \leqslant r < a.$$

If $r = 0$, then b is a multiple of a and, hence, $(a, b) = a$. If $r \neq 0$, then any divisor of a and r is also a divisor of qa and r (Law 1 of Section 3.2) and, thus, is also a divisor of $qa + r = b$ (Law 2 of Section 3.2). Furthermore, if $g|a$ and $g|b$, then $g|r$ (Law 1 and Law 3 of Section 3.2). Hence, the gcd of a and b is the same as the gcd of a and r since *any* number that divides a and b also divides a and r and vice versa. In symbols, this is the fact that $(a, b) = (r, a)$. The advantage is that the number b has been replaced by a smaller one, r. Repeated application of this fact provides a step-by-step procedure for getting the gcd of two numbers. We illustrate with $a = 112$ and $b = 259$.

$$\begin{array}{r|l} & \;\;2 \leftarrow q \\ 112 & \overline{259} \\ & \underline{224} \\ & \;\;35 \leftarrow r \end{array} \qquad \begin{array}{r|l} & \;\;3 \leftarrow q \\ 35 & \overline{112} \\ & \underline{105} \\ & \;\;\;7 \leftarrow r \end{array} \qquad \begin{array}{r|l} & 5 \leftarrow q \\ 7 & \overline{35} \\ & \underline{35} \\ & \;0 \leftarrow r \end{array}$$

According to our observations above, we have

$$(112, 259) = (35, 112) = (7, 35) = 7$$

The last equality stems from the fact that 35 is a multiple of 7. The sequence of remainder r generated by this process is a collection of decreasing nonnegative integers and, hence, a remainder of $r = 0$ must be reached after a finite number of steps.

This procedure for calculating the gcd of two numbers is known as the *Euclidean algorithm*. The number of divisions, or steps, in the Euclidean algorithm can be large; in fact, for any integer $n \geqslant 1$ there are pairs of numbers for which the algorithm takes n steps. The example above has only three steps, but the algorithm applied to 178 and 288 requires ten steps.

Lamé, a French mathematician of the nineteenth century, proved that the number of steps required in the Euclidean algorithm is never more than five times the number of digits in the smaller number. It can actually take this number of steps for certain numbers. For example, the algorithm for 8 and 13 should take no more than five steps by Lamé's theorem since the smaller number has only one digit. Applying the algorithm, we get

$$
\begin{array}{r}
1 \leftarrow q \\
8\,\overline{)13} \\
\underline{8} \\
5 \leftarrow r
\end{array}
\qquad
\begin{array}{r}
1 \leftarrow q \\
5\,\overline{)8} \\
\underline{5} \\
3 \leftarrow r
\end{array}
\qquad
\begin{array}{r}
1 \leftarrow q \\
3\,\overline{)5} \\
\underline{3} \\
2 \leftarrow r
\end{array}
\qquad
\begin{array}{r}
1 \leftarrow q \\
2\,\overline{)3} \\
\underline{2} \\
1 \leftarrow r
\end{array}
\qquad
\begin{array}{r}
2 \leftarrow q \\
1\,\overline{)2} \\
\underline{2} \\
0 \leftarrow r
\end{array}
$$

that is,

$$(8, 13) = (5, 8) = (3, 5) = (2, 3) = (1, 2) = 1$$

so the full five steps are necessary.

The smallest pairs of numbers that produce a given number of steps in the Euclidean algorithm are actually successive numbers of the Fibonacci numbers 1, 1, 2, 3, 5, 8, 13, 21, 34, 55, 89, 144, ... This sequence, named in memory of an Italian mathematician of the Middle Ages, is generated by the fact that each entry from the third onward is obtained by adding the two preceding members of the sequence. If $l(a, b)$ denotes *the number of steps* in the Euclidean algorithm for a and b, then

$$l(1, 2) = 1, \quad l(2, 3) = 2, \quad l(3, 5) = 3, \quad l(5, 8) = 4, \quad l(8, 13) = 5$$

and so on, where we have used successive pairs of Fibonacci numbers.

Fibonacci numbers have a mystic quality and occur frequently in the enumeration of several natural phenomena as well as in certain mathematical problems. For example, in plant growth the number of leaves in regular intervals of the plant's height often form a Fibonacci sequence, the number of petals in various species of flowers form such sequences, and the number of ancestors of a male bee (see the exercises) can be arranged in a Fibonacci sequence.

Actually, Fibonacci discovered the sequence to be an answer to a population model he had for rabbits: A pair of rabbits one month old are too young to reproduce, but suppose that in their second month and every month thereafter they produce a new pair of rabbits. If each pair of rabbits produced contains one of each sex, if each pair begins the reproduction cycle two months after birth, and if no rabbit dies, how many pairs of rabbits will there be at the beginning of each month? If we start with a pair of newborn rabbits, then the number of pairs of rabbits at the beginning of each month is 1 initially, 1 at the beginning of the second month, 2 at the beginning of the third month since the initial pair has now reproduced, 3 at the beginning of the fourth month since the initial pair (only) has reproduced again, 5 at the beginning of the next month since the initial pair and their first-born both reproduce, and so on. The answers form the sequence 1, 1, 2, 3, 5, 8, 13, . . . of Fibonacci numbers.

Exercises

1. Find the indicated gcd by the Euclidean algorithm:

   a. (352,480) b. (178,288) c. (387,301)
   d. (213,46493) e. (323,475) f. (55,89)

2. A male bee has only one parent, his mother of course, whereas a female bee has both a father and a mother. How many bees are there in each generation of ancestors for a male bee if we trace back 3, 4, 5, 6, 7, and 8 generations?

3. Suppose you are on an isolated South Sea Island beach with two unmarked containers, one of which holds exactly 4 qts and the other exactly 9 qts. Using only these two containers, can you measure any number of quarts from 1 through 13?

4. Suppose the containers in Exercise 3 hold in quarts:

   a. 3 and 6; b. 6 and 13; c. 4 and 7.

   List, in each case, all the possible number of quarts that can be measured using only these two containers.

5. Let $0 < a < b$ be two integers. Suppose that the containers in Exercise 3 measure exactly a qts and b qts. Under what conditions on a and b would it be possible to measure any number of quarts of water from one to $a + b$ using only these containers?

6. Show that $l(13,21) = 6$, $l(8,13) = 5$, and $l(5,8) = 4$, where $l(a,b)$ is the number of steps in the Euclidean algorithm.

7. Show $l(ca,cb) = l(a,b)$ if c is any counting number.

8. If $l(a,b) = 3$ and $b > a$, prove that $b \geqslant 5$ and $a \geqslant 3$. Find an analog of this when $l(a,b) = 4$ and $b > a$. (Hint: Work backwards in a three (four) step Euclidean algorithm, selecting the numbers involved in each step to be as small as possible.)

9. There is a game, called Euclid, based upon the Euclidean algorithm. Let a and b be counting numbers with $b > a$. Each of two players moves in turn according to the following rule: The player must reduce the larger of the two numbers before him to any nonnegative number obtained by subtracting a multiple of the smaller number from the larger. The first person to produce a zero wins. For example, with $a = 3$, $b = 7$, the first move is either to replace 7 by 4 $(= 7 - 3)$ or by 1 $(= 7 - 6)$. The former is a winning position for the first player and the latter is a losing position. Can you develop a strategy for this game? (A. J. Cole and A. J. T. Davie. A game based on the Euclidean algorithm and a winning strategy for it. *Math. Gaz.* 53 (1969), pp. 354–357.)

10. If a two digit number is divided by the sum of its digits what are the largest and smallest possible values for this ratio as a real number?

3.9. Diophantine Equations

One application of the alleged irrelevant mathematics of this chapter is to solve equations such as $2x + y = 11$ for integers. This requires that we find all pairs of integers (x, y) such that $2x + y = 11$. There are practical problems that require such solutions.

Equations like $2x + y = 11$ in which integral solutions are desired are called linear *Diophantine equations* after a Greek mathematician Diophantus who lived around 250 A.D. Such equations arise when dealing with problems concerning objects, such as people, which cannot be fractionalized–what, after all, is half a man, other than a cruel jest? Suppose, by way of illustration, that you have had a party with men and women in attendance. Assume that all the people at the party drank beer and that 11 quarts of that beverage were consumed during the evening. Knowing your friends, you realize that *each male* at the party *drank 2 qts* of beer and *each female drank 1 qt.* The next morning you find that you cannot even remember how many men and women were at your party. Can you, in a sober moment, calculate the number of people of each sex at the party?

In order to solve the problem, let x be the number of men at the party and let y be the number of women. Then according to the consumption assumptions (sic) of the problem $2x$ and y represent the number of quarts of beer consumed by the male and female party people respectively and, hence, $2x + y$ is the number of quarts of beer drunk that evening. By hypothesis this is 11, that is, $2x + y = 11$. Thus, $y = 11 - 2x$. Now x is the number of males at the party and, hence, is a positive integer. We assign values $x = 1,2, \ldots$ and compute the corresponding number of females at the party y for each x by the formula $y = 11 - 2x$. Within the physical limitations of the problem the only pairs of solutions are (1,9), (2,7), (3,5), (4,3), (5,1), where the first number in the pairs is the number of men x and the second is the number of women y. Any other choice of x, for example, $x = 6$, produces a negative y which has no meaning in the stated problem. If you cannot pick the particular pair that represents the actual number of men and women at the party, owing to your condition, you can conclude there was at least one woman and not more than five men in attendance.

This absurd example illustrates the fact that a Diophantine equation has a *set of solutions* in positive integers. The solution set, that is, the set of all pairs of positive integers that satisfy the equation, could be empty, finite (as in the party example), or even infinite. For example, $2x + 4y = 11$ has no solution for x and y that are integers since $2 \cdot x$ and $4 \cdot y$ would be even integers when x and y are integers and, hence, their sum could not be the odd number 11. The solution set for this example is the empty set. The solution set for the equation

$x - y = 1$ is infinite. Indeed, x is just one more than y and there are infinitely many such pairs of natural numbers: (2,1), (3,2), (4,3), (5,4), ...

Linear Diophantine equations can be more complicated than the examples so far illustrate. Consider the equation $7x + 12y = 220$ which we wish to solve for positive integers. Divide throughout by 7, the smaller of the two coefficients 7 and 12 of x and y, to obtain

$$x + \frac{12}{7}y = \frac{220}{7} \quad \text{or} \quad x + y + \frac{5y}{7} = 31 + \frac{3}{7}$$

Collecting the fractional terms, we get

$$x + y + \frac{5y - 3}{7} = 31$$

Since x, y, and 31 are integers, it must follow that $(5y - 3)/7$ is also an integer. (If it were a fraction like $\frac{1}{2}$, when it was added to integers, a nonintegral fraction would be obtained.) Set $(5y - 3)/7 = n$ where n is an integer. Then $5y = 3 + 7n$. Now try various choices of n. When $n = 0$, $5y = 3$ which is of no value since y is to be an integer. When $n = 1$, we get $5y = 10$ so $y = 2$, an integer. For this value of y we have

$$31 = x + y + \frac{5y - 3}{7} = x + y + n = x + 2 + 1 = x + 3$$

Hence, $x = 28$ so (28,2) is a solution of the problem. Next try $n = 2$ so $5y = 17$ which is again of no value; $n = 3$ gives $5y = 24$, also of no value since y is an integer. Next, $n = 4$ gives $5y = 31$, which is of no value. When $n = 5$ the same type of answer is obtained. For $n = 6$, however, we obtain $5y = 45$ so $y = 9$ is a solution. The corresponding x value is 16 and (16,9) is a solution to the problem. Continue this trial-and-error method for other choices of n. You will discover that when $n = 11$ another solution (4,16) arises and that there are no additional solutions.

This method of solution may work, but it is tedious. One way out of the drudgery is as follows. Obtain one solution by the trial-and-error method, for example, (28,2), in the above problem. Then write

$$7 \cdot 28 + 12 \cdot 2 = 220, \qquad 7 \cdot x + 12 \cdot y = 220$$

Subtract these two equations to obtain

$$7(28 - x) + 12(2 - y) = 0$$

which is equivalent to $7(28 - x) = 12(y - 2)$.

Now $12 \nmid 7$ so $12 \mid (28 - x)$ and $7 \nmid 12$ so $7 \mid (y - 2)$. Thus

$$\frac{28 - x}{12} = \frac{y - 2}{7} = n$$

where n is an integer. This gives

$$y = 2 + 7n, \qquad x = 28 - 12n$$

This time every choice of n leads to integers for x and y, and $n = 0,1$, or 2 yields positive integers. The solutions are $x = 28$, $y = 2 (n = 0)$; $x = 16$, $y = 9 (n = 1)$; and $x = 4, y = 16 (n = 2)$. Other choices of n lead to a negative value for x or y.

Other, more systematic, methods can be used to solve these equations [See, for example, Hall and Knight. *Higher Algebra.* New York, N.Y.: Macmillan. 1948. pp 110-113 and pp 284-291.] The method of this section, however, is adequate for most problems and it does have the by-product of reviewing the high school algebra topic of linear equations without repeating high school mathematics. Linear equations are worth reviewing since they are frequently encountered in applications of mathematics.

Finally, you can carefully graph an equation like $7x + 12y = 220$ with a ruler and graph paper. This should indicate which pairs of integers (x, y) are on the graph. They can then be proven to be solutions by substitution. We will not pursue this method at this point since graphing is a topic that is avoided until absolutely necessary in this text.

Exercises

1. A man buys stock in two different companies. If one stock costs \$21 a share and the other stock costs \$31 a share, how many shares of each can he buy if he spends \$1770?

2. A boy opens up a lemonade stand where he sells two different sizes of drinks; one for 5¢ and the other for 7¢ per glass. If he has collected 93¢ by the end of the day, how many customers did he have for each type of drink?

3. Some men and women have a party. If each man eats two and a half cookies and each woman eats one, how many people of each sex attended the party in which 17 cookies were consumed?

4. A car salesman received \$75 commission for each sale of one model car and \$100 for each sale of another model. In a certain period of time he must make a commission of \$3300 to pay his personal expenses. How many cars of each model must he sell to meet his obligations?

5. Suppose $c \mid a$ and $c \mid b$ where c is a counting number larger than one. Show that the Diophantine equation $ax + by = d$ has no solutions in integers if $c \nmid d$. Write down three Diophantine equations that have no solutions in integers.

6. Solve, if possible, each of the following Diophantine equations for positive integers.
   a. $5y = 3x + 25$ b. $12x + 9y = 341$ c. $7x + 3y = 23$
7. How large an integer N is necessary for the equation $18x + 23y = N$ to have natural numbers x and y as solutions?
8. How many ways can you make change for 25¢ if only nickels and dimes are used? How many ways can you make change if you also allow pennies?

3.10. The Pythagorean Theorem

Long before the existence of Diophantus, the Pythagoreans considered solutions in integers for the identity they derived concerning right triangles, known as the Pythagorean theorem. The theorem states that $c^2 = a^2 + b^2$ where a, b, and c are the lengths of the sides of a right triangle with c the longest side (hypotenuse). Interest in this theorem and its converse probably arose centuries before the Pythagoreans because of the social need for measurement of land and because of problems in architecture.

Let us digress from number theory and provide a geometric justification of the theorem, one which perhaps the Pythagoreans, at least, would accept. There are around 400 different proofs of the Pythagorean theorem available and one of these is attributed to James Garfield, the former U.S. president.

Let a, b, and c be the lengths of the sides of a right triangle with c the length of the hypotenuse. Consider Figure 3.4, the second and third parts being squares of equal area.

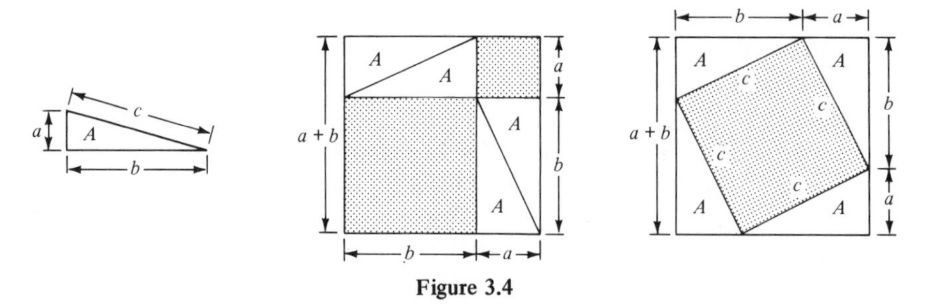

Figure 3.4

The first square contains four copies of the triangle A, and the area of the uncovered region (shaded) is a^2 (from the upper right corner) plus b^2 (from the lower left shaded corner.) In the second square, which also contains four copies of A, the shaded region is a square of area c^2. Since the triangles are all of the same size (congruent) and since the squares are also of the same size, the shaded regions in each case have the same area. Thus, $c^2 = a^2 + b^2$ which is the Pythagorean theorem.

The converse of the Pythagorean theorem, namely, if $c^2 = a^2 + b^2$ for a triangle A with sides of length a, b, and c, then A is a right triangle, is also true. One proof is by the side-side-side argument in geometry, which states that two triangles are congruent (same size and shape) if each side of one triangle has length equal to the length of a side of the other triangle. Now let $c^2 = a^2 + b^2$ for the triangle A. Construct a right triangle with legs of length a and b. The hypotenuse, by the Pythagorean theorem, has square equal to $a^2 + b^2$. Since $a^2 + b^2 = c^2$, it follows that each side of the triangle A has length equal to that of a side of the right triangle just constructed. This implies that triangles are congruent; in particular, corresponding angles are equal, and A is a right triangle.

EXAMPLE. Suppose you wish to purchase a ladder to reach the roof of your 20-ft house. If, for safety reasons, you plan to rest the base of the ladder 8 ft from the wall as in Figure 3.5, how long a ladder should you buy? Let $a = 8$,

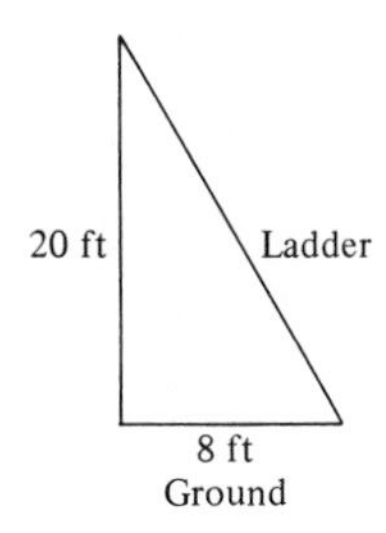

Figure 3.5

$b = 20$ and apply the Pythagorean theorem to obtain $c^2 = a^2 + b^2 = 64 + 400 = 464$. Thus $c = \sqrt{464} \approx 21.5$† so a 22-ft ladder should suffice.

One problem that interested the Pythagoreans in connection with their theorem was the determination of triplets a, b, and c, each of which is a counting number, such that $a^2 + b^2 = c^2$. For example, $a = 3$, $b = 4$, and $c = 5$ have this property. This particular example was known and used by many early societies (Babylonians, Egyptians, and Indians) well before the time of Pythagoras. In fact, a number of other so-called Pythagorean triplets appeared on a recently unearthed Babylonian clay tablet which has been dated in the period 1900–1600 B.C.

The problem of determining Pythagorean triplets of integers is similar to that of solving linear Diophantine equations. In order to attack it, notice first that if two of the three counting numbers a, b, or c are even integers, then so must be the third number. Indeed, squares, sums, and differences of even integers are

†The symbol "≈" means approximately equal.

even. We shall, therefore, initially search for solutions of $a^2 + b^2 = c^2$ in which c and a, say, are odd numbers.

Since c is the length of the hypotenuse of the triangle, $c > a$. Set $c = p + q$ and $a = p - q$ where p and q are counting numbers with $p > q$. This can always be accomplished by taking $p = (a + c)/2$ and $q = (c - a)/2$. ($a + c$ and $c - a$ are both even since, as we have assumed, a and c are odd integers.) Now, by algebra (or by Figure 3.6, if you wish) we have

$$c^2 = (p + q)^2 = p^2 + 2pq + q^2$$
$$a^2 = (p - q)^2 = p^2 - 2pq + q^2$$

Subtracting these two quantities, we obtain $c^2 - a^2 = 4pq$. For us to be certain that this is a perfect square, b^2, we demand that $p = P^2$ and $q = Q^2$, where P and Q are counting numbers with $P > Q$. Then $b^2 = 4P^2Q^2$ or $b = 2PQ$ whereas $a = p - q = P^2 - Q^2$ and $c = p + q = P^2 + Q^2$.

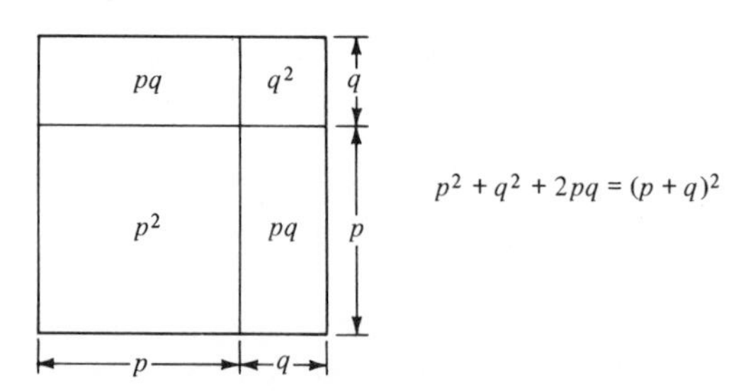

Figure 3.6

The formulas $a = P^2 - Q^2$, $b = 2PQ$ and $c = P^2 + Q^2$ do provide integer solutions for $c^2 = a^2 + b^2$. Indeed, let P be any counting number larger than Q. We can determine the corresponding a, b, and c from the above formulas. For example, $P = 2$ and $Q = 1$ gives $a = 4 - 1 = 3$, $b = 2 \cdot 2 \cdot 1 = 4$ and $c = 4 + 1 = 5$; $P = 7$ and $Q = 5$ gives $a = 49 - 25 = 24$, $b = 2 \cdot 7 \cdot 5 = 70$, and $c = 49 + 25 = 74$. A partial list of Pythagorean triplets obtainable by this method is shown in Table 3.1.

TABLE 3.1

a	3	5	7	15	9	21	11
b	4	12	24	8	40	20	60
c	5	13	25	17	41	29	61

You can obtain additional solutions from any given solution by multiplying a, b, and c by a common number. For example, $a = 3$, $b = 4$, and $c = 5$ satisfies $c^2 = a^2 + b^2$. Therefore, $a = 6$, $b = 8$, and $c = 10$ also is a solution, and so is $a = 9$, $b = 12$, and $c = 15$. In fact, $a = 3x$, $b = 4x$, and $c = 5x$ for any counting number x satisfies $a^2 + b^2 = c^2$ since an x^2 factor appears in each term and, hence, can be factored out of the sum to obtain

$$a^2 + b^2 = 9x^2 + 16x^2 = (9 + 16)x^2 = 25x^2 = c^2$$

One consequence of this is that there are infinitely many distinct Pythagorean triples.

Fermat considered a generalization of the Pythagorean result and asked if there are counting numbers a, b, and c such that $a^n + b^n = c^n$ when n is an integer larger than two. He claimed to have a proof, although one was never published, of the fact that there are *no* such choices of a, b, and c for *any* $n > 2$. For more than three hundred years, mathematicians have tried, without full success, to find a justification of this statement of Fermat. (It has been called Fermat's "last problem" since it was found among his notes after his death.) Fermat did prove, however, that $a^4 + b^4 = c^4$ has no solution among the counting numbers and, more recently, it has been shown that $a^n + b^n = c^n$ has no counting numbers as solutions for all n less than 25,000. Using inductive reasoning, one might believe this is enough cause to state the result in general. But mathematicians are deductors and thus are not satisfied by such arguments!

Exercises

1. A man drives due west from a town at 30 mph and another drives north from the town at 40 mph. If both leave at the same time from the same intersection in the town, how far are they from each other at the end of two hours?

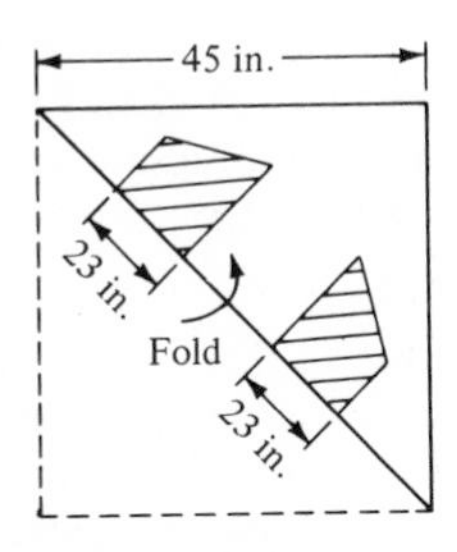

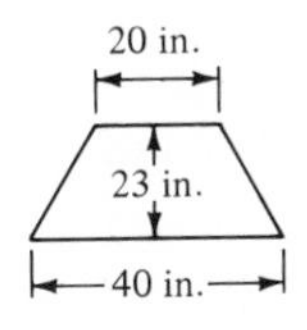

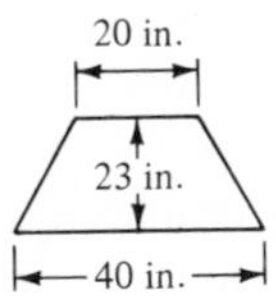

Figure 3.7

2. A dressmaker has to cut a skirt on the bias from a 45-in. width of wool. This means he cuts two identical pieces (front and back) from a triangle obtained by folding the left hand corner of the material to the opposite side. In order to make a skirt, does he have enough material folded to cut out two pieces if the dimensions are as shown in Figure 3.7? (The figure is a regular trapezoid.)

3. Find each choice of P and Q, where $a = P^2 - Q^2$, $b = 2PQ$, and $c = P^2 + Q^2$, which produces the Pythagorean triples in Table 3.1.
4. Use Fermat's result that there is no solution in integers for $a^4 + b^4 = c^4$ to prove that there is no solution in integers for $a^8 + b^8 = c^8$. Can you generalize the conclusion?
5. In each triplet of Pythagorean numbers, it is known that one is divisible by 3, one by 4, and one by 5. (It may be the case that the same number is divisible by two or more of these divisors.) Check this fact for the numbers in Table 3.1.
6. The dual of Fermat's conjecture is to solve the identity $n^a + n^b = n^c$ for integers a, b, and c when n is a given natural number. Prove there are solutions when $n = 2$. This is the only case for which solutions exist. Try $n = 1$, 3, and 4.
7. A right triangle, each side of which is of integral length, will be called a Pythagorean triangle. For each natural number $n \geqslant 3$ there is a Pythagorean triangle with one leg (not the hypotenus) of length n. Find the corresponding Pythagorean triples for the numbers n from 3 through 12.
8. Use the formulas of this section for Pythagorean triplets to obtain a Pythagorean triangle with leg n when n is even. For odd $n > 1$ the Pythagoreans had the solution

$$(n^2 - 1)/2, \quad n, \quad \text{and} \quad (n^2 + 1)/2$$

9. Not every natural number can be the length of the hypothenuse of a Pythagorean triangle. For example, no number that is a power of two can be such a hypothenuse.
   a. Show by enumeration that two and four are not hypothenuses of Pythagorean triangles.
   b. If a is odd, $a^2 - 1$ is divisible by eight (Section 3.3). Use this result to prove $64 = a^2 + b^2$, where a and b are odd, has no solutions for a and b.
   c. If c is a multiple of four, show $c - 2$ is two times an odd integer.
   d. Show that there are no solutions of $c = a^2 + b^2$, where c is a power of two, and a and b are odd natural numbers.
10. One assumption used in obtaining a formula for Pythagorean triples was that p and q can be replaced by perfect squares P^2 and Q^2, respectively, when $b^2 = pq$. This need not be the case, for example, $p = 2$ and $q = 18$ have a product $pq = 36$ which is a perfect square. Find a hypothesis on p and q such that $pq = b^2$ implies p and q are themselves perfect squares.

3.11. Two Finger Counting

Early cultures of man developed number notations which generally proved inefficient and cumbersome as mathematics, with its multitude of applications, advanced. The limitation of the Roman system of writing numbers, used by

Europeans for centuries, is now all too familiar to young students throughout the Western World. A significant step forward for man was the adoption of the Hindu-Arabic number notation that is in common usage today.

The current decimal notation is a clever combination of the ten digits 0, 1, 2, 3, 4, 5, 6, 7, 8 and 9, and positions of the digits relative to a fixed point, the decimal point. A number like 475.2 means we have 4 hundreds, 7 tens, 5 ones, and 2 one tenths. All numbers are expressed by the Hindu-Arabic notation in terms of powers of 10, the *base* of our number system, and the digits 0 through 9, which by their location in this representation define a particular power of ten. For example,

$$39{,}714 = 3 \times 10^4 + 9 \times 10^3 + 7 \times 10^2 + 1 \times 10 + 4$$

There is nothing magical about 10. Other bases can be, and were, advocated and used in the history of man. The selection of 10, however, is likely based on the fact that man has 10 figures which he used in calculating, just as small children do today. Some early cultures counted on both fingers and toes; perhaps this explains why some people adopted a base of 20 instead of 10. Even today certain groups advocate, with good justification, a selection of the base 12 to simplify decimals, although the advantage of any change in this area is undoubtedly highly outweighed by the inconvenience that implementation would cause. Imagine the chaos if everyone in the world had to learn new addition and multiplication tables!

An awareness of different bases for number notation has recently appeared in our society perhaps owing its existence to the so-called "New Math," or to the wide use of the computer. With the "New Math," it was, and likely still is, hoped that students would better understand their numbering system by an exposure to numbers written in various bases.† With computers the base 2, or a base related to 2 like the hexadecimal (16) system, has certain engineering advantages over the standard representation of numbers. Indeed, the arithmetic done by a computer is accomplished by means of what we shall loosely call "switches" with just two states, like "open" and "closed."

With the base two, we have only two digits, 0, and 1, and all numbers are written in a collapsed form of the expression representing the number in terms of sums of powers of two. For example, consider 101011 as a number in the base 2.

First, recall that a number in the base 10 can be expressed as a sum of digits multiplied by powers of 10. For example, 4573 can be written as

$$4 \times 10^3 + 5 \times 10^2 + 7 \times 10 + 3$$

†In order to fully appreciate this point, one should hear Tom Lehrer sing his song entitled *New Math*, Reprise 6179, recorded July, 1965.

A similar representation occurs in the base 2 except that (a) there are only two digits (0 and 1) instead of ten, and (b) the powers of the base 10 are replaced by powers of the base 2. Thus, 101011 in the base 2 is the same as

$$1 \times 2^5 + 0 \times 2^4 + 1 \times 2^3 + 0 \times 2^2 + 1 \times 2 + 1 = 32 + 8 + 2 + 1 = 43$$

in the base 10.

How do we write 1 + 1 in the base 2? Well, one plus one is two in any base but two takes the form 10 in the binary system. Thus, 1 + 1 = 10 in the base 2. Similarly 10 + 1 = 11, but 11 + 1 = 100 since 3 has the form 11 in base 2 while $4 = 2^2$ so 4 has the form 100 in base 2. We can generalize from these examples and conclude that we can "carry" just like we did in addition for the base 10, only it is easier in base 2. Similarly, you can multiply, subtract, and divide in base 2 by analogy with what is normal for the base 10. A few calculations will convince you of this fact.

	Base 2	Base 10	Base 2	Base 10
Carried →	1111			
digits	101011	43	110	6
	+1101	+13	×11	×3
	111000	56	110	
			110	
			10010	18

The numbers 111000 and 10010 in base 2 are easily converted to the base 10. For 111,000 is equivalent to the base 10 number

$$1 \times 2^5 + 1 \times 2^4 + 1 \times 2^3 + 0 \times 2^2 + 0 \times 2^2 + 0 = 32 + 16 + 8 = 56$$

and 10,010 is equivalent to the number

$$1 \times 2^4 + 0 \times 2^3 + 0 \times 2^2 + 1 \times 2 + 0 = 16 + 2 = 18$$

It is also quite easy to convert a number from the base 10 to the binary system. Consider, for example, the number 23. We have by repeated division by 2 that

$$23 = 2 \times 11 + 1; \quad 11 = 2 \times 5 + 1, \quad 5 = 2 \times 2 + 1, \quad \text{and} \quad 2 = 2 \cdot 1 \qquad (3.3)$$

By substitution of each expression into its predecessor, if there is one, we obtain

$$\begin{aligned} 23 &= 11 \times 2 + 1 = (5 \times 2 + 1) \times 2 + 1 = 5 \times 2^2 + 1 \times 2 + 1 \\ &= (2^2 + 1) \times 2^2 + 1 \times 2 + 1 = 2^4 + 1 \times 2^2 + 1 \times 2 + 1 \end{aligned}$$

which is 10111 in the base 2. Each of the operations in Eq. (3.3) is simply an application of the division algorithm, where the divisor is always the number base 2. We can shorten the process by using ordinary short division as follows:

```
        Remainder                              Remainder
   2| 1 --- 1                               2| 1 --- 1
   2| 2 --- 0 <-   23 or (10111)2           2| 3 --- 1
  2|  5 --- 1 <-                          2|  6 --- 0   48 or (110000)2
 2|  11 --- 1 <-   remainder              2| 12 --- 0
2|   23 --- 1 <-                          2| 24 --- 0
                                         2|  48 --- 0
```

where $(10111)_2$ and $(110000)_2$ are notations for the numbers 23 and 48 in the base 2.

Arithmetic would be much easier if we had all been born with two fingers! Some of the exercises will perhaps convince you of this. However, each benefit has cost factors; for instance, it would take many lines of print to express a number equal to the total budget of the Federal Government.

Exercises

1. Carry out the indicated operations for each problem in the base 2.

   a. $\begin{array}{r} 1101101 \\ +\ 11011 \\ \hline \end{array}$ b. $\begin{array}{r} 1101101 \\ -110111 \\ \hline \end{array}$ c. $\begin{array}{r} 11011 \\ \times\ 101 \\ \hline \end{array}$ d. $1011101 \div 11$

2. What is each of the above problems written in base 10 notation? Use this to check your previous answers.
3. Convert each of the following to binary notation.

   a. 164 b. 135 c. 54 d. 31 e. 62
4. Find a rule for divisibility by 10 and 100 in the base 2.
5. Can you find a divisibility rule for 11 in base 2.
6. How many digits are in the base 2 equivalent of a million?

3.12. Nim for Real

The "traditional" (the origin appears unknown) game of Nim is similar to the game in Chapter 1 with the following difference: Piles of sticks are laid out in advance. In a turn each of the two players must select a single pile and take at least one stick from this pile. He can, however, pick up as many sticks as he wishes from the pile, even the entire pile. Turns are alternated and the person who picks the last stick loses. There is a strategy, based on the binary system, for this more complicated game of Nim which was discovered by C. Bouton, an American, around 1900.

In describing the strategy of Bouton, it might be well to proceed inductively from simple examples to suggest general results. We begin with three piles consisting of 1, 2, and 3 sticks, respectively. As a first step we record, in binary notation the number of sticks in each pile and add these numbers as if they were in base 10:

11 (3 sticks)
10 (2 sticks)
1 (1 stick)
22 (sum in base 10)

Next, we define a "losing" position to be one in which *each* digit in this sum is even, and otherwise call the position a "winning" one. The reason for the names, as they themselves suggest, is that there is a strategy which enables a player who finds the piles in a winning position to win. After each turn, therefore, a determination of the position of the remaining piles should be made according to the above rule.

In order to develop the strategy, assume a player faces a losing position, that is that the columns of binary digits each add up to even numbers in base 10. Any move affects only a single row of binary digits since the player must select sticks only from a single pile, and each row is the number of sticks in one of the piles, written in binary notation. Now the removal of one stick from a pile changes at least the last digit in the binary count of the number of sticks in the pile. If the last digit was 1 it changes to 0, and if it was 0 it changes to 1. In either case, the last column's sum is changed from even to odd since no other entry in that column changes when you are permitted to pick from only one pile. If two sticks are removed from the pile, then the second from last column's sum changes from even to odd since the second from last binary digit is changed from 0 to 1 or from 1 to 0 in the row representing this pile while the other rows (piles) are unchanged. Similarly, if $3 = (11)_2$ sticks are removed from a pile, then the last two columns change sums from an even to an odd number. In this way it is found that removing any number of sticks from a single pile changes a losing position to a winning one.

Suppose next a person faces a winning position, that is that the sum of the binary digits in at least one column is odd. By removing enough sticks from the largest pile in the collection at this point you can always change the sum so that each column has an even sum. In this manner, you pass to your opponent a losing position.

The winning strategy, therefore, is to always begin with a winning position and to leave your opponent with a losing position *except when this strategy leaves him with an even number of one stick piles*. In the latter case, you should pick up enough sticks to leave an odd number of one stick piles for your opponent. A few sample games as in Figure 3.8 should suffice to illustrate the strategy.

Number of Sticks

	Base 2	Base 10												
	11	(3)	he	10	(2)	you	10	(2)	he	1	(1)	you	1	(1)
Game I	10	(2)	→	10	(2)	→	10	(2)	→	10	(2)	→	0	(0)
	1	(1)		\	(1)		0	(0)		0	(0)		0	(0)
Total	22			21			20			11			1	

he → you → he → you →

	11	(3)	he	1	(1)	you	1	(1)	he	1	(1)	you	1	(1)
Game II	10	(2)	→	10	(2)	→	1	(1)	→	1	(1)	→	0	(0)
	1	(1)		1	(1)		1	(1)		0	(0)		0	(0)
Total	22			12										

Two single stick piles

	11	(3)	he	11	(3)	you	10	(2)	he	10	(2)	you	0	(0)
Game III	10	(2)	→	10	(2)	→	10	(2)	→	1	(1)	→	1	(1)
	1	(1)		0	(1)		0	(1)		0	(0)		0	(0)
	22			21			20			11				

Figure 3.8

Why does the strategy work? You can lose if your opponent (a) has a single multistick pile *or* (b) has an even number of single stick piles. The first case can never arise by using the strategy since a single pile could not have a column that is even. Case (b) is directly avoided. Hence, you should win if either you start from a winning position or he starts from a losing position.

Exercises

1. Play a 1-2-3 stick game of Nim with a friend and use the winning strategy. (The notation above means three piles, one stick in the first, two in the second and three in the third pile.)
2. Play a 1-3-5-7 stick game of Nim with a friend. Who should start the game if you wish to play the winning strategy?
3. Show that, if your opponent starts a 1-1-1-1-1-1 stick game of Nim, he will force you to lose even though this is a "losing" position.

4. Who should start each of the following games of Nim?
   a. 2-4-6-8-10 b. 1-3-5-7-7-11-13
5. If there are only two piles in a game of Nim, there is an easy strategy which enables you to win. Can you find it?

References

In addition to the references already cited in the text, the following provide excellent supplementary reading.

1. M. Richardson. *Fundamentals of Mathematics*. New York, N.Y.: Macmillan. 1941.
2. I. Barnett. Some ideas about number theory. *National Council of Teachers of Mathematics*. 1961.
3. P. Herwitz. The theory of numbers. *Scientific American*. July, 1951.
4. P. Davis. Number. *Scientific American*. September, 1964.
5. M. Kline. *Mathematics, a cultural approach*. Reading, Mass.: Addison-Wesley. 1962.
6. G. Gamow. *One two three . . . infinity*. New York, N.Y.: New American Library. 1946.
7. M. David and R. Hersh. Hilbert's 10th Problem. *Scientific American*, November, 1973, pp. 84–91.

For a number theory type solution to a common puzzle see:

8. T. Brown. A note on "Instant insanity." *Math. Magazine* 41 (1968), 167–169.

Chapter 4 A Pictorial Applied Mathematics

4.1. Introduction

Our previous chapter was devoted to quantitative relationships whereas this chapter is devoted to the more geometric, spatial relations in mathematics. The principal topic is what is now called "graph theory," although the name may lead to the false impression that we intend, at this point, to discuss analytic geometry and the plotting of functions. The graphs that we are about to study, however, are simple geometrical figures consisting of points and "lines" connecting the points. The word "line" is used in a general sense in that certain continuous (unbroken) arcs are permitted instead of just the straight lines of Euclidean geometry.

In either the plane or space geometry of Euclid, one seeks properties of figures that remain unchanged when the figures are translated (moved along a straight line segment), rotated, or reflected (mirror images). Geometry, in general, can be defined as a study of properties that are invariant (unchanged) under a certain specific collection of transformations. It is the type of transformations that tie down the geometry. The geometry is Euclidean if the transformations are reflections, translations and rotations. Other collections of transformations lead to other geometries.

If one permits the transformations to include the deformation of plane objects by bending lines, straightening arcs, or, in general, stretching figures, we obtain a very general and comparatively new mathematical "geometry," called *topology*. There are invariants in this subject which, at least initially, will appear rather basic and somewhat obvious. For example, a circle divides the plane into two regions (an inside and an outside), and so does any plane figure that can be deformed into a circle by our topological transformations. See Figure 4.1. This fact may seem either very easy or very hard to prove, depending on whether or not you attempt to write down an actual proof!

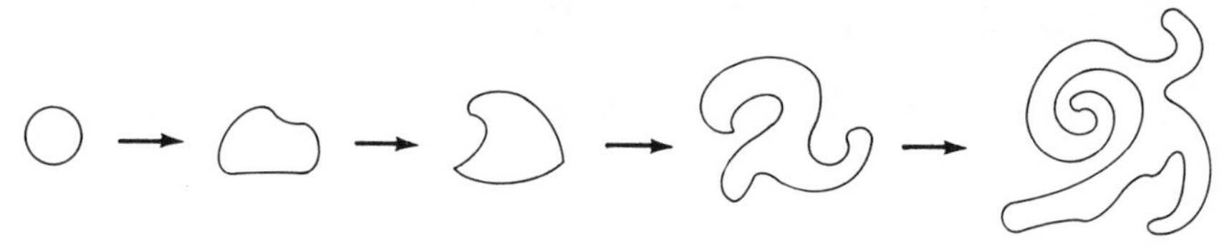

Figure 4.1

In dealing with graphs we shall permit such topological transformations. When one graph can be deformed into another by these transformations, the two graphs are called *equivalent*, just as triangles are called congruent in ordinary geometry if you can translate, rotate, or reflect one to superimpose it on the other. Figure 4.2 contains a set of graphs, all of which are equivalent in the new sense.

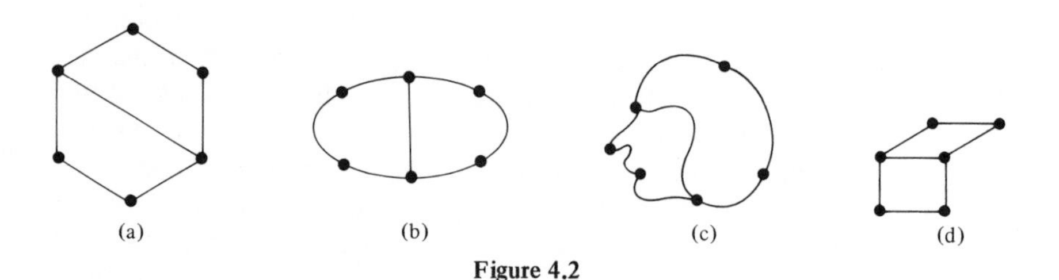

Figure 4.2

There are many general results in graph theory. A few of these are included in this chapter. The primary thrust of the discussion, however, is toward applications rather than theory. The subject has already made its appearance in such diverse fields as psychology, economics, chemistry, and biology. It is hoped that this brief encounter with graphs and their applications will enable you to discover a few applications of your own.

When we apply the theory, the graph is a model of some situation or problem. Consider a trivial little problem, namely, list all the divisors of 12. No graph theory is necessary! They are 1, 2, 3, 4, 6, and 12. Suppose, however, we build a diagram for this problem. We represent each divisor of 12 by a point. A path is a line joining two (or more) of the points. If two points (numbers) are connected by a path, the left-most point represents a number that is a divisor of the number associated with the other point. For example, a path joins 2 and 6 in (a) of Figure 4.3 since 2 is a divisor of 6.

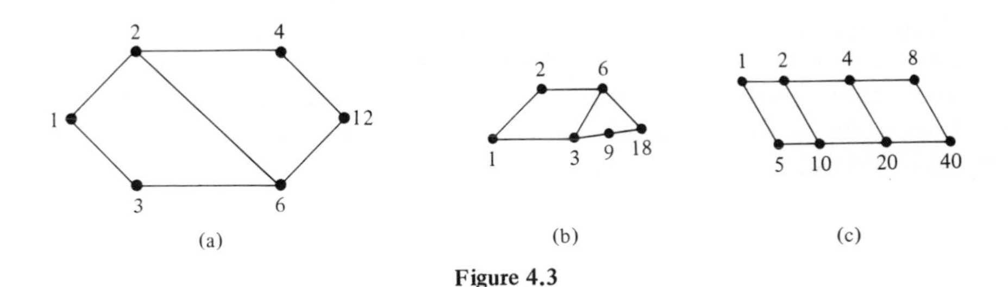

Figure 4.3

Also, there are two paths, 2-6-12 and 2-4-12, joining 2 with 12. Each path indicates 2 divides 12.

Once you understand the particular interpretation that has been placed on the points and lines, the graph is a compact way to convey information. To paraphrase an old adage, one graph is worth a thousand words. The graphs (b) and (c) of Figure 4.3 provide information about divisors of 18 and of 40 without any necessity of additional discussion.

In order to further illustrate the uses of a graph, consider the very ancient *shipping problem*. A man (*m*) has a small boat in which he wishes to transport a tame wolf (*w*), a sheep (*s*), and an open box of cabbage (*c*) across a river. His boat, however, can carry only one of these items, in addition to himself, on each trip. Furthermore, he cannot, for obvious reasons, leave the cabbage with the sheep or the sheep with the wolf. How should he proceed?

We analyze the alternatives by means of a graph. The labels on the points indicate which of the group are on the initial shore. A line segment represents a trip from one shore to the opposite shore. The left end point of each line segment is the content of the original shore before the trip represented by the line. The right end point of the segment represents the contents of the original shore after the trip has been completed. The graph is given in Figure 4.4.

This graph has a first line indicating that the man has transported the sheep, his only possible first move, across the river leaving the wolf and cabbage (*wc*). He then comes back (*mwc*) to join the group and has the option of next transporting either the wolf or the cabbage, as indicated by the lines emanating from the point

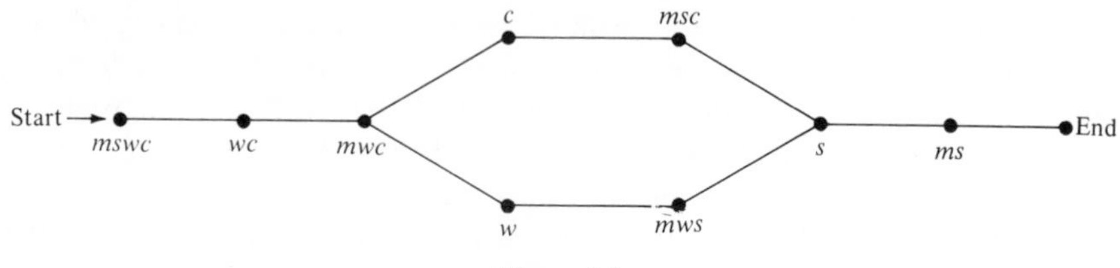

Figure 4.4

mwc. Thereafter, he must return with the sheep since he cannot leave either the sheep with the cabbage or the wolf with the sheep on the far shore. He next leaves the sheep (*s*) and transports the other item and finally comes back to pick up the sheep (*ms*) and leaves nothing. Perhaps the graph aids the reasoning which leads to the solution of this problem. In any case, it illustrates how a problem might be transcribed into a graph.

Exercises

1. Construct a graph, similar to those in Figure 4.3, for the divisors of 16, 24, and 30. (The graph for 30 can be drawn without crossing segments at points other than those corresponding to divisors of 30. However, you can also ignore such common points which have no meaning for the problem at hand.)
2. There are three distinct TV dinners–ham, steak and turkey–and three men–Tom, Dick and Harry–who are to select one dinner each. Their first and second preferences are as follows: Tom–steak, ham; Dick–ham, steak; Harry–steak, turkey. Let the people and the dinners be points on one graph representing their first preferences and on a second graph representing their second preferences. Psychologically which graph represents the most stable situation? Act as an arbitrator and assign the dinners to the individuals in some fair manner.
3. Use a graph to solve the problem in Exercise 6, Section 2.5.
4. Let A be a set with three elements. Draw a graph containing all subsets of A in which two sets are connected by a line segment if the one at the left end point is a subset of the one at the right end point.
5. Suppose Tom, Dick, and Harry, of exercise 2, are competing for young ladies, Ms. Steak, Ms. Ham, and Ms. Turkey. The men's preferences are the same as for the dinners. However, the ladies have the following preference orders: Ms. Ham–Harry, Dick; Ms. Steak–Harry, Tom; Ms. Turkey–Tom, Dick. Is there any way to arrange dates between these men and women such that no man and woman are scheduled for a date with each other if *each* has someone else in the group he/she would prefer to date?

4.2. Stuff on Graphs

We begin our formal study of this subject with the following "salesman problem" which, in an abstract sense, is a "practical" problem. Joe Swindle works for Crummy Toys, Inc. The company tells him to hit all the small towns in Florida along seven major routes joining Pensacola, Jacksonville, Orlando, and Miami. He has to pick up toys in each of the four cities in Figure 4.5 and sell

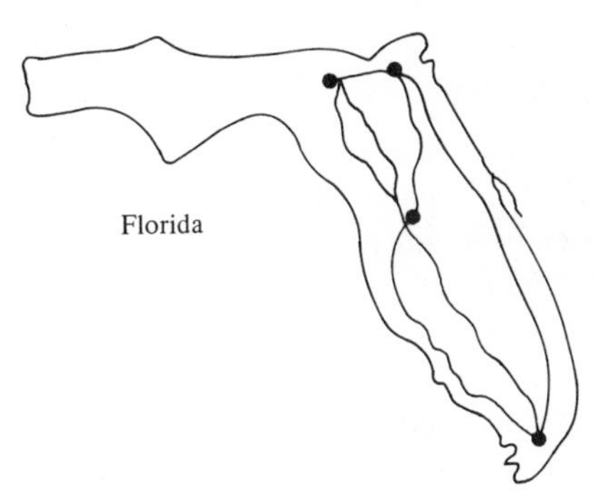

Figure 4.5

them in the towns along the routes joining the cities. But he has very crummy toys and the people who buy them would not treat him kindly should he ever return to their town. Can Joe cover all the routes without, for obvious reasons, returning along any route along which he has sold toys if he can start at any city on the map? The answer is negative. Some attempts to solve the problem should help convince you of this fact.

A solution of this particular problem can be encompassed into a general solution. The basic ideas for the solution and for the foundations of the recent subject of graph theory, date back to 1735 when Leonhard Euler (1707–1783), a Swiss mathematician of the first magnitude, provided a broad result that was initiated by a problem much like (in fact, mathematically equivalent to) our salesman problem. In order to discuss the general approach, we first need some definitions.

A *graph* is any finite set of points together with a finite number of arcs joining these points. The points are called *vertices*, and the arcs are termed *edges*. A graph is *connected* if it is possible to travel from any vertex to any other by moving along edges. (See Figure 4.6.)

A vertex is called *odd* if there are an odd number of edges emanating from it; otherwise it is even. The number of edges emanating from a vertex is called the *order* of the vertex.

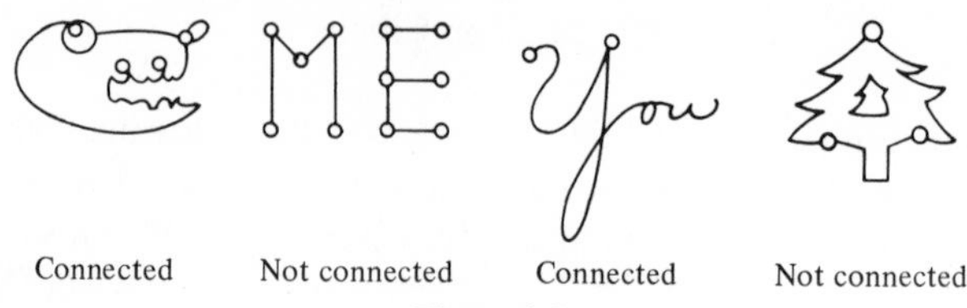

Figure 4.6

THEOREM. (Euler) Let A be a connected graph. If the graph A can be drawn without lifting the pencil from the paper and without tracing any edge more than once, then A has at most two odd vertices.

Proof: Except for the initial vertex, where we start the drawing, and the final vertex, where we finish it, our path must enter each of the other vertices by one edge and leave the vertex by another edge. Thus, if the graph can be drawn, each vertex other than the initial and final ones must be of even order. This means that only the initial and final vertices can be of odd order and, hence, that there are at most two odd order vertices.

With this theorem you can now conclude that the salesman problem has no solution since there are four odd vertices (the cities) in Figure 4.5.

The converse of Euler's theorem is also valid although we shall not attempt to prove it. There are, instead, some simple consequences of the above argument. First, if a vertex has odd order, then the drawing of the graph must begin or end at this vertex. Reread the above proof and show this is the case. Second, there is no graph with exactly one (or any odd number) odd vertex. Third, in drawing a graph with no odd vertices you will end at the initial vertex, and any vertex can be selected as the initial vertex. All these results follow by arguments like the one used to prove the theorem.

Exercises

1. In Figure 4.7 which can you trace without lifting the pencil from the paper and without tracing any edge more than once?

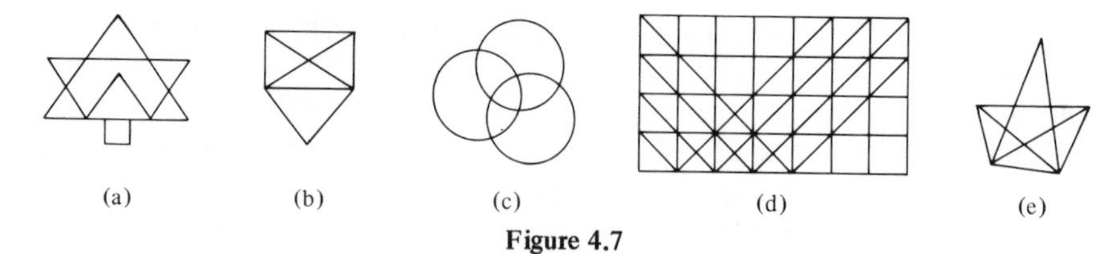

Figure 4.7

2. Euler's original discussion pertained to a problem that the people of the town of Königsberg (now Kaliningrad) suggested: The town was located in

Prussia where two branches of the Pregel river met. There were seven bridges which crossed the river as in Figure 4.8. Could a person cross every bridge once and only once on a continuous walk through the city? Show that the answer is "no." Recently a new bridge was added and it is now possible to take such a walk. Where might it have been placed?

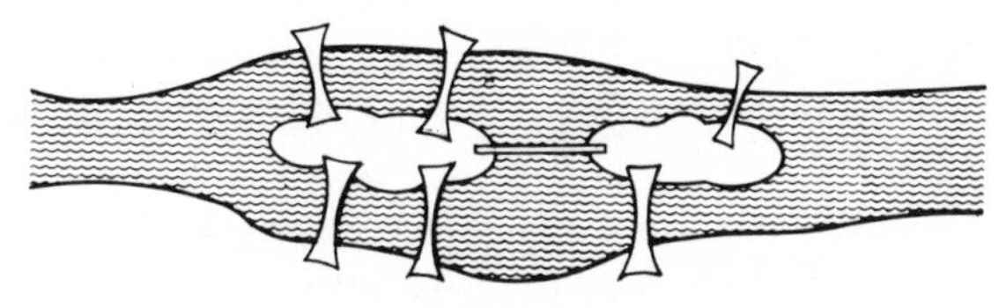

Figure 4.8

3. Given a graph, count the number of edges emanating from each vertex and add these results. The answer is twice the number of edges. Why?
4. Use Exercise 3 to prove that there *must* be an even number of odd vertices in any graph.
5. Suppose Joe Swindle sells toys only in the four Florida cities. Can he find a route so that he *never* passes through the same city twice? (He does not have to use all the routes and he can start at any of the four cities.)
6. Which of the graphs in Figure 4.9 has the property that there is a path which passes through each vertex once and only once? (You do not have to draw the figure. Just trace paths which take you through each vertex once and only once.)

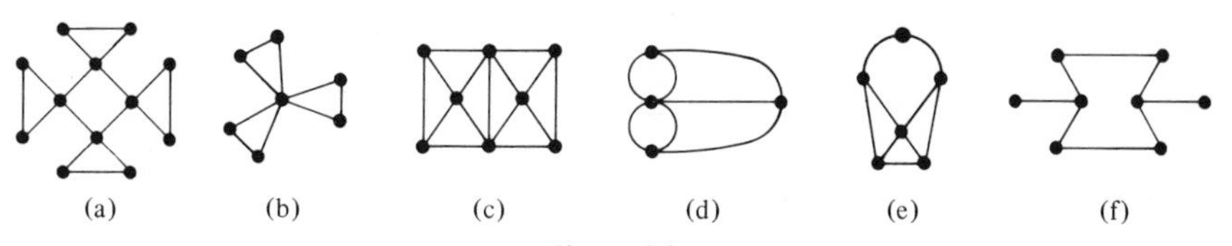

Figure 4.9

7. The problems described in the last two exercises are examples of a problem suggested by the mathematician W. R. Hamilton around 1850. The path is now called a *Hamilton Path* in a graph. No general result like the one for Euler's problem has been discovered. Show that there are Hamilton Paths for some graphs with four odd vertices.
8. If, in a Hamilton path, we can join the last vertex in the path to the initial vertex, then the path is called a Hamilton circuit. Which of the figures in Exercise 6 have Hamilton circuits? (A European mathematician, G. A. Dirac, proved in 1952 that a graph with n vertices ($n > 3$), where each vertex has order at least $n/2$, *always* has a Hamilton circuit.)

9. What is the converse of Euler's theorem? Assuming Euler's theorem and its converse, how many times must you lift your pencil to draw a connected graph with four odd vertices? With six? Can you generalize this result? (Hint: Work from examples; no proof is requested.)

10. A small airline has its central quarters in city A and routes each of its planes through cities B, C, and D. The distances between the cities in miles are:

$$AB = 200, \quad AC = 220, \quad AD = 260, \quad BC = 180, \quad BD = 190, \quad CD = 210$$

Find the shortest round trip for a plane beginning at city A and stopping in all the other cities.

11. Show by a few examples that if a graph has no odd vertices, you can trace it starting at any point and without retracing any edge. The initial point is also the terminal point. Prove this result.

4.3. Duals and Houses

A problem related to, but different from, the ones we have already considered in graph theory is illustrated by the following statement: Draw a continuous path through each edge of the graph (a) in Figure 4.10 such that you pass through each edge once and only once and such that you do not pass through any vertex.

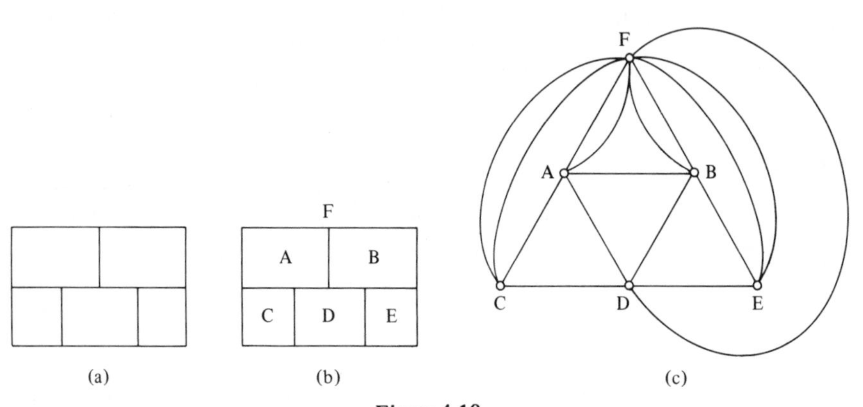

Figure 4.10

This problem can be reduced to the type of graph discussed in Section 2 by the following method. Label the regions in (a) as in Figure 4.10(b). Replace each of the six regions in (b), namely the regions A, B, C, D, E, and the exterior F, by a point. Then replace each edge by a path joining those points which correspond to the regions bounded by the edge. For example, the left vertical line of A bounds A and F and so does the top horizontal line of A. Each of these edges becomes an edge joining point A and point F in the corresponding graph, called the *dual* graph. The other edges are drawn likewise, and the dual graph of (a) is the configuration (c) of Figure 4.10.

The original problem can now be restated in terms of the dual graph. Indeed, passing through a side bounding two regions is equivalent to tracing in the dual graph the edge that joins the points corresponding to these regions. Hence, if we do not wish to pass through any edge of (a) more than once, then we should not *retrace* any edge in the dual graph (c). In this way one soon discovers that the original problem is equivalent to drawing the dual graph without retracing any lines.

Now the original question becomes one which Euler's theorem can be invoked for an answer. The dual graph has vertices A, B, C, D, E, and F; A, B, D, and F are odd vertices whereas C and E are even. In fact, the order of each of these vertices is, from figure 4.10(c), A - 5, B - 5, C - 4, D - 5, E - 4, F - 9. The dual graph cannot, therefore, be drawn without lifting the pencil from the paper since there are more than two odd vertices. This implies that the original problem has no solution since lifting the pencil in drawing the dual graph is equivalent to breaking the path through the edges of the original figure.

It is not difficult to construct the dual graph of a given graph. Perhaps some practice at this task is necessary. The pairs in Fig. 4.11 are dual graphs. (Why?)

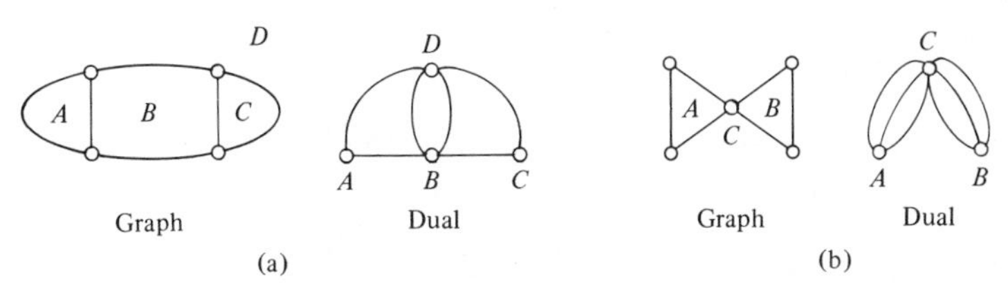

Figure 4.11

Note that each dual graph in Figure 4.11 has no more than two odd vertices; in fact, (a) has none. Hence, each of the original figures has the property that you can find a continuous path which passes through each edge once and only once and which passes through no vertex. Solutions are indicated by Figure 4.12.

A variation of the preceding problem is the one of finding a path which enables a person to walk through each door of a house once and only once. If we denote doors by parallel line segments in a floor plan of a house the answer can be obtained from the dual graph. Here each room, as well as the exterior of the house, is replaced by a point, and these points are connected by paths if and only if there is a door between the corresponding rooms or room and exterior. Figure 4.13 has a dual (b) of the house whose floor plan is (a). The dual graph can be drawn without retracing an edge by starting at either of the points E or D and ending at the other. Indeed, all the other vertices in the dual are of even order.

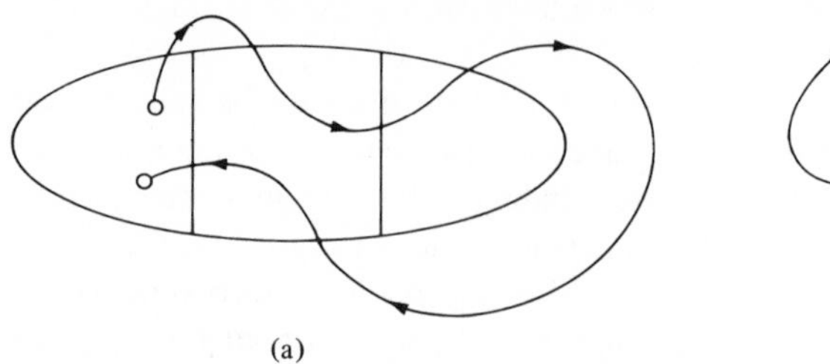

(a)

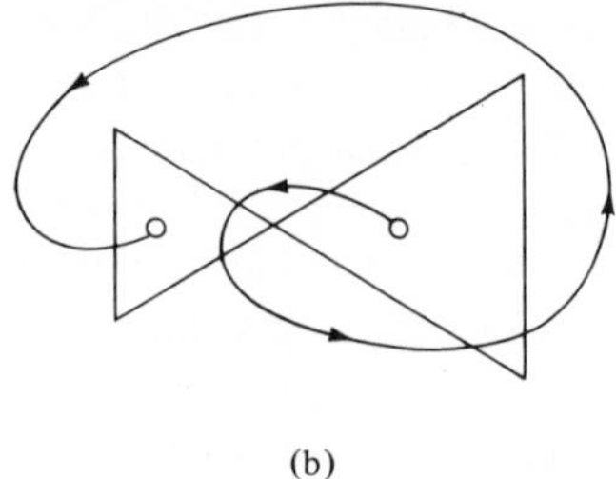

(b)

Figure 4.12

It should be observed that the actual construction of the dual graph is not necessary for solving the problem. One can count the order of the points in the dual graph corresponding to room A in Figure 4.13, for example, by counting the number of doors in room A. The order of the dual point of the exterior F is the number of doors in the house which enable a person to enter the house.

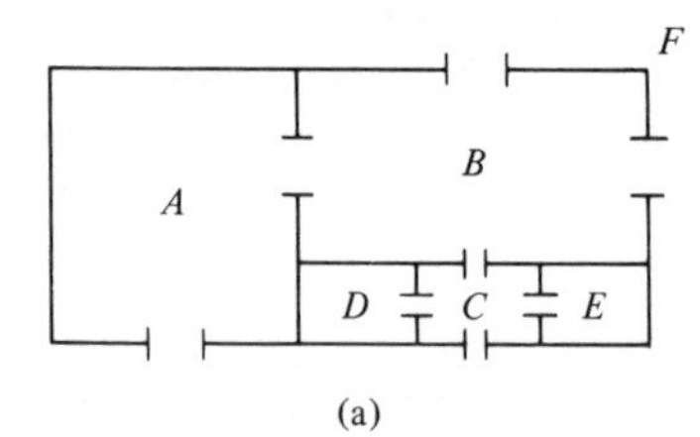

(a)

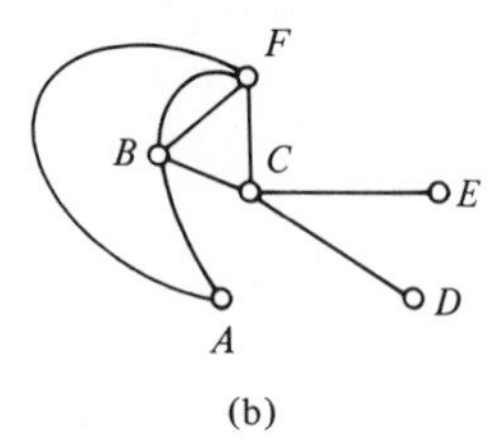

(b)

Figure 4.13

Exercises

1. Can a person walk through each door once and only once in Figure 4.14? Draw the path if he can.

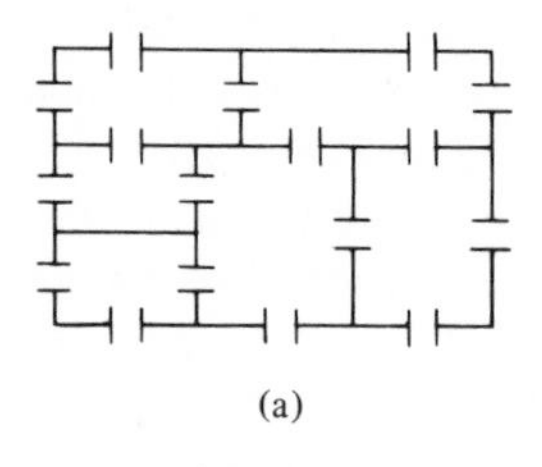

(a)

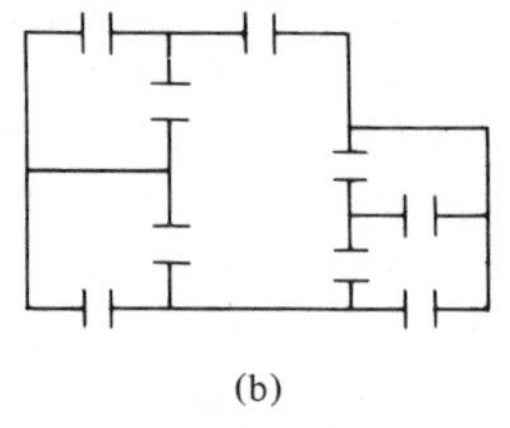

(b)

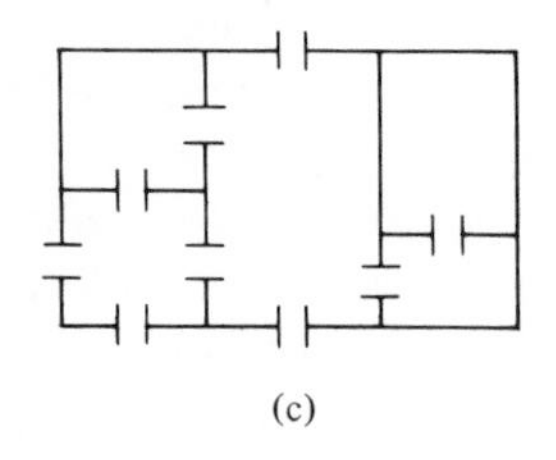

(c)

Figure 4.14

2. Draw the floor plan of your house and determine if you could walk through each door once and only once. (Use only doors to rooms.) How many additional doors (minimum number) would be needed so that you could walk through each door once and only once?

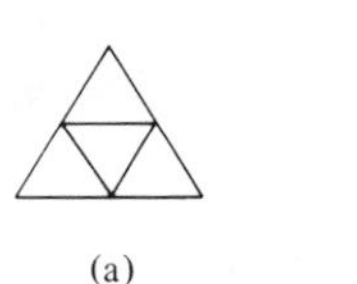
(a)

(b)

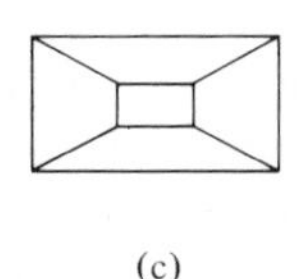
(c)

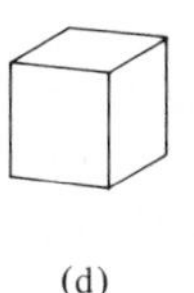
(d)

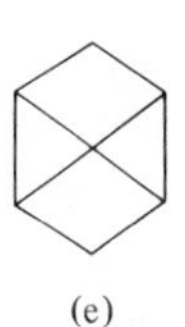
(e)

Figure 4.15

3. For which of the graphs in Figure 4.15 can you draw a continuous line which passes through each line segment exactly once? Draw the path whenever it is possible.

4. Find the dual of each of the graphs in Figure 4.16.

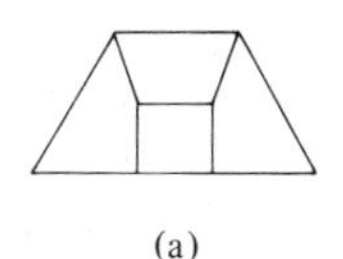
(a)

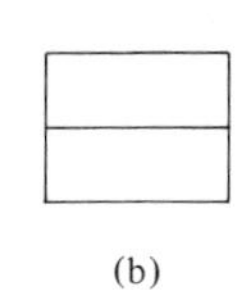
(b)

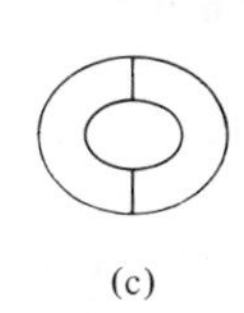
(c)

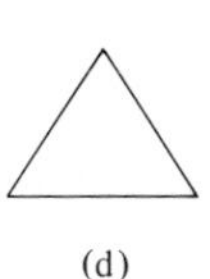
(d)

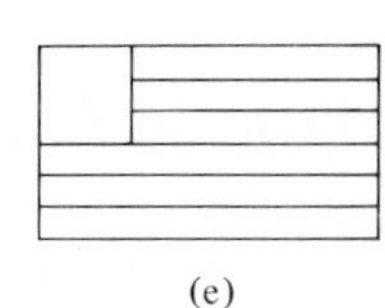
(e)

Figure 4.16

5. What is the fewest number of doors that a person must pass through more than once in each of the houses of Exercise 1 if he passes through each door of the house starting from a position outside the house?

4.4. PERT Charts

Graphs have been used to solve a number of industrial and environmental problems in recent years. One such application is the so-called PERT (Program Evaluation and Review Technique) chart that originated in the development of the Polaris submarine.

In order to illustrate the technique, suppose that in the production of new automobiles some of the major tasks of the construction can be accomplished simultaneously, and that some cannot be started until certain other tasks are completed. We will oversimplify and list these tasks as: 1. Building the frame; 2. Installing the engine and the controls; 3. Installing the fuel tank and lines; 4. Completing the body. Assume that task 1 takes 3 days, task 2 takes 10 days, task 3 takes 2 days, and, finally, that the last task takes 1 day. Moreover, the priorities that must be observed in initiating the task are that task 2 requires the completion of 1, and task 3 requires the completion of task 1, whereas task 4 cannot be started until 1, 2, and 3 are completed. We now graph these tasks using line segments to represent the time period during which the task is performed. The tasks are placed from left to right on the graph according to the priority list with tasks that can be accomplished simultaneously placed in parallel (Figure

4.17 (a)). Broken lines indicate the passage from one task to another. We did not place a line from task 4 to task 1 since task 4 cannot be initiated upon completion of 1; task 4 requires completion of task 2 and task 3 before it is begun. Diagram (b) in Figure 4.17 lists the days required to complete the various tasks in the graph.

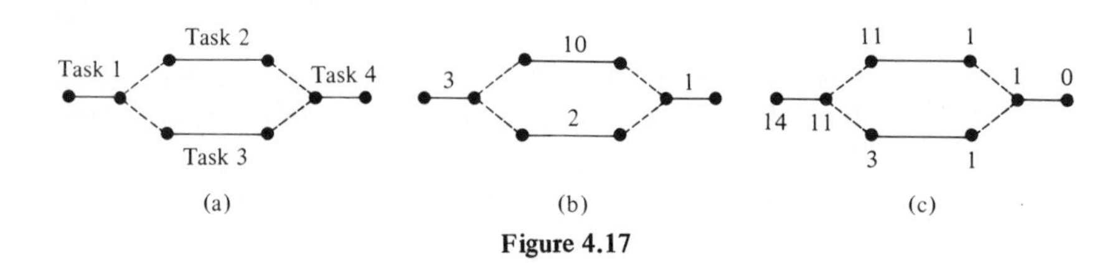

Figure 4.17

Next, starting at the right end point in the graph of the entire project, we assign numbers to the end points of each segment in the graph according to the following rule: The number assigned to each point is the number of days needed to complete the remainder of the project. For example, it takes one more day to complete the project after tasks 2 and 3 have been completed; hence, 1 is assigned to the left end point of the last segment since one day is needed to complete the body (task 4). The right end point of the task 2 segment is 1 since it takes one day to complete the project after task 2 is accomplished. The left end point of this task is assigned an 11 since it takes 10 days to complete task 2 and one more to finish the automobile. The assignment for the end points of task 3 is similar. Should you encounter a branching point, you always assign the maximum number of all the tasks that are fed from this point. For example, task 1 in Figure 4.17(c) feeds tasks 2 and 3 so we assign 11 to its right end point since 11 is the maximum of 11 and 3. Since task 1 itself takes three days, we add this to the 11 to obtain the number for its left end point, namely, 14. After doing all of this, the number at the beginning of the first task is the minimum time required to complete the project. The *critical path* is that path in the graph which requires this amount of time to complete, for example, the upper path indicated by arrows in Figure 4.17(c). This particular path is of interest to the project manager since any delay in the tasks along this path will definitely delay the completion of the new automobile. A delay in task 3, unless prolonged beyond eight days, has no such effect. (Why can task 3 be postponed up to *eight* days without delaying the entire project?)

We next turn to a more complex example, given by a chart.

TASK T	1	2	3	4	5	6	7	8	9
Days needed	3	1	1	2	1	4	2	1	2
Priorities	0	0	2	2, 3	1, 2, 3	1, 2	4	4, 5	4, 5, 6, 7, 8
Implied Priorities	0	0	2	3	1, 3	1, 2	4	4, 5	6, 7, 8

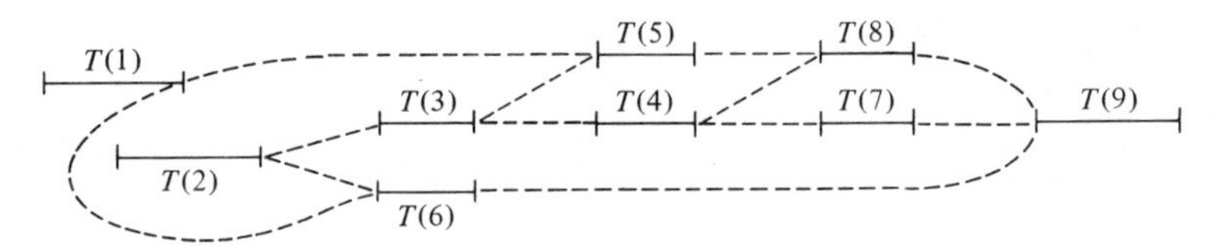

Figure 4.18

The last row eliminates redundant priorities. For example, task 8 requires 4 and 5, and task 9 requires 8, and, hence, 9 automatically requires 4 and 5. The PERT chart, after these redundancies have been removed, is Figure 4.18.

The number PERT chart is next constructed as in Figure 4.19, where for il-

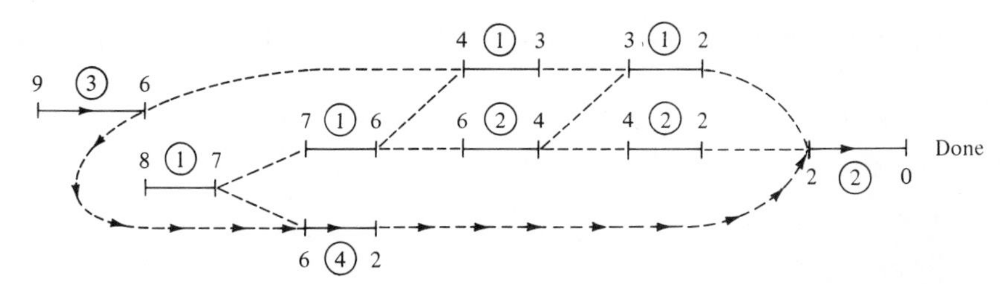

Figure 4.19

lustration the number of days needed to complete a task is circled. One starts the chart at the zero point on the extreme right and adds the circled number to obtain the other end point. When a branch occurs this number is passed on to each end point of the line segments leading to the branch. The *maximum* of *all* the left end points of segments connected with a right end point is the number associated with the latter point. This process continues until each end point of the segments has a unique number assigned. Since we have two starting points for this project, task 1 and task 2, the critical path is the path taking the *most* time. Hence, the critical path begins with task 1 since we are nine days from completion once this task is initiated as opposed to the eight days needed when we begin task 2. The critical path is indicated by arrows in the last figure. Any delay in tasks 1, 6, or 9 will, therefore, postpone the completion of the entire project.

Exercises

1. Consider the example in the text represented by Figures 4.18 and 4.19.
   a. What affect does a one day delay in completion of task 4 have on the project?
   b. Suppose task 8 is delayed three days before it is begun. Does this affect the total time needed to complete the project?

c. If task 7 were delayed four days, show that the project would take 12 days, instead of 9, to complete.

2. Construct a PERT chart for the project consisting of obtaining a B.A. in mathematics if the following chart indicates the minimum requirements for the degree. Locate the critical path and calculate the minimum time necessary.

Course	Qtr.	Prereq.	Course	Qtr.	Prereq.
1. Calculus	3	none	6. Numerical Analysis	2	3, 4
2. Calculus V	1	1	7. Math. Statistics	3	3, 4
3. Linear Algebra	2	1	8. Number Theory	1	1, 2
4. Advanced Calculus	3	1, 2	9. Geometry	1	1
5. Real Numbers	1	2, 3	10. Senior Colloquium	1	4, 5, 6

Suppose only 1, 3, and 4 are required. Write a PERT chart for these requirements.

3. Build a PERT chart for whatever subject you are majoring in.

4. Construct a PERT chart for the following project:

Task	1	2	3	4	5	6	7
Requires for completion		1	1	3	3, 4	2, 3, 4	5, 6
No. of days to complete	10	2	3	2	4	5	3

What is the minimum time necessary to complete the task, and what is the critical path?

5. In preparing a meal there are often tasks that can be done simultaneously. Build a PERT chart for the preparation of your favorite holiday meal.

4.5. Euler's Formula and Map Coloring

In the first section of this chapter, mention was made of invariants in geometry. There is an invariant in graph theory, discovered by Euler, which has proved to be useful in answering certain questions. If we call the region enclosed by the edges of a graph a *face* and also count the exterior of a planar graph as a face, then we can establish the formula of Euler.

THEOREM. Let V be the number of vertices, E the number of edges, and F the number of faces (including the exterior) of a planar connected graph. Then $V - E + F = 2$.

This is a result that is quite independent of the complexity of the planar graph. For example, each of the figures in Figure 4.20 has $V - E + F = 2$.

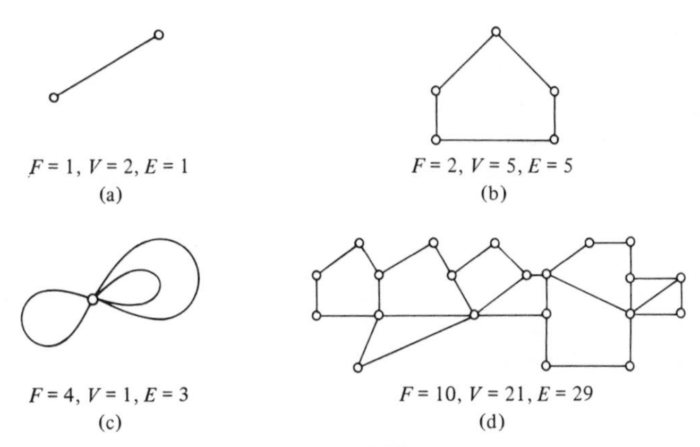

Figure 4.20

Proof: The proof is by construction. A graph is made up of loops, as in Figure 4.20(c), or else it has an edge with two distinct vertices, as in Figure 4.20(a). If it has only loops with a common vertex, each loop bounds a face, so the number of edges, E, is equal to the number of these faces. But the exterior is also a face. Hence, $F = E + 1$ so $F - E = 1$. Adding $V = 1$ gives $F - E + V = 2$ in this case.

Suppose that there is an edge with two vertices in the graph. Then we begin drawing the figure step-by-step as follows. We either (1) pick a new vertex and join it to an old vertex by a simple (nonintersecting) edge that does not cross the existing edge of the graph or (2) join the two existing vertices by a new simple path not crossing the existing edge or (3) add a loop to one of the vertices so that this loop does not cross the existing edge. Each of these steps is illustrated in Figure 4.21.

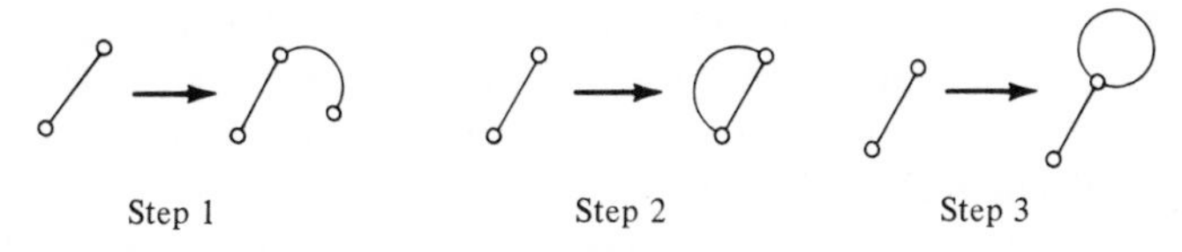

Figure 4.21

In each case $V - E + F = 2$ for the new figure since the following things happen for these steps:

Step 1. Adding a vertex also adds an edge. Hence, there is no change in the formula since E is subtracted from V.

Step 2. Adding an edge also adds a face, and this does not change $V - E + F$.

Step 3. Addition of a loop (an edge) also adds a face and again $V - E + F$ is unchanged.

Now to draw the entire graph we need only apply steps 1, 2, and 3 in a finite number of times. But each application leaves $V - E + F$ unchanged. Since it was initially 2, it must be 2 after the graph is completed. Hence, $V - E + F = 2$ for all planar connected graphs.

Figure 4.22 illustrates the step-by-step application of the above procedure in drawing the graph on the extreme right.

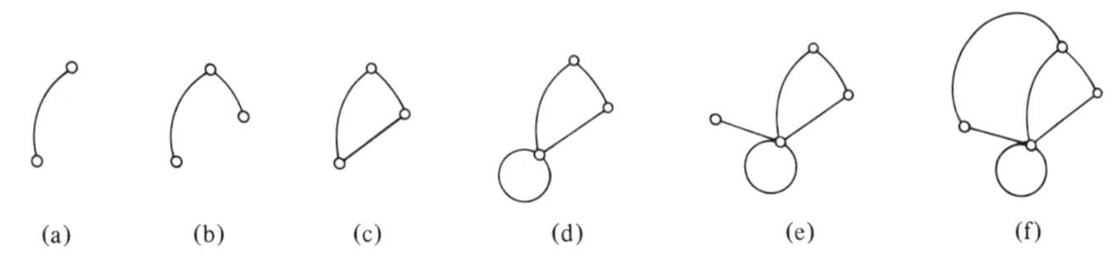

Figure 4.22

One application of the formula of Euler is in map coloring. A British student, Francis Guthrie, conjectured around 1850 that it would take no more than four colors to color a map so that countries with common borders (edges) would be contrastingly colored. This problem was brought to the attention of many leading mathematicians of the day and no justification was found. A "proof" was reported in 1879 by a lawyer, Kempe, and the problem was believed to be solved until 1890 when it was noted that there was a fundamental error in the proof. The method of Kempe, however, could be adjusted to verify that no more than five colors are necessary in the map coloring problem proposed by Guthrie. Until this day it has never been proved or disproved that more than four colors are needed to color a map.†

There is an application of the construction in the proof of Euler's theorem to the design of printed circuits. For many modern electrical devices, a circuit is pressed onto a flat surface and the printed wires are not insulated. Hence, wires may touch only at specified points, called terminals. For example, if we have four terminals and each must be connected by wires in the pressed circuit, we cannot use the circuit in Figure 4.23(a) since the wires A and B cross. However, the four can be connected to each other, as indicated in Figure 4.23(b), without crossing wires.

For a more complicated example, consider the six terminals 1, 2, 3 and *a, b, c* as shown in Figure 4.24(a). Can we find a printed circuit which joins each numbered terminal with each lettered terminal? When treated as a graph, there

†The justification of the fact that five colors is sufficient is long and is omitted from this text. Euler's Formula plays an important role in the proof. For a complete and elementary discussion, see E. L. Spitznagel, Jr. *Selected Topics in Mathematics*. New York, N.Y.: Holt, Rinehart and Winston. 1971. Chapter 5.

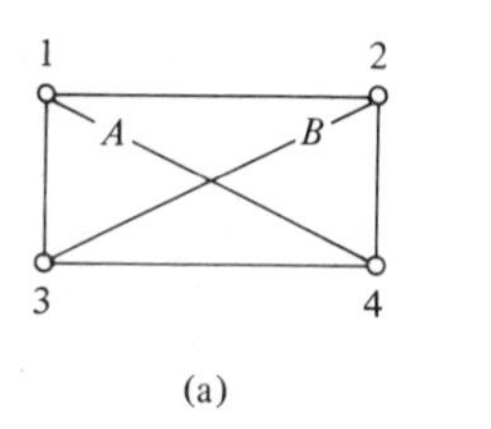

(a)

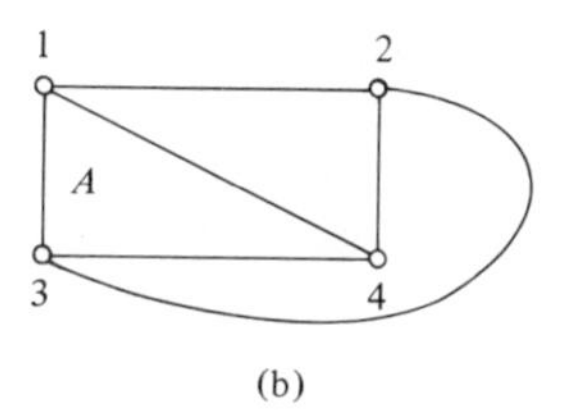

(b)

Figure 4.23

are six vertices V and nine edges E. By Euler's formula, there must be $F = 2 - V + E = 2 - 6 + 9 = 5$ faces. Now it is clear by construction that each face must be bounded by four or more edges. (There are no connections between numbered vertices or between lettered vertices.) Thus, if we list the number of edges bounding each face and add these together, we have a total that is greater than or equal to $4 \times 5 = 20$. Now each edge separates two (and obviously only two) faces, one on the left and the other on the right (with any orientation). Hence, in our total of 20 or more we have counted each edge twice, once when it bounds one face and then again when it bounds the other. This means there must be $20/2 = 10$ or more edges. But this is a contradiction since $E = 9$. It follows that the construction is impossible.

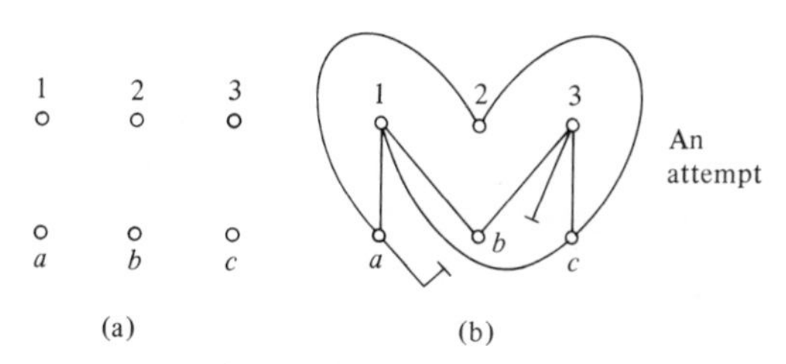

Figure 4.24

Exercises

1. Build models of all graphs with $E = V$ when $V = 2$, 3, and 4.
2. Can you build a printed circuit with each of four numbered terminals connected to each of a pair of lettered terminals?
3. Can you build a printed circuit with each of five terminals connected to all the others?
4. Suppose there is a graph in which (1) every vertex has three edges, (2) every edge has two ends, and (3) each edge is shared by two faces. Show that this implies that one face has at most five bounding edges.
5. What is the value of $V - E + F$ if we total the edges, vertices, and faces for three disconnected graphs? For four? Generalize.

6. Suppose two points are represented on a sheet of paper. Draw an arc joining these points. When this arc and all subsequent arcs are drawn a new vertex is chosen on the arc. Then a new edge is drawn either joining existing vertices or else forming a loop so that (a) no edge crosses itself or another edge in the graph and (b) no vertex has more than three edges emanating from it. Prove, using Euler's formula, that there are 7 vertices and 10 edges in the graph when the instructions can no longer be carried out. Figure 4.25 is an example. (This can be played as a game, called "Sprouts,"† in which players alternate turns. The last person who is able to draw an arc wins the game. In this case the first to draw always wins.)

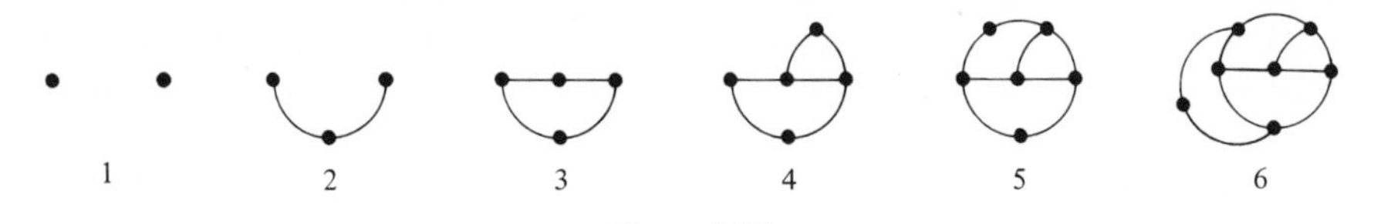

Figure 4.25

4.6. Is the Two Party System the Greatest?

In recent years graph theory has proved fruitful in forming mathematical models in the social sciences. We will attempt to illustrate this by creating a model for groups of human beings working together.

PROBLEM: A group of people, such as the U.S. Senate, is required to make decisions on certain major issues. Each pair of individuals has favorable, unfavorable, or neutral relations. By a favorable relation we shall mean that they have similar views and can work together. It is an unfavorable relation if they have divergent views and naturally oppose each other in their work. If the relationship is neither favorable nor unfavorable it is termed neutral. (They may have similar views but cannot work together.)

Assume that there are relationships of each type in the group. What type of division occurs in the group?

MATHEMATICAL TOOLS

We represent by a vertex (dot) each person in the group and join the vertices by edges if there is a relationship, other than a neutral one, between the persons corresponding to the vertices. If the relationship is favorable, we assign a "+" to the edge and assign a "–" when the relationship is unfavorable. For example, some of the possible graphs with a three person group are shown in Figure 4.26.

Note in (d) that all get along and in (c) that none get along. In (e), Mr. *a* and Mr. *c* have a neutral relationship whereas Mr. *b* gets along with both of them.

†M. Gardner. Mathematical Games. *Scientific American*. July, 1967. pp. 112–115.

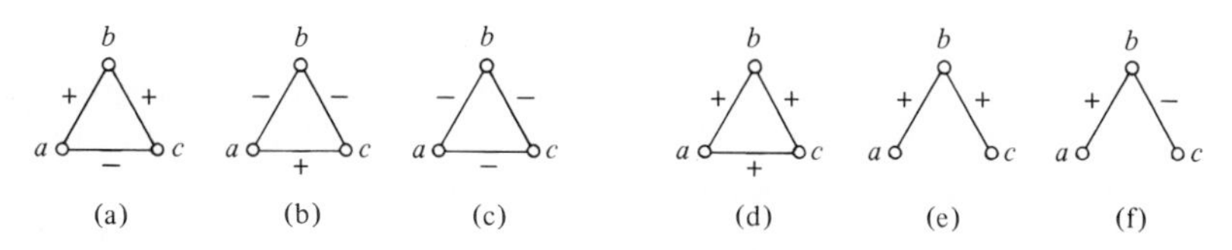

Figure 4.26

We now define a cycle to be any path in a graph that begins and ends at the same point and does not pass through any vertex more than once.

For instance, $a - b - c - a$ is a cycle in Figure 4.27 and so is $c - d - e - c$. But $a - c - d - e - c - a$ is not since this path passes through c twice. A cycle is *positive* if it has zero or an even number of negative signs and *negative* if it has an odd number of negative signs assigned to the edges in the cycle. The *abca* and the *cdec* cycles in the figure are positive.

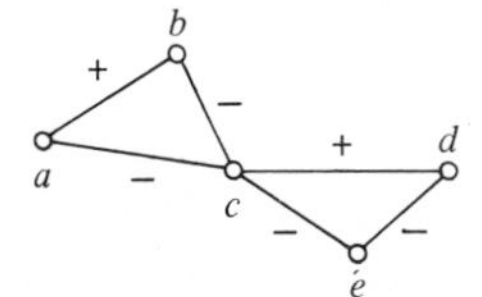

Figure 4.27

What type of cycle corresponds to a group of people who, if divided, are clearly divided into camps of opposition. For example, consider the three person situations in Figure 4.26. (a) Here a and b work well together but are faced with conflicting relations with c. It would be hard for such a relation to be productive. The cycle is unstable since, if b works with both a and c, they will resist b's efforts because they themselves do not work well together. (b) This graph corresponds to a balanced (stable) situation since a and c will work together against b. The camps of opposition are well defined. (c) Obviously no progress can be made here at all! The situation is volatile. (d) All three are happy with the relationship represented here. (e) and (f) are not cycles but both are stable relationships.

The cycles in Figure 4.27 were positive whenever the situation represented by the graph was stable. If a cycle was negative, a change in the relationship of at least two persons would be necessary before the opposing camps could explicitly be identified. These observations serve to motivate the following definition. (Harary, F. and Cartwright, D. Structural Balance. *Psychological Review*. vol. 63, pp 277–293).

A graph is *balanced* if every cycle in it is positive.

With all these definitions, we are now ready to prove a theorem concerning group interaction. Consider the U.S. Senate with its (essentially) two party structure. Assume that the people in each party have a favorable or neutral relationship with each other and that all persons have a neutral or negative relationship with individuals of the opposite party. Is the graph associated with the Senate then balanced? Well, consider any cycle in the graph. This cycle can have a negative edge only if we cross over from one party to the other. Since a cycle must end at the starting point, there must be an even number of these cross overs and thus the graph is balanced. The Senate is, according to these assumptions, a "stable" group of individuals.

We have, in fact, proved with the above argument a theorem which can be applied to other organizations.

THEOREM. Let the set of all vertices in a graph be partitioned into two disjoint sets, A and B (perhaps one is empty), such that:

a. If two vertices in the same set have an edge joining them, this edge is positive.
b. If two vertices in different sets have an edge joining them, this edge is negative.

Then the graph is balanced.

The converse of this Theorem is also true. Thus, although we will not prove it, every balanced graph has the partitioning of the vertices as stated in the Theorem. In particular, if the U.S. Senate corresponds to a balanced graph then we must have a two party system (party A and party B). Before you become too satisfied with this conclusion, remember that one party might have no members!

EXAMPLES: The graph for two six man committees is given in Figure 4.28. Do the committees' members have a stable working relationship?

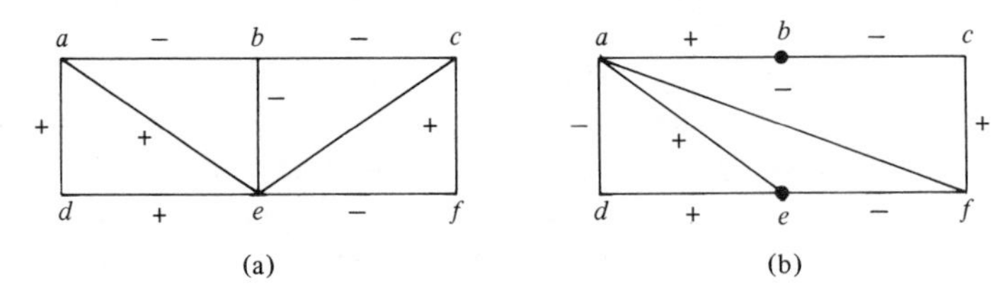

Figure 4.28

Case (a). $A = \{a, e, d, c\}$. $B = \{b, f\}$. The graph is balanced and a majority are in set A. The opposition is well defined. When any issue comes to a vote, the opinion of Group A will prevail.

Case (b). This is not a balanced graph since *adea* is a negative cycle. We cannot partition the graph as in the theorem. This committee would not work well.

The legislative body in many states must correspond to an unbalanced graph.

Exercises

1. Consider the graphs in Figure 4.29. Which are balanced?

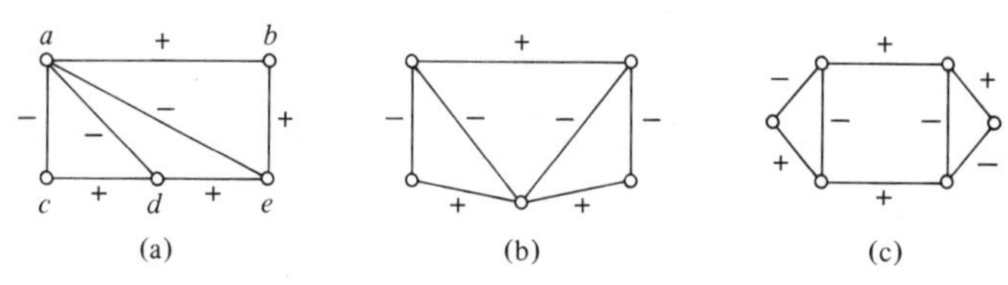

Figure 4.29

2. Which of the graphs in Figure 4.30 represents the best working relationship for a five person committee? (None are balanced, so the meaning of "best" is perhaps subjective.)

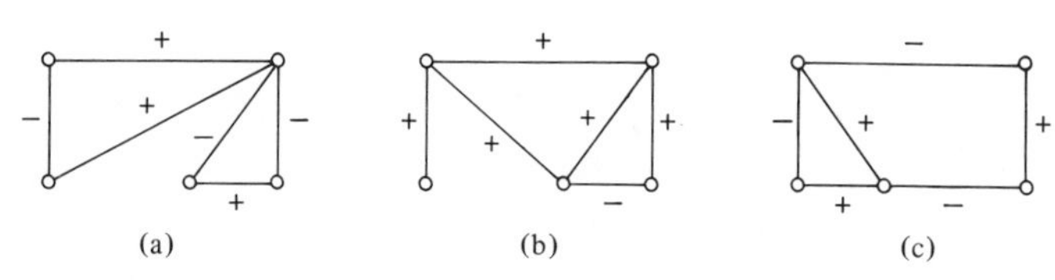

Figure 4.30

3. Can you build a map of a town in which every street is one way, (except dead-end streets and single bridges) and yet a person cannot travel from a certain point in the city to any other point? Denote with a "+" one of the directions of a one-way street and denote the opposite direction with a "–." Then the first part of this problem corresponds to a signed graph. Can you build a map that meets the conditions in the first sentence of this problem and that corresponds to a *balanced* graph?

4. At any party, some of the guests know each other while others are unacquainted. *If the party is attended by at least four people, then there must be at least three people who are acquaintances or at least three people who are mutual strangers.* Represent each of the four persons at the party by points in a graph. Use green lines to connect acquaintances and red lines to connect strangers. Three acquaintances appear as a green triangle (with vertices representing the people) and three strangers as a red triangle. Verify the underlined statement for a number of special cases with these colored graphs.

5. Draw a graph for six such people at a party and connecting people who know each other by green lines and those who are unacquainted by red lines. (Use the pattern in Figure 4.31. What you decide is the color of the lines.) There should then be *at least* two triangles in the figure of the same color. (There are at least two sets of three mutual acquaintances or mutual strangers.)

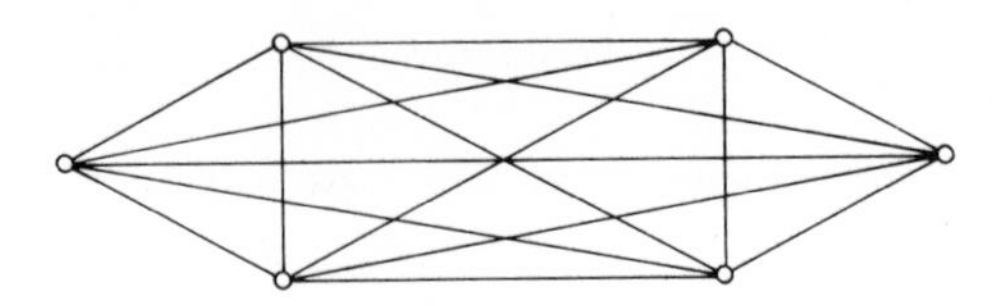

Figure 4.31

(Frank Harary. *The two-triangle case of the acquaintance graph. Mathematics Magazine*. 45 (1972), 130–135)

4.7. Who Can't Tell the Trees from the Forest!

A *tree* is defined as a connected graph that contains no cycles. A forest is any collection of trees. For example, Figure 4.32 is a forest, according to this definition, with three trees. (It must be late autumn or winter!)

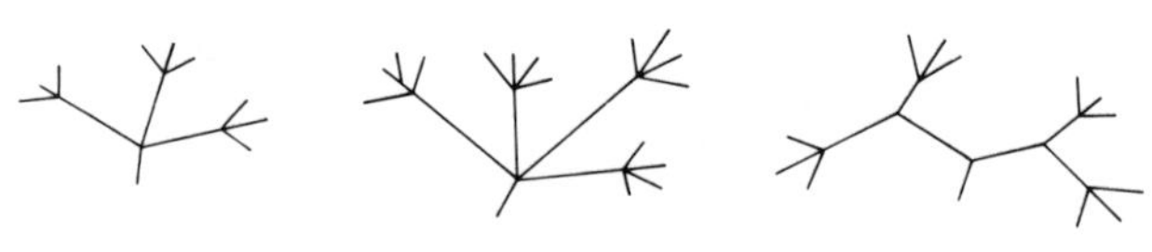

Figure 4.32

Tree graphs are useful for counting purposes. For example, if you are taking four courses this quarter, how many possible grade reports can you receive? Checking the college bulletin, suppose one finds that there exist possible grades, A, B, C, D, F, I, and W. The associated tree would have four initial branches, one for each course, and each of these subdivides into seven branches, one for each possible grade. By counting, there are 28 possible grade reports, as in Figure 4.33.

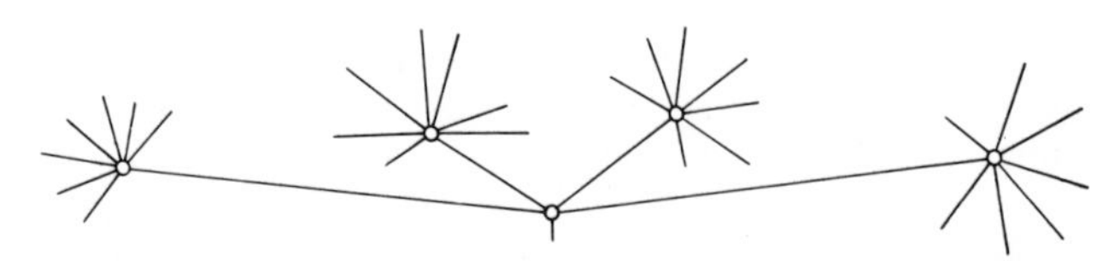

Figure 4.33

The counting can be much more complicated. Consider a game of Nim in which each player can pick up one to three sticks in a turn from a single pile. If there are five sticks in the original pile and if the turns alternate, there is a tree which completely describes the game. In order to draw it, start with the first player A and consider each of his moves as the first set of branches (edges), as in Figure 4.34 (a).

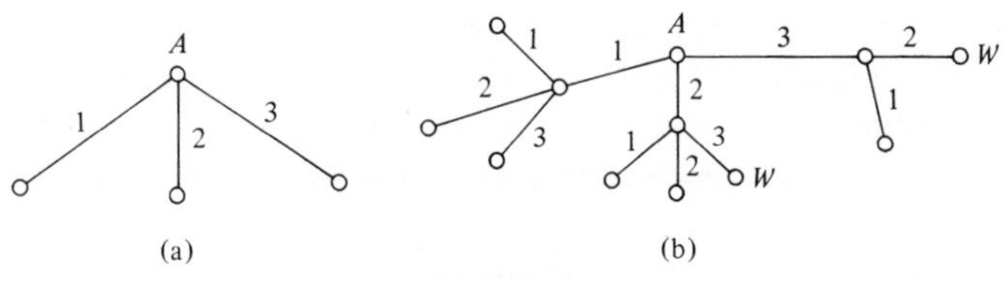

(a) (b)

Figure 4.34

The numbers on the branches indicate the sticks that A has removed on his first turn. Now for each of these moves, draw branches indicating all possible moves of the second player, B, as in Figure 4.34(b). Note that if A initially picks up 2 sticks and B follows with 3, then B loses since he picked up the last of five sticks. We indicate this move by a W (win for A) on the graph. The same type of situation occurs if A draws 3 and B follows by 2. In all other cases the game continues through at least one more turn.

If this process of drawing the tree continues until there are no additional moves along any branch, the complete tree graph of the game is obtained (Figure 4.35).

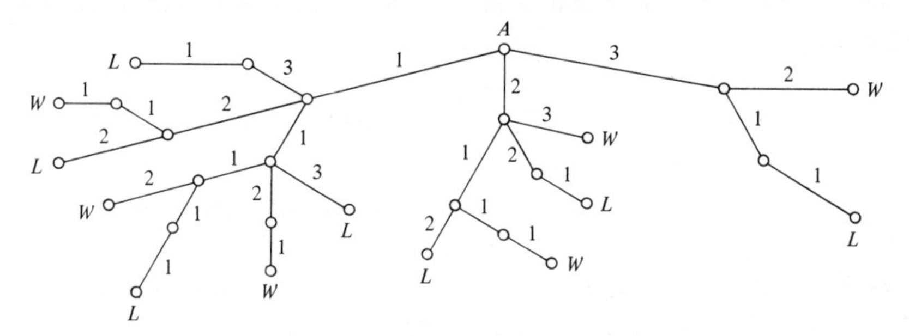

Figure 4.35

The graph indicates that each initial move of A could lead to a win as well as to a loss. Note, however, that the second player, B, by suitably selecting his first move, can always force A down a losing branch of the tree.

Even a simple game, such as the one in the last illustration, has a complicated tree. Yet, if there is no other method to attack a problem, trees, or at least partial trees, can serve as guides for the proceedings.

There are some general principles about trees that can be useful. One is the following.

THEOREM. A tree with n vertices has $n - 1$ edges.

Proof: A tree with one edge has two vertices so the result is certainly true in this case. If we add a branch (edge), it must emanate from one of the existing vertices but it also adds a new vertex. Thus a new edge and one new vertex has been added and the number of edges remains one less than the number of vertices. In general, when a branch is added to an existing tree, only one new vertex and one new edge is added. This implies the number of edges always remains one less than the number of vertices since this is the way things were at the beginning of the drawing.

As an application, consider the problem of connecting a number cities A, B, C, ... by an interstate highway. The costs $c(AB)$, $c(AC)$, $c(BC)$, ... for constructing the road between pairs of cities are known. The graph representing the cheapest connecting network must be a tree, called the *economy tree*. Indeed, if there was a cycle in the graph, one road could be removed and the cities would still be connected. Now, if there are n cities in the original collection, the economy tree must, by the theorem, contain $n - 1$ links.

In order to build the economy tree, the first step is to connect those two cities which have the cheapest link. If a number of cities have links of minimal cost, then any of these links can be used to initiate the tree. In each step thereafter we add the cheapest remaining link to the cities already in the graph. The complete economy tree is built in this fashion.

For example, let A, B, C and D be the cities, and let the cost factors be given by a chart.

AB	AC	AD	BC	BD	CD	Link
100	90	85	110	75	85	Cost
						(thousand dollar units)

We begin with the cheapest connecting road (link), BD. Then either the AD or the CD road (which have equal cost) can be added. Assume that we have added the AD road. The next link is to add the road CD since $c(CD)$ is smaller than the cost of any remaining road between two cities. All the cities are now connected and the economy tree AD, BD, CD has been constructed. (See Figure 4.36.)

Does this process really produce the economy tree? Are we certain that some other tree T cannot be drawn which has a smaller total cost? Assume that there is such a tree. The edges of the constructed tree E were added in a definite order as described in the previous paragraphs. Since T is not E, there must be a first edge, e_1, in E that is not in T. All previous edges are part of T as well as E. Let A and B be the two cities connected by e_1. If we adjoin the edge e_1 to the tree T, then $T + e_1$ has a cycle since we can travel from A to B along the branches in T and then from B to A along e_1 as illustrated in Figure 4.37. Now E has no cycles, since it is a tree. Hence, there is at least one edge in the created cycle, call it e_2, that is not a link in the tree E. Next create a new tree S by removing e_2 from the tree T and adjuncting in its place e_1. The result is indeed a tree since we can still pass from any city to another along the links in S. The tree S is the

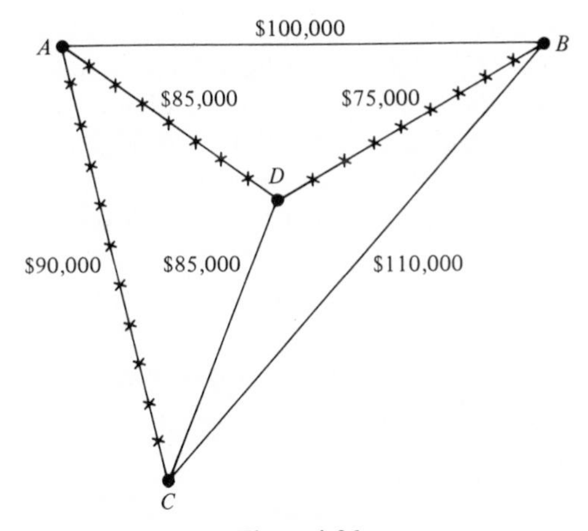

Figure 4.36

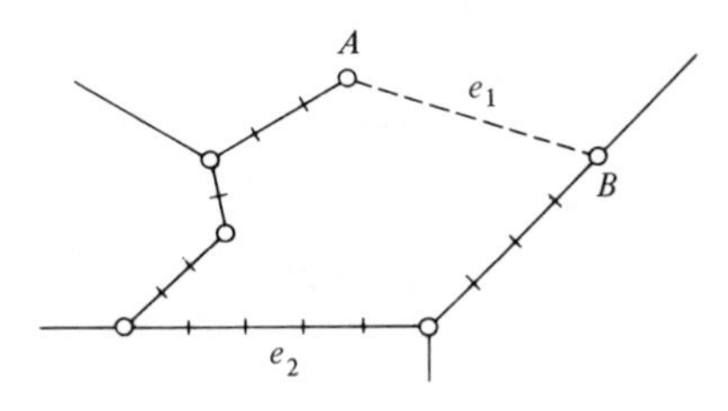

Figure 4.37

same as T, except that edge e_1 has been replaced by e_2; in symbols

$$S = T + e_1 - e_2$$

where "+" and "-" mean adjuncting and removing an edge, respectively. If we consider the cost of the new tree S, we have, by the method of calculating cost, that

$$c(S) = c(T) + c(e_1) - c(e_2)$$

If T is a tree with smallest cost, then $c(S) \geqslant c(T)$ so $c(e_1) - c(e_2)$ must be a nonnegative number if the inequality is to hold. Thus, $c(e_1) \geqslant c(e_2)$. But by the construction of E, e_1 was an edge which when added to the previously used edges of E increased the cost by no more than that obtained by adding any other edge. Now e_2 is an edge that also can be added to the previously used edges of E; hence, $c(e_2) \geqslant c(e_1)$ by the selection process of e_1. The two inequalities imply that $c(e_2) = c(e_1)$, so $c(S) = c(T)$. In this manner, a tree S has been found with minimum cost and having one more edge in common with E than does T.

Continue this process with S replacing T. In a finite number of steps we will obtain a tree, S', identical with E. Each step of the process adds one more edge that belongs to E. Since $c(T) = c(S) = c(S')$ by the construction process, we have $c(T) = c(E)$ and, hence, the economy tree is one with minimum cost.

For instance, if the trees E and T are as defined in Figure 4.38 the sequence of steps to pass from T to E that is described in the previous paragraphs is the sequence of construction. The cost factor of each link is indicated by the circled number in each diagram.

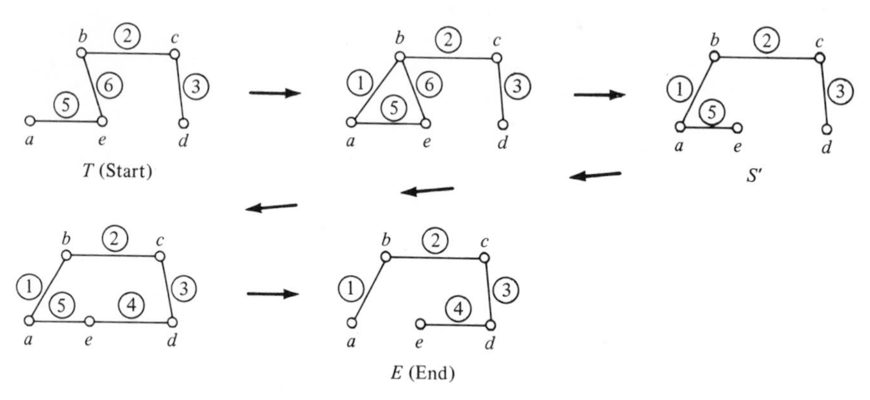

Figure 4.38

This problem is rather long and has "tight" logical steps. It is, however, an example of considerable practical value and the method of proof is similar to many arguments needed by a practitioner of the art of applied mathematics.

Exercises

1. Build a tree for a game of Nim with three piles of 1, 2, and 3 sticks, respectively.
2. Draw five points on a sheet of paper and find the path of shortest length that connects all five of these points. (Use a ruler to measure the distance between the points.)
3. Suppose the telephone company wishes to link four cities, A, B, C and D, by wires. The cost of the wire is proportional to the distance between the cities and the latter is given in miles by the following chart: $AB = 110$, $AC = 200$, $AD = 75$, $BC = 150$, $CD = 210$, $BD = 140$. Build an economy tree for this problem.
4. Show that there are two economy trees for the previous problem if $AB = 110$, $AD = 75$, $AC = 150$, $BD = 160$, $BC = 150$, and $CD = 200$.
5. If (a) in Figure 4.39 is the economy tree, parallel the steps in the text that

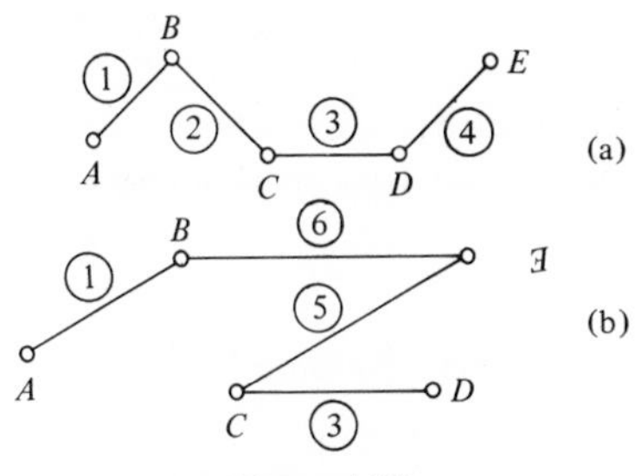

Figure 4.39

enable us to pass from (b) to (a) by adding a link from (a) and removing a link from the cycle so formed in (b). (Also see Figure 4.37.)

6. Draw a tree diagram for the following variation of the simplified game of Nim: There are 5 sticks. Each player can pick from 1 to 3 sticks in a turn. After his initial turn, the player must select either one more or one less stick than he did on his previous turn. The player must select the remainder of the pile and lose if he is required to select in his turn as many or more sticks than are in the pile. Prove that the person who moves second in this two-person game can always win.

7. Draw a tree diagram for the game in Exercise 6 if six sticks are in the initial pile. Show that the second player can always win.

8. Label with letters each vertex of a connected graph. Can you label each edge uniquely to correspond to the label of one of its vertices? If the graph is a tree, this can be done as in the first example in Figure 4.40.

   a. Construct a few trees and attempt to develop a method for naming each edge uniquely in terms of one of its vertices.

   b. Find such a label for each edge in Figure 4.40(b).

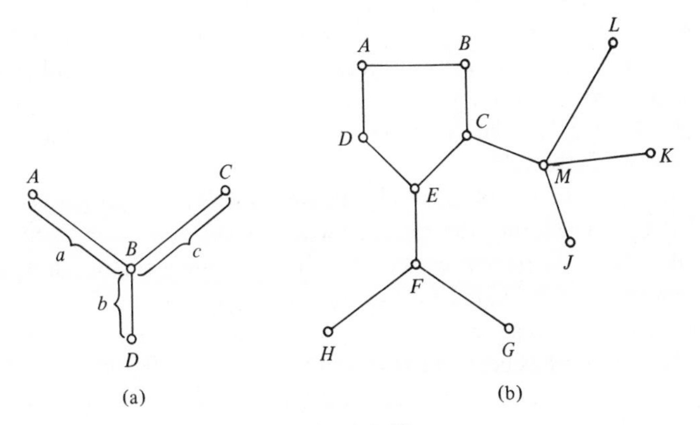

Figure 4.40

c. Can you construct a simple example which shows that this labeling process cannot be accomplished for all graphs?
d. Finally, from a conjecture as to when this task can be accomplished. (This type of problem can arise when a city attempts to name streets the same as squares from which the streets emanate.)

4.8. Platonic Solids

The Greek philosopher Plato believed in the existence of four elements–fire, earth, air, and water–each of which was composed of atoms. The atoms were certain geometric solids, for example, earth atoms were cubes and fire atoms were tetrahedrons. These assumptions were undoubtedly influenced by the Pythagorean school with the result that there are only five "regular" solids. Euclid discussed these "Platonic" solids and proved that exactly five such bodies exist in what appears to be a climax of his *Elements*.

This section is devoted to a graph theory proof for the existence of, at most, five Platonic bodies. We begin the study with a few definitions.

A *regular polygon* is a plane figure with all sides the same length and all angles equal. For example, the equilateral triangle and the square are regular polygons and, in fact, the only regular polygons with three and four sides, respectively. See Figure 4.41 for other examples.

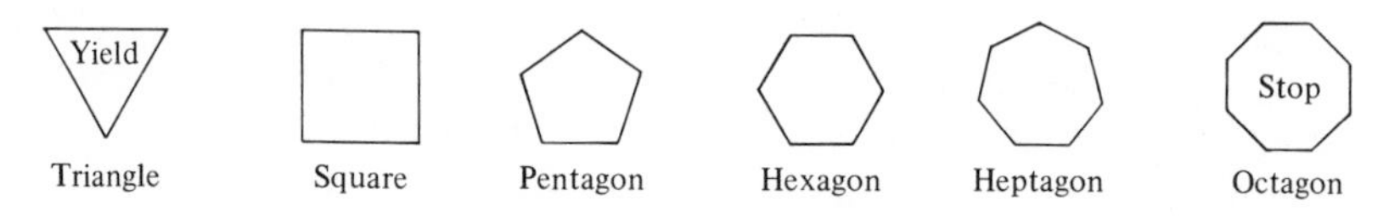

Figure 4.41 Regular polygons

A *polyhedron* is a solid figure with faces that are polygons.

A *regular polyhedron* is a solid figure with faces that are all congruent to the same regular polygon and with vertices that have the same number of edges emanating from them. The regular tetrahedron, which has faces congruent to an equilateral triangle, and the cube are examples of regular polyhedrons. Other examples are in Figure 4.42.

The formula of Euler applies to solid figures as well as planar graphs. This can be justified by considering the polyhedra as hollow and made from a flexible material like clay. By removing one face, the remaining figure can be flattened into a planar graph as indicated for the cube in Figure 4.43.

The flattened figure is a planar graph and hence by Euler's formula $V - E + F = 2$, where the exterior has been counted as a face. But the flattened figure has the same number of edges and vertices as the original solid figure since none were removed when the face was removed. If we think of the exterior in the planar

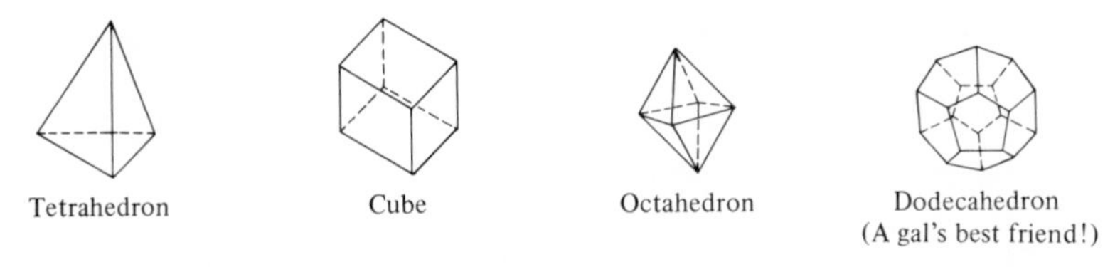

Figure 4.42 Regular polyhedrons.

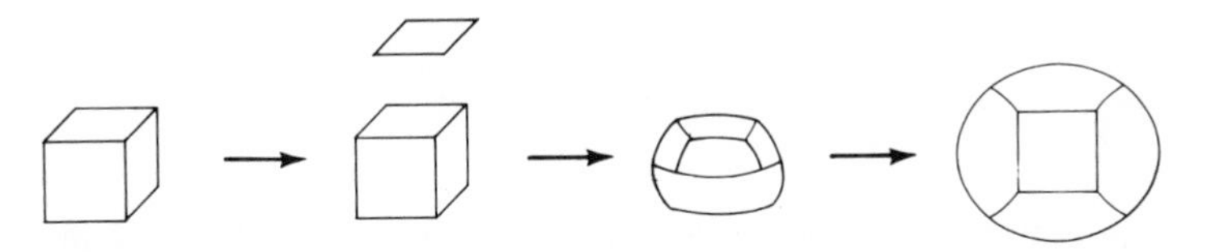

Figure 4.43

graph as corresponding to the removed face of the solid figure we have that $V - E + F = 2$ for the original solid. Thus, the number of vertices minus the number of edges faces is always two for any polyhedron which can be flattened into a plane when omitting one face.

We now turn to a proof of the classical result that there are only five regular solids. Assume that we have a polyhedron with each face bounded by $r(>2)$ edges and each vertex having $n(>2)$ edges emanating from it. Since each face has r edges, there must be no more than rF edges. But each bounds exactly two faces so each edge has been counted twice, that is, $2E = rF$. Furthermore, each vertex has n edges emanating from it and each edge has two vertices. Hence, nV must be a count of the edges in which each edge has been included twice. We conclude $2E = nV$ and $2E = rF$. Substitute $E = nV/2$ and $F = 2E/r = nV/r$ into Euler's formula to obtain

$$2 = V - E + F = V - \frac{nV}{2} + \frac{nV}{r}$$

or, after multiplication through by $2r$,

$$4r = 2rV - nrV + 2nV = (2r + 2n - nr)V$$

All these numbers must be positive so

$$2r + 2n - nr > 0 \quad \text{or, equivalently,} \quad nr - 2r - 2n < 0$$

Notice that $(n - 2)(r - 2) = nr - 2r - 2n + 4$ so the expression above can be re-written as

$$0 > nr - 2r - 2n = nr - 2r - 2n + 4 - 4 = (n - 2)(r - 2) - 4$$

that is, $(n-2)(r-2)<4$. We now try cases:

n	r	$(n-2)(r-2)$	V	E	F	Name
3	3	1	4	6	4	Tetrahedron
3	4	2	8	12	6	Cube
3	5	3	20	30	12	Dodecahedron
3	6	4				
4	3	2	6	12	8	Octahedron
4	4	4				
5	3	3	12	30	20	Icosahedron
5	4	6				

There are only five possible regular solids and, although we have not proved it, each of these solids can be constructed. Four of the five solids appear in Figure 4.42. The icosahedron has 20 triangular faces ($r = 3$) with five sides ($n = 5$) joining each vertex. It is an exercise in art for you to draw the solid.

Exercises

1. Examine the list of regular solids and form a definition of the dual of a regular solid so that the cube and octahedron are dual, the dodecahedron and icosahedron are dual, and the tetrahedron is self-dual.
2. For which of the regular polyhedron can an ant traverse all the edges once and only once if it starts at a vertex?
3. Which of the regular polyhedron have Hamilton circuits?
4. A semiregular polyhedra is a solid with faces congruent to one of two or more regular polygons and the same number of edges emanating from each vertex. Prove that there are no more than 13 semiregular solids. (Each of the 13 solids exists and are called Archimedean solids.)

References

1. S. K. Stein. *Mathematics, The Man made Universe*. San Francisco, Calif.: Freeman and Company. 1969.
2. O. Ore. *Graphs and Their Uses*. New York, N.Y.: Random House. 1963.
3. H. Steinhaus. *Mathematical Snapshots*. New York, N.Y.: Oxford Univ. Press. 1969.
4. M. Gardner. *Mathematical Games*. Scientific American, April, 1975. (It should be noted this is the April 1 issue!)

Chapter 5 Anyone Can Count?

5.1. One, Two, Three, . . .

We deal so frequently in daily life with small collections of items that it is likely that most of our counting is performed by direct enumeration. When faced with a large number in a collection, it is natural for us to use an indirect method of counting. For instance, how many dots are in Figure 5.1? Certainly you do not intend to count the dots successively from one to 48; instead, you notice that there are six rows of eight dots each and obtain the total by multiplication. An indirect method has been used to count the dots.

This chapter is concerned with principles and methods of indirect counting. The fact of the matter is that the underpinning of the subject is essentially the idea used to count the dots in Figure 5.1. Indeed, this primary principle, often called "the rule of products," can be paraphrased as follows: "If there are n of these items and each contains m of those things, then there are $n \times m$ of those things in all." If there are six rows and each row contains eight dots, then there are $6 \times 8 = 48$ dots in all. Simple it is, indeed! Let us take a closer look.

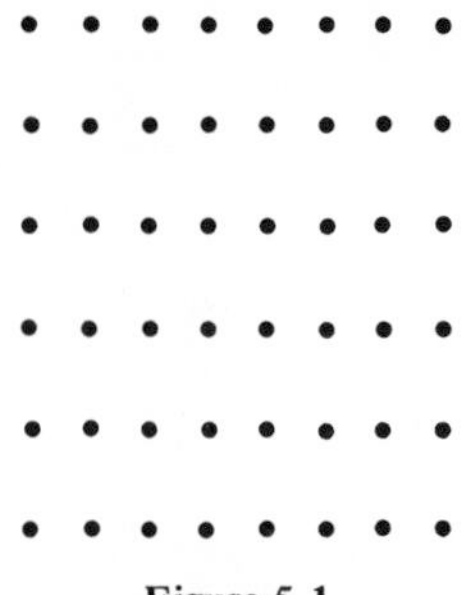

Figure 5.1

5.2. A Basic Counting Principle

There are many experiments that are intended to display the learning ability of animals. One such is to place a small animal such as a rat at the entrance of a T-shaped maze and to place food at the end of the left arm of the maze, as in Figure 5.2. By repeating the experiment the small animal eventually learns that food can be found in the left arm and, after some trials, will turn to the left consistently when placed at the entrance.

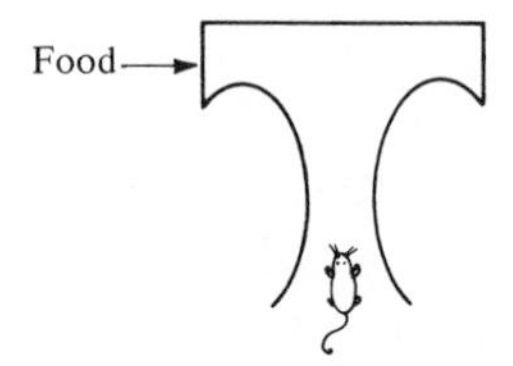

Figure 5.2

Suppose that this experiment is performed three times. What are all the possible paths that the rat can take in the three trials? In each trial he can turn either left or right. If he should turn right (R) on the first trial, he can turn left (L) or right on the second. The tree diagram of Figure 5.3 shows all paths that the rat can take in three trials.

There are eight possible distinct outcomes. They are listed in the columns of Table 5.1.

If there were 20 trials, instead of three, the tree diagram would be tedious to draw. Yet, a little analysis would enable us to determine the number of possible paths for the rat to take. Indeed, he has two choices, R or L, in the first trial. With each of these choices, there are two choices for a path in the second trial. Thus, there are $2 \times 2 = 4 = 2^2$ choices for two trials, namely LL, LR, RL, and RR as seen from the diagram. Now with each of these four, there are two choices for the third path, that is, $4 \times 2 = 8 = 2^3$ choices of paths in all. When

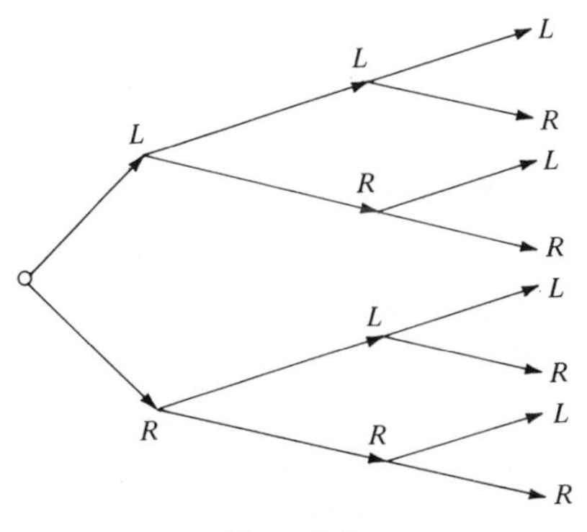

Figure 5.3

TABLE 5.1

Trials	Outcomes							
1	*L*	*L*	*L*	*L*	*R*	*R*	*R*	*R*
2	*L*	*L*	*R*	*R*	*L*	*L*	*R*	*R*
3	*L*	*R*	*L*	*R*	*L*	*R*	*L*	*R*

L = left turn, *R* = right turn

there is a fourth trial, there are $8 \times 2 = 16 = 2^4$ choices of paths. By repeating this process we can show that there are $2^{20} = 1{,}048{,}576$ paths possible in twenty trials.

The ideas in this example illustrate the general principle for counting the number of outcomes of an experiment.

RULE OF PRODUCTS

Suppose there are two separate steps in a selection where the first step can be made in m different ways and the second can be made in n different ways. Then there are $m \cdot n$ ways to make the selection.

For example, if a restaurant offers four different kinds of desserts and five different choices of beverages, then there are $4 \times 5 = 20$ ways in which you can order a dessert and drink. (Draw a tree diagram for this problem.) If, in addition, there are 10 main course items on their menu, then there would be $20 \times 10 = 200$ ways to select a main course, dessert, and beverage. In other words, we can use the above principle in case there are more than two steps in a selection by applying the principle, as stated, two steps at a time. Rather than discuss our problems two steps at a time, it is more convenient just to remember this and write the solution briefly as $4 \times 5 \times 10 = 200$.

EXAMPLE. Suppose you have a choice in a restaurant of 8 soups or salads, 10 main dishes, 4 desserts, and 5 drinks. How many different ways can you select a full meal?

Solution: $8 \times 10 \times 4 \times 5 = 1600$ ways.

EXAMPLE. In a multiple choice test, there are 5 possible answers for each question. Suppose you take a test with 4 such questions. In how many different ways can you fill in the answer sheet?

Solution: There are 5 ways to answer each question and there are 4 questions. Hence, there are 5 ways to answer the first question, 5 ways to answer the second, 5 for the third, and finally 5 ways to answer the last question. Thus, there are $5 \times 5 \times 5 \times 5 = 625$ ways to complete the test.

Suppose there are 20 questions in the multiple choice test. Then, by inductive reasoning as above, there are 5^{20} ways to answer the test. This is approximately 95,370,000,000,000 ways.

EXAMPLE. In the four question multiple choice test, how many of the 625 ways to answer the questions will produce a zero grade?

Solution: For each question there should be 4 incorrect answers and exactly one correct answer. Thus there are 4 ways to answer each question incorrectly. With 4 questions this gives $4 \times 4 \times 4 \times 4 = 256$ ways to answer all questions incorrectly. Notice, however, that there is only *one way* to answer all questions correctly.

EXAMPLE. There are 5 air lines with flights from Cincinnati to Chicago. In how many ways can a man make a round trip between these cities if (a) he can take any air line each way or (b) he must take a different air line each way.

Solution: (a) $5 \times 5 = 25$; (b) $5 \times 4 = 20$. There are only four options for the return trip since the man must take a different air line than the one selected for the initial phase of the trip.

The exercises of this section are not difficult. Each should be thoroughly understood before you proceed to the next section.

Exercises

1. There are five finalists for the Miss Ohio contest, namely, Miss Cincinnati, Miss Dayton, Miss Cleveland, Miss Lima, and Miss Columbus. In how many ways can the judges select a winner and a first runner-up? Draw a tree diagram for this problem.
2. In how many different ways can a visitor to Tisney Universe take in one of the 8 attractions of the day of Morrowland, one of the 16 attractions of

Frightfulland, one of the 10 attractions of Skyland, and one of the 4 attractions of Never-Neverland? (The order in which the attractions are enjoyed is to be ignored.)

3. Pizza comes in many ways–plain, with bacon, with peppers, with onions, with mushrooms, with sausage, with anchovies, or with pepperoni. It also comes in three sizes, with extra cheese and with extra sauce. How many different ways can you order a pizza? (No two or more way orders, please.)
4. Thirty-five students take a true-false test with five questions. Show that at least two students must have identical papers if each student answers all questions. Is this conclusion valid if they need not answer all questions?
5. a. In how many ways can we mail four letters in three mailboxes?
   b. In how many ways can we mail three letters in four mailboxes?
6. Three people arrive in a town where there are four hotels.
   a. How many ways can they take up quarters in these hotels, with each in a different hotel?
   b. What is the answer if there are four people?
   c. Suppose that the condition that they be in different hotels is dropped. What then are the answers to questions a. and b.?
7. a. A secretary has typed four letters to four different people and has addressed an envelope for each letter. She puts one letter in each envelope. In how many ways can she assign the letters to envelopes incorrectly?
   b. In how many ways can the secretary send each of the four individuals a letter addressed to someone else?
8. A man has five ties. He plans to wear a different tie each day of his five-day work week. For how many years can he select his ties so that he wears the ties in a different sequence each week? (Assume he has two weeks vacation per year and that he is never absent from work otherwise.) Suppose next that he has seven ties. What is the answer now?
9. How many subsets are there of a set with 10 elements? (Hint: Each element in the set is either in a given subset or not in it.)
10. How many automobiles have to be licensed in a state before it must relinquish its system of starting each license plate with two letters from the the alphabet excluding I and O, and following them with four numerals?

5.3. Jetsam

To develop a skill with counting methods, we recommend reading the example and, immediately thereafter, proceeding to the indicated exercise. The exercise is similar to the example and the solution should be readily obtained. After this task is completed there are additional exercises that you should find instructive and challenging.

EXAMPLE. How many 3-digit numbers greater than 424 can be formed from the digits 2, 4, 5, and 6? No digit is to be used more than once.

Solution: In order that the number be greater than 424, the first digit must not be two. Indeed, the 3-digit number is in the two hundreds range (<424) if its first digit is 2. Thus, there are 3 choices for the first digit, namely, 4, 5, or 6. The second digit can be any of the remaining numbers so there are 3 choices for the second digit. We now have used two numbers so the last digit must be selected from the unused two. In all, there are $3 \times 3 \times 2 = 18$ ways to form the desired numbers. They are listed below:

425	426	452	456	462	465
524	526	542	546	562	564
624	625	642	645	652	654

EXAMPLE. How many of the integers in the preceding example are even?

Solution: There are 14, as you can obtain by counting in the above chart. Let us, however, find a mathematical method for doing this counting.

First method. It is, in fact, easier to obtain the odd numbers since these are the numbers that end with a 5 and there are fewer such numbers in the list. (If a numbers ends with a 2, 4, or 6, it obviously is even.) Now how many odd numbers are there in the collection? There is one way to fill in the last digit, namely with a 5. There are two ways to select the first digit, namely from 4 or 6, since the number is to be greater than 424. We now have used two of the numbers, 2, 4, 5, and 6, so a choice of two remains for the middle digit. Thus, there are $2 \times 2 \times 1 = 4$ odd integers in the collection. The number of even integers is then the total less the number of odd, that is, $18 - 4 = 14$.

Second method. We could compute the number of even entries in the collection directly. Indeed, if the first digit is a 4, then to be even the last digit must be 2 or 6—two choices. The middle digit is any one of the remaining 2 giving us $2 \times 2 = 4$. (See the previous list.) In the same way we conclude:

$$\text{First digit} = 5 \rightarrow 2 \cdot 3 = 6 \text{ ways}$$
$$\text{First digit} = 6 \rightarrow 2 \cdot 2 = 4 \text{ ways}$$

The total $4 + 6 + 4 = 14$ is the number of even integers in our collection. (Now proceed to Exercise 1.)

EXAMPLE. There are eight men in a room and each man shakes hands once with the other men in the room. How many handshakes are made?

Solution: With eight men, each performs seven handshakes. It appears that there are $8 \cdot 7 = 56$ handshakes in all. But notice that each handshake has been counted twice. For, we counted Mr. *A* shaking hands with Mr. *B* and then Mr. *B* shaking hands with Mr. A, etc. Hence, the number of handshakes is $56/2 = 28$. (See Exercise 2.)

Another solution. Line the men up in a row. The first, Mr. A, shakes hands with the seven men to his left. Then the second man, Mr. B, say, shakes hands with the six men to his left. Mr. C, the third man, shakes hands with the five men to his left, etc. In this process we are not counting handshakes twice and the total number of handshakes is

$$7 + 6 + 5 + 4 + 3 + 2 + 1 = 28$$

The last two illustrations show that there is often more than one way to reason to the answer. Just hope each way gives the same answer. (Try exercise 2.)

EXAMPLE. In how many different ways can three people arrange themselves at a round table?

Solution: Where the first person sits does not really matter. Once he sits down the second person sits in either the chair to his right or the first chair to his left—two choices. The last persons sits in the one remaining chair. There are thus two ways for three people to arrange themselves at a round table. Once you sit down, what matters is which of the others is to your left. (Figure 5.4). Exercise 3 is similar to this example.

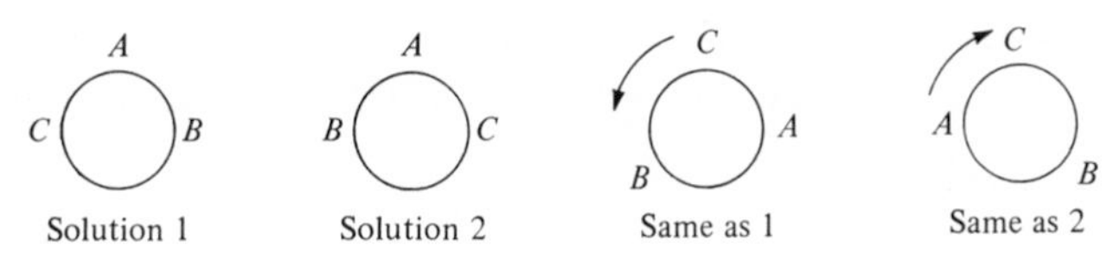

Figure 5.4

EXAMPLE. a. In how many different ways can a group of five children be arranged in a row of five chairs?

b. Suppose three are girls and two are boys. In how many ways can they be arranged in this row of chairs if the boys must sit together and the girls must sit together?

c. Suppose only the girls must sit together? What is the answer now?

Solutions: a. The first child can sit in any of the five chairs, the next child can sit in any of the remaining four chairs, etc. There are $5 \times 4 \times 3 \times 2 \times 1 = 120$ different arrangements of the children in the row.

b. If the boys and the girls are to be segregated the boys must be at one or the other end of the row. (Figure 5.5) With each of these two choices, there are two ways for the boys to sit down. Indeed, the boy named Peter, say, can sit either in the end seat or the seat next to the end. The girls can be seated $3 \times 2 \times 1 = 6$ ways in the three remaining seats. Thus, there are $2 \times 2 \times 6 = 24$ ways to seat the children within the limitation of b. in the example.

Figure 5.5

c. In how many ways can a block of three adjacent seats be picked from the five seats? The answer is the same as the number of ways you can place three adjacent "x's" in five spaces. This can be solved directly: $\underline{x}\ \underline{x}\ \underline{x}\ \underline{\ }\ \underline{\ }$, $\underline{\ }\ \underline{x}\ \underline{x}\ \underline{x}\ \underline{\ }$, or $\underline{\ }\ \underline{\ }\ \underline{x}\ \underline{x}\ \underline{x}$. There are $3 \times 2 \times 1 = 6$ ways to seat the girls in the reserved three seats and $2 \times 1 = 2$ ways to seat the boys in the remaining seats. Hence, there are $3 \times 6 \times 2 = 36$ ways to arrange the groups as specified. (See Exercise 5.)

We close this section with a counting problem, the solution of which runs contrary to intuition. Although no large sample statistical test was conducted, we suspect that most people would select the hearts in the next example.

EXAMPLE. Two cards are selected from a standard deck of playing cards. It is better to bet (even odds) that they will both be hearts or to bet that exactly one of them is a black ace?

Solution: We assume that the first card is *not* replaced before drawing the second. Thus, we have 13 ways to draw the first heart and only 12 ways to draw the second heart (since the first one is missing). In total, there are $13 \times 12 = 156$ ways to draw two hearts without replacement.

The case for drawing exactly one black ace is different. We can succeed in this event in two *different* ways. First, we could draw a black ace on the first drawing and then follow this by drawing something *other than* a black ace (50 such choices). Hence, there are $2 \times 50 = 100$ ways to draw a black ace first followed by a card that is not a black ace. Second, we could draw something other than a black ace on the first turn and draw a black ace on the second turn. This can happen in $50 \times 2 = 100$ ways also. Now each of these two cases provides exactly one black ace. The first case (100 ways) gives a black ace followed by some other type of card and the second case (100 ways) provides a non-black ace followed by a black ace. Together these two independent ways of success yield 200 cases. Hence, the black ace choice is the better bet! (See Exercise 7.)

Exercises

1. a. Using only the digits 1, 2, 3, 4, and 5, how many three-digit numbers with distinct digits can be constructed?
   b. How many of these are even?
   c. How many are larger than 300?
   d. How many even three-digit numbers can be formed from these digits if the numbers can contain repeated digits?
2. In a friendly group of four women and three men, each woman greets all others with a kiss on the cheek. How many such kisses occur when they gather for a party? (The men use handshakes with one another.) We count the greeting of two women as one kiss.
3. In how many different ways can five girls arrange themselves in a ring for a May Day dance?
4. How many different ways can you put four keys on a key chain? (Caution: The answer is not six.)
5. There are two works of three volumes each and two works of two volumes each. In how many ways can the 10 books be placed on a shelf so that volumes of the same work are *not* separated?
6. A certain club has 50 members, 35 men and 15 women.
   a. How many different ways can they select from their membership a president and a vice president?
   b. Suppose the president must be a woman. In how many ways can the two officers be selected?
   c. Suppose either the president or the vice president must be a woman, but not both. In how many ways can the club now fill the positions?
   d. In how many ways will both officers of the club be of the same sex?
7. Three cards are drawn in succession from an ordinary deck of playing cards. It is more likely that one be the ace of spades than that all three be red cards?
8. In how many ways can at least one club be selected when two cards are drawn from a deck of playing cards?
9. A family of four (two adults) and a family of five (two adults) buy a row of nine seats for a local football game.
   a. In how many ways can they be seated if the families are to stay together?
   b. In how many ways can they be seated such that an adult separates every two children?
10. Three union officials, three management representatives, and two government arbitrators are to settle a labor dispute. In how many ways can they arrange themselves at a round table if the members of each group are to be next to each other?

5.4. Permutations and Combinations

Suppose that a certain club, at a meeting, selects from its membership two individuals to represent the club at some national meeting. Suppose, furthermore, that the club elects a president and vice-president. Ignoring politics, they have twice as many ways to fill the two administrative offices as compared with sending two members to a national meeting. Indeed, if Pat and Joe are to be the representatives at a meeting, it really does not matter who was picked first since both are to attend the event. However, the order is important in selecting the officers of the club. Pat, as president, and Joe as vice-president is quite different than Joe as president and Pat as vice-president.

This example attempts to illustrate the basic difference between what mathematicians call a permutation and a combination. The former is defined as a set of r objects drawn *in a certain order* from a set of n objects. A combination is a set of r objects drawn from a set of n objects. The order is not considered in a combination.

EXAMPLE. Consider the set of four letters $\{a, b, c, d\}$. How many ways can we draw two of these letters? (Combination) The order is not important so drawing (a, b) is the same as drawing (b, a). The number of two letter combinations is six, namely,

$$ab \quad ac \quad ad \quad bc \quad bd \quad cd$$

If the problem is to draw *in a certain* order two letters from the set $\{a, b, c, d\}$, then ab is a different case than ba. In fact, for each of the six combinations there are two permutations, namely, ab, ba; ac, ca; ad, da; bc, cb; bd, db; cd, dc; twelve cases in all.

Initially, at least, it is easier to compute the number of permutations than combinations for a problem. For example, how many ways can we draw three letters in a certain order (permutation) from $\{a, b, c, d\}$? The first letter can be drawn from $\{a, b, c, d\}$, so there are four cases; the second letter is drawn from the remaining three letters and the last letter is drawn from the remaining two. Thus, there are $4 \cdot 3 \cdot 2 = 24$ ways to draw three letters in order. They are as follows:

$$\begin{array}{cccccc} abc & bac & bca & cba & cab & acb \\ abd & bad & bda & dba & dab & adb \\ bcd & cbd & cdb & dcb & dbc & bdc \\ acd & cad & cda & dca & dac & adc \end{array}$$

Now, in the selection of three letters from the set $\{a, b, c, d\}$ if the order is not to be considered (combination) then there are only four cases. For, each row in the above list represents just one combination since each row uses the same triplet of letters. Notice that if we know the number of combinations, four in this case, we can compute the number of permutations by rearranging the letters in each combination in all possible orders. There are $3 \cdot 2 \cdot 1 = 6$ ways to arrange three letters in all possible orders. Since there are four combinations and six ways to arrange each into all possible orders, there are $4 \cdot 6 = 24$ permutations. It is generally not difficult to pass from the number of combinations in a problem to the number of permutations.

The number of permutations of n objects taken (in a certain order) r at a time is denoted by P_r^n. Notice that the first object can be selected from any one of the n objects, the second from any of the remaining $n - 1$ objects, the third from any of the remaining $n - 2$ objects, etc. In picking the rth object, the selection is made from $n - r + 1$ remaining objects. (Check when $r = 1$, $r = 2$, and $r = 3$. We always subtract the r from n and add 1.) This argument proves that

$$P_r^n = n(n - 1)(n - 2) \cdots (n - r + 1) \tag{5.1}$$

the product of the r integers $n, n - 1, \ldots, n - r + 1$. For example, $P_2^5 = 5 \cdot 4$, the product of two integers starting with 5 and decreasing by 1 to get the next factor. $P_4^{10} = 10 \cdot 9 \cdot 8 \cdot 7$, starting with 10 and decreasing by 1 for the next factor until we have four factors. Similarly, $P_7^{25} = 25 \cdot 24 \cdot 23 \cdot 22 \cdot 21 \cdot 20 \cdot 19$, and there are seven factors.

The number of combinations of n object taken r at a time is denoted by C_r^n. What is more natural?

As illustrated above, we can pass from the C_r^n answer to P_r^n in a simple manner. Indeed, the r objects in a combination are not arranged in a particular order since order is not important for this counting. If we consider the order of these r objects we can arrange them in $r \cdot (r - 1) \cdot (r - 2) \ldots 2 \cdot 1$ ways. For, the first object can be selected from the set of r, the second from the remaining $r - 1$, and so on until we have used up the entire set of r objects. Therefore, we have

$$C_r^n \cdot r(r - 1)(r - 2) \cdots 2 \cdot 1 = P_r^n \tag{5.2}$$

since the number of combinations times the number of ways to arrange the r objects in each combination is the number of permutations. Using the notation $r!$ (r factorial) for the product $r \cdot (r - 1) \ldots 2 \cdot 1$, we write Eq. (5.2) as

$$C_r^n = P_r^n / r! \tag{5.3}$$

Hence,

$$C_3^5 = P_3^5/3! = (5 \cdot 4 \cdot 3)/(3 \cdot 2 \cdot 1) = 10$$

and

$$C_4^{10} = P_4^{10}/4! = (10 \cdot 9 \cdot 8 \cdot 7)/(4 \cdot 3 \cdot 2 \cdot 1) = 210$$

If you know the permutation, it is easy to obtain the combination by Eq. (5.3).

One way to guide your memory is to notice that the combination formula has the same number of factors in the numerator as in the denominator. The denominator is the factoral of this number and the numerator is a product of successively decreasing integers starting with n. For instance, C_4^8 has four factors in both the numerator and denominator; the numerator is thus $8 \cdot 7 \cdot 6 \cdot 5$ and the denominator is $4! = 1 \cdot 2 \cdot 3 \cdot 4$. (We counted one as a factor in the last product. If you do not like it, invent your own method to remember combinations.)

With the factorial notation at hand, we can also write Eq. (6.1) in a compact manner. For, $(n - r)! = (n - r)(n - r - 1)(n - r - 2) \ldots 2 \cdot 1$ so

$$\frac{n!}{(n-r)!} = \frac{n \cdot (n-1) \cdots (n-r+1)(n-r)(n-r-1) \cdots 2 \cdot 1}{(n-r)(n-r-1) \cdots 2 \cdot 1}$$

By cancellation of the last $n - r$ factors in the numerator with the corresponding factors in the denominator, we obtain P_r^n as given in Eq. (5.1), that is,

$$P_r^n = \frac{n!}{(n-r)!} \qquad (5.4)$$

In view of Eq. (5.3) we also have

$$C_r^n = \frac{n!}{(n-r)!\,r!} \qquad (5.5)$$

Thus,

$$C_5^{10} = \frac{10!}{5!5!}, \qquad C_4^{12} = \frac{12!}{8!4!}, \qquad C_9^{30} = \frac{30!}{21!9!}$$

The formula (5.5) has an interesting and practical symmetry. It shows, for example, that $C_5^{15} = C_{10}^{15}$ by the commutative law for the product of integers. In general,

$$C_r^n = C_{n-r}^n \qquad (\text{Define } C_0^n = 1 \text{ and } 0! = 1 \text{ so } C_n^n = C_0^n)$$

Thus, the number of combinations of 100 objects taken 98 at a time is the same as the number of combinations of 100 objects taken 2(= 100 − 98) at a time. The latter is easier to calculate and is $C_2^{100} = 100 \cdot 99/1 \cdot 2 = 4950$.

EXAMPLE. There are five finalists in the Miss Ohio contest, namely, Miss Cincinnati, Miss Dayton, Miss Cleveland, Miss Lima, and Miss Columbus. In how many ways can the judges pick a winner and the first and second runners-up?

Solution: The order is important, at least to the contestants, so this is a permutation problem. There are $P_3^5 = 5 \cdot 4 \cdot 3 = 60$ different ways to make the selection.

EXAMPLE. Suppose Ohio can send three ladies to the Miss America contest. In how many ways can the judges pick three from the five finalists of the previous example?

Solution: This is a combination problem and its solution is $C_3^5 = 5!/(3!2!) = (5 \cdot 4)/(1 \cdot 2) = 10$.

EXAMPLE. The student senate contains five representatives from each class (freshman, sophomore, junior, and senior). How many committees can be formed containing three seniors, two juniors and one of each of sophomores and freshmen?

Solution: We can select the seniors in $C_3^5 = 10$ ways. We can select the juniors in $C_2^5 = 10$ ways. The sophomore and freshman can be selected in $C_1^5 = 5$ ways each. Thus there are $10 \cdot 10 \cdot 5 \cdot 5 = 2500$ different ways to select the committee.

EXAMPLE. A baseball team has 11 players. In how many ways can the manager field a team (nine players) if each individual can play any position?

Solution: $C_9^{11} = C_2^{11} = (11 \cdot 10)/(1 \cdot 2) = 55$ ways.

How many ways can he arrange the batting order for each of these combinations?

Answer: $P_9^9 = 9! = 362{,}880$.

EXAMPLE. How many four letter "words" can be formed from the letters in *MISS*?

Solution: Any permutation of all four letters is called a "word" in this problem, even though it may not be in the dictionary. In order to illustrate the methods, let us label the two S's in the problem as S_1 and S_2. Then there are $P_4^4 = 4 \cdot 3 \cdot 2 \cdot 1 = 24$ ways to select four letters from M, I, S_1, and S_2. But the distinction between the S's was artificial and each word is, therefore, counted twice. For example MIS_1S_2 and MIS_2S_1 are both *MISS*. Hence, the answer to the problem is $24/2 = 12$.

How many five letter words can be formed from *MISSS*? If we could distinguish among the S's, namely an S_1, S_2, or S_3, there are $P_5^5 = 5! = 120$ ways. But three letters are alike so we have counted each word $3! = 1 \cdot 2 \cdot 3 = 6$ times. For example, *MISSS* is obtained from $MIS_1S_2S_3$, $MIS_2S_1S_3$, $MIS_2S_3S_1$, $MIS_3S_1S_2$, $MIS_3S_2S_1$, or $MIS_1S_3S_2$. Thus, the solution to the problem is $120/3! = 20$. This illustrates a general rule, which states that when a letter is repeated r times in a word, then one should divide the answer obtained by assuming the repeated letters are different by $r!$ to obtain the number of distinct words formed from the letters.

EXAMPLE. How many different seven letter "words" can be formed from the letters *AAACHUU*?

Solution: The answer would be 7! if the letters were all distinct. Since this is not the case, we adjust for the repetition and obtain 7!/(3!) (2!) = 420.

Exercises

1. a. In how many ways can you select four library books from a group of 12?
   b. In how many different ways can you read four books drawn from a set of 12?
2. If any arrangement of four different letters is a "word," how many words can be formed from our alphabet using no letter more than once in any word? (Do not list them even if you are not the timid type!)
3. How many seven digit phone numbers can be formed if the first digit cannot be zero? How many of these have distinct digits?
4. Find the numerical value of each of the following:
   a. $(1000)!/(998)!$ b. P_5^7 c. P_3^7 d. P_3^{10} e. C_5^7
   f. C_3^7 g. C_{98}^{100} h. C_3^{10} i. $7!/(3!3!2!)$
5. a. How many five card hands can be dealt from a deck of 52 playing cards?
   b. How many of these hands have five cards of the same suit?
6. How many different 11 letter words can be formed from the word MISSISSIPPI?
7. How many different "words" can be formed from the letters of MOON if the words can be of any length from one to four letters?
8. a. There are six campers and two tents, each of which holds three persons. The campers agree to remain away from home as long as they can arrange the sleeping assignments such that each group of three never occupies a tent together more than once. How long must they remain camping?
   b. Assume that there are three two-man tents. Now how long must the campers remain out if no two people are to occupy the same tent for more than one night?
9. How many committees, including one-man committees, can be formed by selecting the members from a group of eight men?
10. Each time 4 objects are selected from a collection of 10 objects, 6 objects are left behind. When only 3 are selected, there are 7 left. Generalize these observations and show that they imply

$$C_r^n = C_{n-r}^n$$

Chapter 6 Why Take a Chance?

6.1. The Origin

The success of the axiomatic approach in the physical sciences encouraged, since the seventeenth century, attempts to produce laws for the behavior of man and his society. For the most part, however, its application has met with failure since the complexities of the problems in these areas defies the discovery of clear, deterministic, general axioms. Successful techniques for providing scientific methodology in social and biological problems were uncovered in the nineteenth century. They are the mathematical theories of probability and statistics. The mathematical theories, nevertheless, were initiated more than a century before their ultimate importance to the social sciences was widely noticed.

The mathematical theory of probability began when Chevalier deMere, a gambler and amateur mathematician, asked the French mathematician Blaise Pascal (1623-1662) why it was unprofitable to bet even money that at least one double six would come up in 24 throws of two dice. Pascal contacted Fermat (1601-1665) concerning this and related questions, and both mathematicians established some of the foundations for the general theory in 1654. Although probability theory may have now outgrown its gaming origin, the basic notions are still most easily stated in terms of some game of chance.

One of the more important applications of probability was by Gregor Mendel (1822–1884), abbot of a monastery in Moravia. With carefully planned experiments, Mendel studied the effects of cross-fertilization of two strains of peas, one yellow and one green. Mendel interpreted the results of his study in terms of what are now called genes and showed how the outcomes of cross-fertilization could be assigned probabilities. The science of genetics is an outgrowth of Mendel's work.

We initiate the formal study by first introducing the classical definition of probability due to Fermat and Pascal and later turning to a more general axiomatic approach. As an application we include a brief introduction to some laws of heredity that essentially owe their existence to Mendel.

6.2. Probability

Little in life is certain; most occurrences have more than one possible outcome. The probability of a particular outcome for a future occurrence is a real number between 0 and 1 which measures the likelihood of this outcome happening. For instance, the weatherman reports the probability of rain tomorrow is 30% (=.3). If you flip a quarter, there is a 50% (=.5) chance that it falls with Washington's face showing.

Whenever you have a way of counting the total number n of distinct outcomes of an experiment, then the probability of a particular outcome is $1/n$ provided each outcome is as likely to occur as any of the others. For example, in tossing a quarter we assume that there are two and only two distinct outcomes, namely, a head (Washington's) or a tail (the eagle's). We also assume that the coin is balanced so that it does not favor either a head or a tail. With these assumptions, the probability of a head is $1/2$.

Notice, however, that the assumptions are not really valid; the coin could stand upright on its edge and it is questionable whether each face is balanced since they are stamped in different ways. We will ignore these relatively minor difficulties and replace the actual coin by a mathematical model of a coin, one which satisfies our assumptions. Like other mathematical entities, our mathematical model has no real world existence; nevertheless, it is closely, although not perfectly, approximated by most coins. A study of the behavior of the conceptual coin, therefore, translates into quite satisfactory information about the real coin. And the model coin is much easier to study!

The set of all possible outcomes of any experiment, or occurrence, is called the *sample space*. If we toss two coins, the sample space is HH, HT, TH, and TT, where H and T, respectively, stand for head and tail. Any subset of the sample space is an *event*. For example, obtaining at least one head in tossing two coins is an event since the cases in the sample space which produce the outcome, namely, HH, HT, and TH, form a subset of the sample space. The events which contain only one entry from the sample space are called points of the space, for example, HH is a point and so is HT, TH, or TT. For finite sample space problems the probability $P(A)$ that an event A will occur is, provided that the points in the sample space S are equally likely,

$$P(A) = \frac{n(A)}{n(S)} \tag{6.1}$$

where $n(A)$ is the number of points in A and $n(S)$ is the number of points in the sample space S. For example, if A is the event of obtaining at least one head in tossing two coins, then $P(A) = 3/4$ since $n(A) = 3$, $n(S) = 4$ and the points HH, TH, HT, and TT are equally likely. The last statement follows from the fact that each outcome of the coin tossing experiment is as likely to occur as any of the other outcomes.

EXAMPLE. The sample space in rolling a die is 1, 2, 3, 4, 5, and 6. Each outcome has probability 1/6 if the die is "honest." (Our mathematical dice are always honest. Actually, it is physically possible for the die to stand on an edge although this is unlikely. Also, the holes for numbers in the faces may, ever so slightly, produce a bias in the die.) What is the probability of the event A of having a three or a six turn up in the roll of a single die?

Solution: $n(A) = 2$ since 3 and 6 are the only points from the sample space that comprise the event A. Thus

$$P(A) = n(A)/n(S) = 2/6 = 1/3$$

Sometimes plausible, yet incorrect, probabilities are obtained from formula (6.1) because the condition that the outcomes are equally likely is not met. For example, one can falsely argue that the probability of obtaining two heads when two coins are tossed is 1/3 since there are three outcomes, namely, two heads, two tails, or a head and a tail. The probability of a head and a tail is, in fact, greater than the probability of two heads, so the outcomes are not equally likely. Using a similar false argument we claim that the probability that all three children in a family are female is 1/4. Indeed, there are four cases–no girls, one girl, two girls or three girls. What should the answer be?

If you toss a coin repeatedly, counting the number of heads and the number of tails, then the ratio of the number of heads to the total number of times the coin is tossed may not be exactly 1/2. However, as the number of trials increases this ratio tends towards 1/2. That is, the magnitude of the difference between 1/2 and the experimental ratio generally tends toward zero as the number of trials increases. This phenomenon is called the "law of large numbers" and can be precisely stated and proved for the mathematical model of a coin in addition to many other situations. What is important for us is the existence of the law, rather than its proof, since it provides faith in the practicality of the subject of probability.

In many applications the probability of a certain event is desired and, yet, there is no simple mathematical model as in the case of the coin tossing experiments. If the law of large numbers is assumed, we can estimate the probability of the event by observing the proportion of times similar events have occurred in the past. For instance, suppose it rained in 30% of the recorded cases where weather conditions were similar at the location during the given season to those of today. A prediction of a 30% chance of rain seems reasonable. Is it fair to condemn the weatherman, who makes the prediction, should it rain? Not if he properly interpreted the available data. Expecting to picnic and being fair may not be compatible in case rain falls.

The exercises of this section should primarily provide practice with the counting principles of the last chapter. The introduction of the classical definition (Eq. (6.1)) of probability adds no new complications.

Exercises

1. What reasons would you give to refute each of the following arguments?
   a. Since lightning just struck this spot, it is safe to stand here.
   b. Tomorrow you will be either alive or dead. Hence, you have a 50–50 chance to survive this day.
   c. When I roll this pair of dice, the total will be a number from two through 12. Hence, I have 11 possibilities so my chance to roll a three is one in 11.
   d. Since we have had three children, all girls, the law of averages certainly indicates that it is likely that our next child will be a boy.
   e. Most people call heads much more frequently than tails when a coin is tossed. Hence, when you flip a coin, let the other man call the toss. Indeed, the coin is as likely to fall on tails as heads but he will be calling heads more frequently than tails.
   f. The probability that you will marry in 1984 is 1/2 since there are two possibilities, namely, you will marry or you will not marry. (Even if you are now married, there still are two cases each year!)
2. Toss a thumbtack 100 times. Count the number of times it rests on its head and the number of times it rests on a side. (⊥ ⋋) What is the experimental probability that it rests on its head? (The answer depends on the type of tack used.)
3. What is the probability of obtaining a sum of six in rolling a pair of dice? Build a chart showing the probabilities of all other sums with a pair of dice.
4. Assume that you guess the answer for each of four true-false questions on a test. List the number of ways you can have no correct, one correct, two

correct, three correct, and four correct answers. What is the probability that *at least* two will be correct?

5. An anti-zero population growth couple desire to have three children. If the probability of having a boy equals the probability of having a girl, what is the probability that they have (a) three boys, (b) no boys, (c) at least one boy, (d) two girls and one boy, (e) at most one boy?

6. If there are 15 men and 35 women in a club, what is the probability that, in selecting a committee of two from its membership, the club selects one woman and one man?

7. An urn contains 12 red, 10 white, 15 blue and 3 black balls. If a ball is drawn at random, what is the probability that the ball is (a) red, white, or blue, (b) neither black or white, (c) not red?

8. Suppose two balls are drawn from the urn in succession without replacement. (a) What is the probability they are both blue? (b) What is the probability that at least one is blue? (c) What is the probability that none are blue?

9. The sum of two positive integers is 12. What is the probability that they are 9 and 3?

10. A two-card hand is dealt from an ordinary deck such as in Blackjack. What is the probability that it contains (a) exactly one ace, (b) at least one ace, (c) exactly one face card, (d) two hearts, (e) two queens?

6.3. What Are the Odds?

Gambling odds are often quoted instead of probabilities. For example, one hears that the odds are 5 to 2 against a certain horse in a race. In terms of probabilities, this means that the bookies believe the horse's chance of winning the race is $2/7$ $(7 = 5 + 2)$. Odds of a to b against mean the probability of the event is $a/(a + b)$. Nothing could be simpler!

The reason odds are used instead of probabilities for gambling is tradition and, more important, the odds clearly state the proportions involved in any bet. For example, if the odds are 5 to 2 against, then you must place \$2 on the line to win \$5, or \$4 to win \$10, or \$20 to win \$50, etc. The ratio of the amount won to the amount bet must be equal to $5/2$.

Let S be a sample space containing $n(S)$ equally likely points. A certain subcollection of these points is called the event A. If $n(A)$ is the number of points in A, then $n(S) - n(A)$ is the number of points in S and not in A. The odds in *favor* of A are $n(A)$ to $n(S) - n(A)$. For instance, if two coins are tossed, the sample space has four points, *HH*, *HT*, *TH*, and *TT*. The odds in favor of obtaining two tails are, therefore, 1 to 3 $(= 4 - 1)$.

If only probabilities are given, it is possible to convert to odds. To illustrate, suppose the chance of rain today is quoted at 60%. This means there is a 40% chance it will not rain, and the odds in favor of rain are, thus, 60 to 40. Here the odds should be replaced by the simpler, yet equivalent, odds of 3 to 2. The odds represent proportions and as such can be replaced by proportional numbers in the problem $60/40 = 3/2$.

Exercises

1. There is a gambling wheel with numbers 0, 1, 2, . . . 50 equally spaced around the circumference. You can play a number and receive 50 to 1 odds from the house. You can also play odd numbers in which case you receive one-to-one odds. If you must bet, what would you play? Why?
2. The odds in favor of Percy, Ford, Buckley, Kennedy, and Church being elected president are quoted at some point to be 2 to 5, 1 to 1, 3 to 5, 1 to 2, and 1 to 9, respectively. What is the probability, according to these odds, of each man getting elected? Why should you not believe the cited odds?
3. Except for long-shots, odds are usually quoted in terms of two small positive integers. Why do you think this is the case?
4. Suppose the probability that you will be mugged, should you walk outside your hotel room, is .15. What are the odds that you will be mugged?
5. If 237 out of 300 housewives interviewed preferred a product of your company, what quotation of odds would you make in an "honest" media advertisement describing your survey?
6. In order to determine how a young man feels about the success of his recent marriage, someone offers him a bet of $100 against his $30 that the marriage will not last two years. What can we say about the young man's personal probability concerning the success of the marriage if he asks for better odds?

6.4. Simple-Minded Heredity

The study of biological characteristics that are transferred from parents to their children is the science of genetics. The laws of heredity are based on some biological facts in conjunction, at least initially, with mathematically simple applications of probability theory.

An inherited trait occurs in an offspring by a pairing of genes of a certain type, one contributed by each parent, in bisexual reproduction. The parent has itself a pair of genes for each inheritable characteristic, and these pairs of genes may be of the same type or different types. A reproductive cell of an adult carries *one* of these two genes, with either being as likely to be passed on to the offspring as the other. Furthermore, the reproductive cell carries one of the parent's genes for every inheritable trait.

Consider, by way of illustration, the genes associated with eye color. It is known that a blue-eyed person has appair of similar genes, which we label "*bl*." A brown-eyed person may have a pair of similar genes, which we label "*bn*," or he may have a mixed pair, *bn* - *bl*. Either way the person has brown eyes. Suppose now that two blue-eyed adults have a child. What is the probability that the child will have brown eyes? The answer here is zero. Indeed, each parent will contribute a *bl*-gene to the pair in the child and, hence, it will have a *bl* - *bl* pair which produces only blue eyes.

Suppose that a brown-eyed person marries a blue-eyed person. Could the offspring have blue eyes? The answer depends on the brown-eyed parent's gene pair. For, should this parent's eye color gene pair be *bn* - *bl*, he is as likely to pass on the blue gene as the brown gene. The probability of blue-eyed children thus equals that of brown-eyed children in this case since the other parent only contributes a blue gene. In the event that the brown-eyed parent has a gene combination of *bn* - *bn* there is no hope for a blue-eyed child. One parent supplies the child with a blue gene, but the *bn* - *bn* parent only provides a brown gene. The offspring has, therefore, *bl* - *bn* combination of genes which produces brown eyes.

The fact that this mixture of genes still produces brown-eyed individuals suggests the terminology of a "dominant" characteristic for the brown-eye producing gene and of a "recessive" characteristic for the blue-eye producing gene. Although for some inherited traits one or the other of the gene pair is dominant, there are also characteristics for which no gene of the pair has such a central role.

Probability certainly comes into play with this simple analysis of inheritance. Consider any trait that is inherited by a pairing of any genes, G and g, where G is dominant and g is recessive, as in the brown-eyed, blue-eyed example. A GG and GG pair of parents produce a GG child, definitely! When a Gg and a GG produce children, the g gene is as likely to be passed by the first listed parent as is the G gene. Hence, the probability is one-half that the children produced by the couple are of the Gg type. The Gg child, furthermore, would display the G type characteristic since G is dominant. All children of the Gg and GG parents, therefore, display the G characteristic (Figure 6.1(a)).

The outcome is different when the Gg individuals mate. The offspring can be a gg type by picking up a g gene from each parent. All other combinations of gene pairs, one originating from each parent, produce children that display the dominant G characteristic. In fact, the probability is 3/4 that the children of these parents should have the G trait. (See Figure 6.1(b).)

In addition to the dominant and recessive characteristics, there are traits that are produced by different gene pairs working together. Consider, for example, the color of a certain species of flowers. A red flower, let us assume, is produced from a pair of red genes, and a white flower from a pair of white genes. Should a red gene be paired with a white gene, the color of the flower is neither red nor

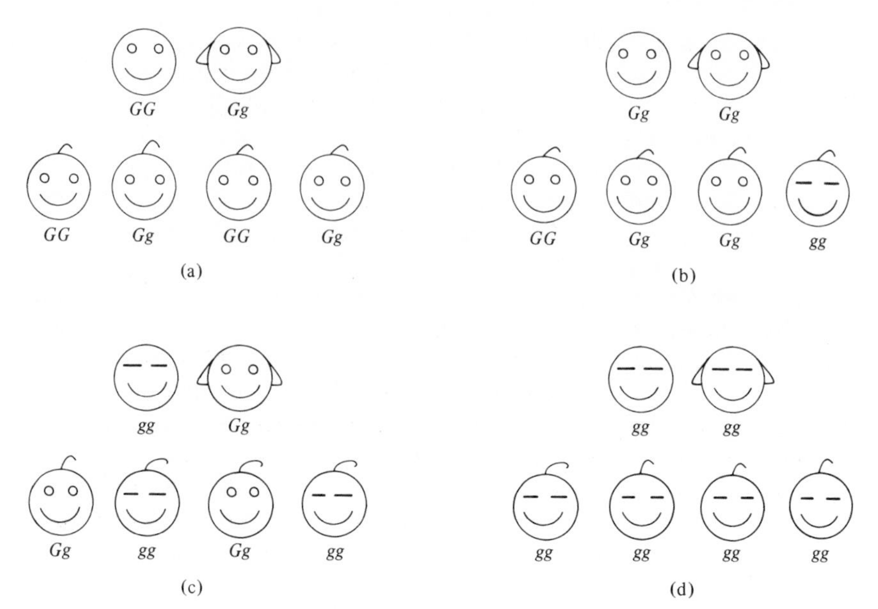

Figure 6.1

white; rather it is the in-between color of pink. Neither the red gene nor the white gene is dominant in this pairing. The chart in Figure 6.2 indicates some of the possibilities when different colored flowers of this species pollenate one another.

When the pollen of a pink flower fertilizes another pink flower, the probability that the seed produces a pink is 1/2, whereas the probability that it produces a white is 1/4 as it also is for a red.

Genes are carried by chromosomes which, like the genes, are paired in a living cell. There are 46 chromosomes in a human cell, each of which carries numerous genes. One pair of these chromosomes determines the sex of the individual as well as certain other characteristics. In a female there are two such chromosomes, called X-chromosomes, that are of the same type. The male has an X-chromosome paired with one of another type, called a Y-chromosome. The reproductive cells of each parent carry one of these chromosomes, with equal likelihood of either. Hence, an offspring always gains an X-chromosome from its mother and an X or a Y-chromosome, with equal probability, from its father. The child is male if the father supplies a Y-chromosome and female if he supplies an X-chromosome. Think of all the queens of the past that were "removed" from their throne since they would not supply the country with a male child. Alas, it would have been biologically more reasonable to replace the king! It was his contribution that determined the sex of their children.

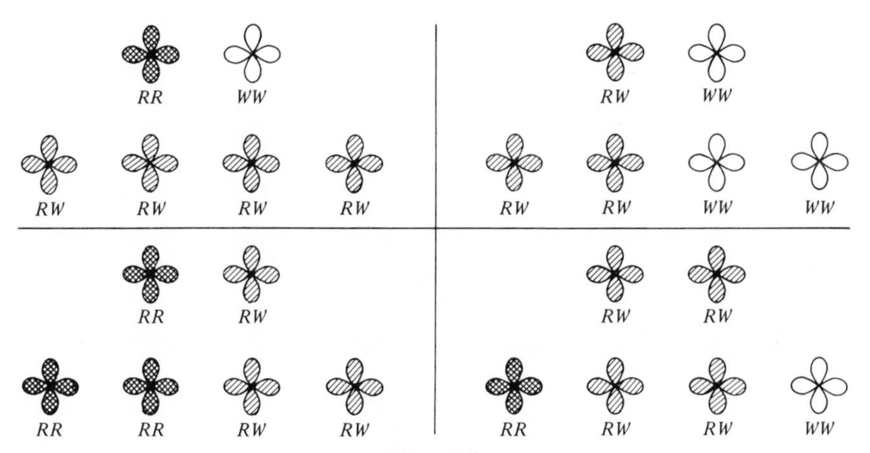

Figure 6.2

Exercises

1. What are the odds that a blue-eyed male married to a brown-eyed female who has a blue-eyed gene will have a blue-eyed child?

2. Assume color blindness is caused by a defect in the X-chromosome. Also assume that a male has the affliction if he has a defective X-chromosome while a female must have both X-chromosomes defective before the problem is noticed.

   a. Prove that a color blind man married to a "normal" (that is, has no defective chromosome) woman cannot have color blind children.
   b. Prove all males produced by a color blind woman and a "normal" man will be color blind. Can any females produced by these parents be color blind?
   c. Prove that a pair of color blind parents will always have color blind children.
   d. Can the children of the parents in a. themselves have color blind children? Why?

3. Skin color is another example of a trait in which there do not appear to be dominant or recessive genes. The problem is more complicated than in the pollenation of flowers cited in the text. There are *two* pairs of genes that determine skin color in each human, call them A-types and B-types. With, for example, the mating of a "black" and a "white" individual, the offspring carry one A-type gene from each parent and one B-type from each parent; none are dominant, so the children are of a skin color that is somewhat between that of their parents (mulattos). See Figure 6.3. Build a set of charts that indicate all possible combinations of genes for the offspring of these children. Assume that their mates are (a) black, (b) white, (c) mulatto.

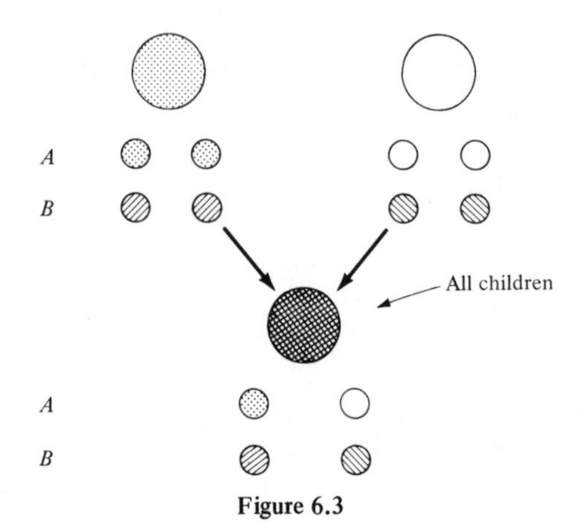

Figure 6.3

4. Suppose that G is a dominant gene of a pair and g is the recessive gene. Assume a GG individual marries a gg type. What is the probability that the grandchildren of this marriage will have the G characteristic? Assume that the mates of their children have probability 1/4 of being GG, 1/4 of being gg, and 1/2 of being Gg types.
5. Hair color genes (with the possible exception of red) follow this pattern of increasing dominance: blonde, brown, black. Build a chart for the hair colors of the offspring of parents with the following gene combinations:
   a. blonde-blonde; brown-blonde
   b. blonde-blonde; black-blonde
   c. blonde-blonde; black-black
   d. blonde-brown; brown-black
   e. brown-brown; black-blonde

6.5. An Axiomatic Approach

A major difficulty in the classical approach to probability, as presented in Section 6.2, is the requirement that the outcomes are equally likely. In a sense we must have a preconceived concept of probability in order to assert that one event is as likely as another. How then can the concept of outcomes being "equally likely" be used in the definition of probability itself? There is a serious charge of circular reasoning which can easily be avoided through an axiomatic approach to probability.

First, a few comments about notation are necessary. We have referred to the totality of points representing all possible outcomes of an experiment as a sample space S. Moreover, an event A was a particular collection of points drawn from S, that is, a subset of S. It is convenient, therefore, to call upon set theory notation in the statements of the laws of probability.

Recall that if A and B are subsets of S, the following symbols can be defined: (1) $A \subset B$ means that all elements in A are also in B. (2) $A \cup B$ is the set consisting of those elements that are in A *or* in B (or both). (3) $A \cap B$ is the set consisting of those elements that are in A *and* in B (in both). (4) If A is a subset of a sample space S, then A' denotes the set of all elements in S that are not in A. For example, if S is the set of all students at this university, A is the set of all students in this course, and B is the set of all students in biology courses, then A' is the set of all students not in this course, $A \cap B$ is the set of all students in both this course and at least one biology course and $A \cup B$ is the set of all students in biology courses or in this course.

Once more we introduce the empty set, written ϕ, to denote a set with no elements. We let ϕ be a subset of every set, again for convenience.

With this notation at hand, let us attempt to state some axioms or laws of probability. Let S be a sample space.

Axiom 1. The probability of an event $A \subset S$ is a nonnegative real number.

If $P(A)$ denotes the probability of event A, then this axiom states that $P(A) \geqslant 0$.

Axiom 2. The probability of S is 1, that is, $P(S) = 1$.

Two events A and B are *mutually exclusive* if they have no points in common. This means $A \cap B = \phi$. For example, A and A' are always mutually exclusive by the definition of A'.

Axiom 3. If A and B are mutually exclusive events in the sample space S, then the probability of one *or* the other events is the sum of the individual probabilities of A and B. In notation, if $A \cap B = \phi$, then $P(A \cup B) = P(A) + P(B)$.

The last axiom agrees with the intuition we built about probability from Section 6.2. Let S be a sample space with $n(S)$ equally likely points, let A and B be two subsets of S, and suppose that A and B have no point of S in common. For example, if the sample space S is the set of all students on this campus, the set A could consist of members of the football team and the set B could consist of members of the women's swimming team. It is more than likely that $A \cap B = \phi$. Now how many persons are either on the football team, or on the women's swimming team? In notation the answer is $n(A \cup B)$. If we know the number of people in A, $n(A)$, and the number in B, $n(B)$, then $n(A \cup B) = n(A) + n(B)$. The number on either team is the sum of the numbers on each team since there is *nobody* on both teams. By this observation and the definition in Section 6.2, we have

$$P(A \cup B) = \frac{n(A \cup B)}{n(S)} = \frac{n(A) + n(B)}{n(S)} = \frac{n(A)}{n(S)} + \frac{n(B)}{n(S)} = P(A) + P(B)$$

Hence Axiom 3 could be "proved" if we accept the classical definition of probability.

If A is an event in the sample space S, then $A \cap A' = \phi$, and $A \cup A' = S$. By Axioms 2 and 3 we have $1 = P(S) = P(A \cup A') = P(A) + P(A')$. Thus the probability of A and not A is one. If, for example, the probability that it will rain tomorrow is .2, then the probability that it will not rain tomorrow is $1 - .2 = .8$ since $P(A') = 1 - P(A)$ by this formula. A less trivial example is the following: What is the probability that you will obtain at least one head in the toss of five coins? If A is the event of getting at least one head, then A' is the event of obtaining no heads, that is, all tails. The probability of this is $P(A') = 1/32$ since there is only one way to obtain all tails and there are $2^5 = 32$ ways for the coins to fall. By the formula,

$$P(A) = 1 - P(A') = 1 - 1/32 = 31/32$$

which is the probability of at least one head.

A consequence of the formula $P(A) + P(A') = 1$ is that $P(A) \leqslant 1$. Indeed, $P(A) = 1 - P(A') \leqslant 1$ since $P(A') \geqslant 0$ by Axiom 1. This and Axiom 1 assure us that every event has probability between 0 and 1. Also, $P(\phi) = 0$ since $\phi = S'$ so

$$P(\phi) = P(S') = 1 - P(S) = 1 - 1 = 0$$

In words, $P(\phi) = 0$ states that an event that cannot occur has probability zero.

EXAMPLE. Suppose that the probabilities that a student will receive an A, B, C, D, or F in this course are, respectively, .12, .38, .40, .08, and .02. What is the probability that a student (a) passes, (b) receives at least a B, (c) receives at least a C, (d) obtains neither an A nor an F?

Solutions: Let A, B, C, D, and F, respectively, represent the events of receiving A, B, C, D, and F grades. (a) If F is the event of failure, then F' is the event of passing the course. Thus, $P(F') = 1 - P(F) = 1 - .02 = .98$. (b) The events A and B are mutually exclusive (only one grade for the course can be received by a student). Therefore, by Axiom 2, we have $P(A \cup B) = P(A) + P(B) = .12 + .38 = .50$ and $A \cup B$ is the event of obtaining an A or B grade. (c) The event of at least a C grade is $A \cup B \cup C$ and A, B, C are mutually exclusive in pairs. Thus, by Axiom 3,

$$P(A \cup B \cup C) = P(A \cup B) + P(C) = P(A) + P(B) + P(C) = .12 + .38 + .40 = .90$$

This illustrates how we might apply Axiom 3 to the case where there are more than two events. It requires $A \cap B = \phi$, $B \cap C = \phi$, and $A \cap C = \phi$, that is, the events are pairwise mutually exclusive. (d) The event cited is $B \cup C \cup D$ so

$$P(B \cup C \cup D) = P(B) + P(C) + P(D) = .38 + .40 + .08 = .86$$

Exercises

1. If the probabilities that you receive grades of A, B, C, D, and F in a course are, respectively, .05, .21, .60, .10, and .04, what is the probability that you obtain an A or a B? What is the probability that you obtain at least a C?
2. Explain why there is an error in each of the following situations:
   a. A friend claims he has even odds of receiving an A in a course and 3 to 1 odds in favor of a B.
   b. A restaurant owner has found that 42 percent of his patronage spend at least \$8 while 32 percent spend at least \$2 on alcoholic beverages. He concludes that the probability that a patron will have at least a \$2 bar bill or will spend at least \$8 is .42 + .32 = .74.
   c. The probability that you will receive exactly one traffic citation this year is .03 and the probability that you will receive *at least* one citation is .02.
3. Which of the following phrases describes two mutually exclusive events?
   a. Picking a queen or picking a heart from a deck of playing cards.
   b. Obtaining a seven or an eleven when rolling a pair of dice.
   c. Wearing brown shoes or wearing black shoes.
   d. Listening to a radio or studying mathematics.
   e. Driving a Dart or driving a Dodge.
4. An urn has 40 balls; 10 red, 12 green, 8 blue, 6 black, and 4 white. What is the probability that a ball drawn at random from this urn is (a) not green, (b) black or blue, (c) red, white, or blue, (d) neither white nor black? (Use Axiom 3)
5. What is the probability of obtaining a sum that is greater than 7 in rolling two dice? What is the probability of rolling a sum less than 7? (See Exercise 3, Section 6.2.)
6. Given the mutually exclusive events A and B for which $P(A) = .3$ and $P(B) = .65$, find (a) $P(A')$, (b) $P(B')$, (c) $P(A \cup B)$, (d) $P(A' \cup B')$, (e) $P(A \cap B)$, (f) $P(A' \cap B')$. What are the odds in favor of A? What are they for the event B?

6.6. What Are the Consequences?

With the axioms at hand we can now *define* the term "equally likely." Indeed, two events A and B are *equally likely* if $P(A) = P(B)$.

Now if a sample space S has three distinct points a, b, and c, and if these points are equally likely, then we can prove that $P(a) = 1/3$. First denote by $\{x\}$ the set containing the single point x. Then the union of the sets $\{a\}$, $\{b\}$, $\{c\}$ is S, that is, $\{a\} \cup \{b\} \cup \{c\} = S$. Also the sets $\{a\}$, $\{b\}$, and $\{c\}$ have nothing in common since a, b, and c are distinct points. Thus, the events $\{a\}$, $\{b\}$, $\{c\}$ are mutually exclusive. By Axiom 2 and Axiom 3, therefore, we have

$$1 = P(S) = P(\{a\} \cup \{b\} \cup \{c\}) = P(a) + P(b) + P(c)$$

Finally, $P(a) = P(b) = P(c)$ since a, b, and c are equally likely. This implies that $1 = 3P(a)$ by substitution and, hence, that $P(a) = 1/3$.

There was no necessity to confine the discussion to just three points. Any finite sample space with equally likely points can be used. A moment's reflection should convince you that the formula (6.1) of Section 6.2 can, in fact, be verified by an argument like the one in the preceding paragraph.

Another basic consequence of the axioms is the formula $P(A \cup B) = P(A) + P(B) - P(A \cap B)$ which is valid for *any* two subsets of a sample space, mutually exclusive or not. In order to illustrate this law, consider the problem of drawing a card from a bridge deck. What is the probability that it is a king or a heart? Since there is a king of hearts in the deck, these are not mutually exclusive events. The probability of a heart is $P(A) = 13/52$, the probability of a king is $P(B) = 4/52$, and the probability of $A \cap B$, a king and a heart, is $1/52$ since there are 13 hearts, 4 kings, and 1 king of hearts in the 52 card deck. Thus, $P(A \cup B) = 13/52 + 4/52 - 1/52 = 16/52 = 4/13$ is the probability of a king or a heart. There are, of course, other ways to compute such probabilities but this method is convenient in many cases.

In order to make the formula plausible let S be a finite sample space with $n(S)$ sample points. If A and B are subsets of S, then $n(A \cup B)$ counts the points in A or in B or in both. Now if we add $n(A) + n(B)$, then the points that are in both A and B have been counted twice. For they were counted in the $n(A)$ as well as in the $n(B)$. Therefore, $n(A \cup B) = n(A) + n(B) - n(A \cap B)$ since we subtract from the sum of $n(A)$ and $n(B)$ the number of points that have been counted twice, namely, $n(A \cap B)$. Dividing the last equality through by $n(S)$, we obtain $P(A \cup B) = P(A) + P(B) - P(A \cap B)$, which is the formula. In order to replace, for example, $n(A)/n(S)$ by $P(A)$, however, it was necessary to assume that the points in the space S are equally likely. A more general justification is in the next set of exercises.

Exercises

1. A survey of married couples in a certain city determined that 21 percent of the husbands voted in the last election, 27 percent of the wives voted, and both voted in 15 percent of the cases. What is the probability that at least one member of a couple voted?

2. In a group of 100 students, 35 are enrolled in mathematics, 50 are enrolled in English, and 40 are not taking either English or mathematics. What is the probability that a student selected at random from this group is taking English and mathematics? What is the probability that he is taking English or mathematics?

3. Given two events A and B for which $P(A) = .25$, $P(B) = .63$, and $P(A \cap B) = .15$, find

   a. $P(A')$ b. $P(A \cup B)$ c. $P(A' \cap B')$
   d. $P(A' \cup B')$ e. $P(A' \cap B)$ f. $P(A \cup B')$

4. Let C and A be subsets of a sample space S. Suppose $C \subset A$. Define $A - C$ to be all points in A that are not in C. Show $P(A - C) = P(A) - P(C)$.

5. Use problem 4 and the fact that for any two subsets A and B of S, the events A and $B - (A \cap B)$ are mutually exclusive, to prove the formula $P(A \cup B) = P(A) + P(B) - P(A \cap B)$.

6.7. Conditional Probability

Suppose that you roll a pair of dice. The probability that you obtain a sum of five is 1/9. If, however, one of the dice has come to rest on a four, then the probability that you obtain a total of five is 1/6 since in this case the second die must show a one in order to have a sum of five. This is an example of *conditional probability*. If A is the event that a total of five is obtained with two dice, and B is the event that the first die is a four, then $P(A \mid B)$ is the probability of A knowing that the event B occurs. Hence, $P(A \mid B) = 1/6$ while $P(A) = 1/9$.

EXAMPLE. The pollution level of 100 new cars is measured for those consistently using either lead-free or regular gasoline. The results are summarized in the table below. What is the observed probability $P(H|F)$ that a car has a high pollution level if it uses lead-free gas?

	H High Poll	L Low Poll	row sum
F lead-free	12	40	52
R regular	13	35	48

Solution: $P(H|F) = 12/52 = 3/13$ since there are 52 cars in the sample that consistently use lead-free gas and twelve of them have a high pollution level. Note also that $P(H|R) = 13/48$ is the probability that a new car has a high pollution level if it uses regular gas. Some other probabilities in this example are $P(H) = (12 + 13)/100 = 1/4$, $P(L) = 3/4$, $P(F) = 52/100 = 13/25$ and $P(R) = 48/100 = 12/25$. The probability that a car has a high pollution level *and* uses lead-free gas is $P(H \cap F) = 12/100 = 3/25$ according to the table. Using the results of the last section we also have $P(H \cup F) = P(H) + P(F) - P(H \cap F) = 1/4 + 13/25 - 3/25 = 13/20$, which is the probability that a car is a high polluter or uses lead-free gas.

We now have two illustrations of conditional probability and, yet, the term has not been formally defined. The definition should not be based on counting; it must be based instead on probabilities so that the axioms of Section 6.5 can be utilized. An analysis of the preceding example shows that

$$P(H \mid F) = P(H \cap F)/P(F) = (3/25) \div (13/25) = 3/13$$

Here the conditional probability is computed directly from other known probabilities. This observation suggests the following formal definition of conditional probability.

DEFINITION. If A and B are events in a sample space S and if $P(B) \neq 0$, then $P(A \mid B) = P(A \cap B)/P(B)$. $P(A \mid B)$ is read "the probability of A given B."

The definition appears artificial at first. It defines a new type of probability $P(A \mid B)$ for a sample space in terms of probabilities for events rather than from any counting principle. This is necessary if we are to proceed from axioms since the latter do not depend on counting. However, if we assume that S is a finite sample space with $n(S)$ equally likely points, then the definition may be made more plausible as follows: If B has happened then only the sample points of A that are also in B could occur. There are $n(A \cap B)$ such points and these points make up the fraction $n(A \cap B)/n(B)$ of B. Thus, we would naturally define $P(A \mid B) = n(A \cap B)/n(B)$. But $n(A \cap B)/n(S) = P(A \cap B)$ and $n(B)/n(S) = P(B)$ so

$$P(A \mid B) = \frac{n(A \cap B)}{n(B)} = \frac{n(A \cap B)/n(S)}{n(B)/n(S)} = \frac{P(A \cap B)}{P(B)}$$

EXAMPLE. What is the probability of picking two defective cars from a lot of 20 automobiles if it is known that five are defective?

Solution: The probability that the first car, selected at random from those on the lot is defective is 5/20, and the probability that the second one is defective, given that the first auto was defective, is 4/19. Hence, the probability of choosing two defective cars is (5/20)(4/19) = 1/19. The event that at least one car selected is not defective has probability 1 - 1/19 = 18/19. By a similar argument, the probability of choosing two cars which are not defective is (15/20)(14/19) = 21/38, and the probability of selecting at least one defective car is 1 - 21/38 = 17/38.

EXAMPLE. What is the probability of selecting exactly one defective car in the process?

Solution: Suppose, on one hand, that the defective car is chosen first. This has probability $(5/20) \cdot (15/19) = 15/76$ since 15/19 is the probability of choosing a good auto given that you have previously selected a defective one. (If you pick a defective car first, the next car must be a good one since we are permitted exactly one defective car in the problem.) On the other hand, you could select the defective car second, an event which has probability $(15/20) \cdot (5/19) = 15/76$ by a similar argument. Choosing a defective auto followed by a good one and a good followed by a defective car are mutually exclusive events. Hence, the probability of one or the other is, by our third axiom, $15/76 + 15/76 = 15/38$. Notice next that picking no defective cars, picking exactly one, and picking exactly two are pairwise mutually exclusive events. Moreover, together they include all possibilities in the sample space S. Thus, the sum of their probabilities should be one. Check to see if this is the case.

In this example we have used the formula defining conditional probability in the form $P(B|A)P(A) = P(A \cap B)$. For example, if A is the event of selecting a defective car first and B is the event of choosing a good car second, then $A \cap B$ is the event of a defective car *and* a good car being selected. There is no temporal order between the events in this last statement. As a consequence, we imply no temporal ordering between the events represented by A and B when writing $P(A|B)$ or $P(B|A)$. We could ask for the probability that the good car is selected first, given that the defective car is selected second. The answer is 15/19. Thus, $P(A|B)P(B) = (15/19) \cdot (5/20) = 15/76$ which also is $P(A \cap B)$. In general, $P(A|B)P(B) = P(A \cap B) = P(B|A)P(B)$.

Exercises

1. From the chart on pollution levels of cars, find the probability that a car is a low polluter or uses lead-free gasoline.
2. There are 200 unemployed workers in a small town. Some of the group have had previous experience and some were skilled workmen, with the actual breakdown in numbers given in the chart below. (The skilled workers with no on-the-job experience were trained at a local trade school.) What is the probability that a worker picked at random from this group is (a) skilled; (b) skilled if he has no previous experience; (c) skilled and has experience; (d) skilled or has previous experience; (e) unskilled or has no experience?

	S	U
P	35	50
N	30	85

P = previous experience
N = no experience
S = skilled
U = unskilled

3. A freshman engineering student has a probability of .8 of passing calculus and a probability of .7 of passing chemistry. If the probability of failing both subjects is .1, what is the probability that a student selected at random from the freshman engineering class?
   a. fails chemistry, given that he failed calculus;
   b. fails calculus, given that he failed chemistry;
   c. fails calculus or chemistry;
   d. fails calculus and chemistry, given that he failed calculus?

4. A mathematics professor knows from past experience that 99% of the students who consistently do their homework pass his course. A student who does not do his homework consistently has a probability of .6 of receiving a passing grade. Ten percent of the students in his class do their homework consistently.
   a. What is the probability that a student will pass and do his homework?
   b. What is the probability that a student will pass and not consistently do his homework?
   c. What is the probability that a student, selected at random, receives a passing grade?

5. Mortality tables are used by insurance companies for the determination of rates. They record the life of a large sample of individuals. Let X = age and $L(X)$ = number of living. The following table is based on 100,000 living at age 10.

X	$L(X)$	X	$L(X)$
10	100,000	35	81,822
15	96,285	40	78,106
20	92,637	45	74,173
25	89,032	50	69,804
30	85,441	55	64,563
		60	57,917

   a. What is the probability that a 10 year old will live to be 50?
   b. What is the probability that a 20 year old will live to be 40?
   c. What is the probability that a 40 year old will live to be 60?
   d. What are the odds that a 20 year old will live to be 60?

6.8. Independent Events

Two events are independent of the occurrence if one is not affected by the occurrence of the other. Formally A and B are independent if $P(A \mid B) = P(A)$. Getting a total of six in each of two successive rolls of a pair of dice is an example of independent events. Being intoxicated and having an automobile accident when driving are usually not independent (dependent) events.

Since $P(A \mid B) = P(A)$ for independent events A and B, we conclude from the results of the last section that $P(A \cap B) = P(A \mid B)P(B) = P(A) \cdot P(B)$. Thus the probability of A *and* B is the product of the probability of A by that for B. For example, if the probability that a passenger has a bomb on a flight from Cincinnati to Chicago is .0001, then the probability that two passengers have a bomb on the flight is (.0001)(.0001) = .00000001. (Moral: Always carry a bomb on the flight!) In the Apollo flights they have a backup system since the probability of failure of both systems is considerably smaller than failure of one. If an item has a 5% chance of failure, then the probability that two such items fail is

(.05)(.05) = .0025, which is quite small. It is likely to be cheaper to have a backup system than to attempt to perfect one system so that its failure probability is 1/4 of a percent.

The probability of *not* rolling a pair of sixes with two dice is 35/36. The probability that two successive rolls of the pair of dice will not produce a pair of sixes is $(35/36)^2$ since the second trial is independent of the result in the first trial. With three successive rolls the probability that you do not obtain a pair of sixes in any roll is $(35/36)^3 \approx .919$. The complementary event of observing at least one pair of sixes in three rolls of the dice has probability 1 - .919 = .081. How many times should you roll the dice before you have approximately an even chance that a pair of sixes will (or will not) appear? We calculate our way to an answer. With 15 rolls the probability of no double sixes is $(35/36)^{15} \approx .655$ which is still too large. For 25 rolls, $(35/36)^{25} \approx .494$ and this is as close to one half as any power of 35/36. This calculation shows that there are approximately even odds of getting double sixes with 25 rolls of the dice.

The chart below computes the probability that you do *not* obtain a pair of sixes in n rolls of the dice. n is the number of rolls and p is the probability that a pair of sixes is *not* obtained.

n	2	3	10	15	20	23
p	0.945	0.919	0.754	0.655	0.569	0.523

n	24	25	30	40	50	60	100
p	0.508	0.494	0.429	0.324	0.244	0.184	0.060

Exercises

1. Which of the following describe independent events?
   a. Being unemployed and being a college graduate;
   b. Being a teacher and being a musician;
   c. Being a student and being for Jackson;
   d. Being a mathematician and knowing basic algebra;
   e. Wearing black shoes and wearing brown shoes;
   f. Getting seven in two successive rolls of a pair of dice;
   g. Two mutually exclusive events.
2. What is the probability of drawing from an ordinary deck two aces in succession if
   a. the cards are replaced before each drawing and
   b. the cards are not replaced?
3. What is the probability of *not* obtaining a six in three rolls of a single die? How many rolls are necessary before this probability is less than 1/2?

4. If A and B are independent events and if $P(A) = 1/3$ and $P(B) = 1/4$, what is $P(A \cup B)$?
5. A field of 80 percent pure red flowers and 20 percent pure white flowers is planted. If cross-fertilization of a red and a white flower produces a pink flower, what percent of the first generation of the planted flowers will be red? What percent will be white? What percent will be pink?
6. Consider the second generation of the flowers in problem 5. Show that the percent of white, pink, and red flowers, respectively, are exactly the same as the percents in the first generation.

6.9. Birthday Problem

Suppose that three people play a game in which each person secretly writes a digit (0 through 9) on a piece of paper.

EXAMPLE. What is the probability that at least two recorded digits are the same?

Solution: It is easier to consider the complementary event A' than the event A that at least two persons record the same digit. Indeed, there is only one case to consider for the event A' that all three persons select distinct digits. The event A has two cases: (1) exactly two have recorded the same digit, and (2) all three persons record the same digit.

Now suppose the first person has selected a digit. The second person can record any *other* digit so he has probability 9/10 of getting a different digit. The third person must record a digit different from the other two and his probability of doing this is 8/10. Thus, $(9/10)(8/10) = 72/100$ is the probability that the three record different digits. Now $1 - 72/100 = 28/100$ is the probability that *at least* two record the same digit.

EXAMPLE. Suppose that there are four people at the party. What then is the probability that they all record different digits?

Solution: By reasoning as in the previous problem, the probability that they record different digits is $(9/10) \cdot (8/10) \cdot (7/10) = 504/1000 = .504$. A fifth person at the party would make this probability $(9/10) \cdot (8/10) \cdot (7/10) \cdot (6/10) = .3024$. Thus, with five people, the probability that at least two record the same digit is $1 - .3024 = .6976$ which is much larger than 1/2. Yet, many people would intuitively accept the probability to be 1/2 since there are 10 digits and 5 people involved in the game.

The birthday problem is similar to this game. How many people are necessary in a group before the probability is greater than half that at least two have birthdays on the same day of the year? Since, ignoring leap year, there are 365 days in the year, most people intuitively suspect the number to be around 150. Actually the answer is 23. The reasoning is as follows. Suppose there are three people. Select one. The probability that the next person selected has a birthday different from the first is 364/365. The third person has a birthday different from the other two and this has a probability 363/365 of occurrence. Hence, the probability that the three have different birthdays is $(364/365) \cdot (363/365) \approx .992$. The probability that at least two have the same birthday is approximately $1 - .992 = .008$. If we add another person to the group, the probability that all four have different birthdays is $(364/365)(363/365)(362/365) \approx .984$. A fifth person added to the group gives a probability of $(364/365)(363/365)(362/365)(361/365) \approx .973$ that they have different birthdays. We can keep adding people to the group until we obtain a probability that they have different birthdays of less than a half. This happens when there are 23 people in the group. The table below gives some results for various size groups.

TABLE 6.1

Number	$P(A)$	$P(A')$	Number	$P(A)$	$P(A')$
6	.960	.040	23	.493	.507
10	.883	.117	24	.462	.538
15	.747	.253	30	.294	.706
20	.589	.411	35	.186	.814
21	.556	.444	40	.097	.903
22	.524	.476	45	.059	.941

$A \equiv$ The people have different birthdays.
$A' \equiv$ At least two people have the same birthday.

Remark. The mathematical model in this problem assumes that being born on one day is equally likely as being born on any other day. The fact is, however, this is not entirely valid. More people are born in August and September than in other months. The variation is not sufficient, however, to seriously affect the approximation of the model.

Exercises

1. How many tosses of a coin are necessary before the probability that at least one head is obtained is greater than .99?
2. How many people must be in a group before there is a probability greater than 1/2 that at least two were born in the same month?
3. What is the probability that three people secretly record different letters selected from A, B, and C? What is the answer if there are five letters?
4. What is the probability that you will roll at least one seven in four tosses of a pair of dice?
5. What is the probability of drawing four cards, each of a different suit, from an ordinary deck of 52 playing cards?

6.10. Bayes Rule

There are times when the ultimate outcome of an experiment or event is dependent on the occurrence of various intermediate stages. Suppose, for example, that Mr. Rich and Mr. Poor are campaigning for governor of a state, and that the probabilities for each to be elected are .55 and .45, respectively. Furthermore, the probability that there will be a graduated tax increase should Mr. Rich be elected is .3, whereas this probability is .9 should Mr. Poor get elected. What is the probability that there will be a graduated tax increase?

In order to answer the question, we first introduce some notation. Let A be the event of a graduated tax increase occurring and let B be the event that Mr. Rich is elected governor. Then B' is the event that Mr. Rich is not elected, that is, that Mr. Poor is elected governor. Now in set notation

$$A = (A \cap B) \cup (A \cap B')$$

since this partitions the set A into two disjoint sets, namely, those points of A that are also in B, $A \cap B$, and those points of A that are not in B, $A \cap B'$. By Axiom 3, therefore, we have

$$P(A) = P(A \cap B) + P(A \cap B') \tag{6.2}$$

From the discussion in Section 6.5 we know that $P(A \cap B) = P(A \mid B)P(B)$ and $P(A \cap B') = P(A \mid B')P(B')$. Thus we obtain, upon substitution of these results into formula (6.2), that

$$P(A) = P(A \mid B)P(B) + P(A \mid B')P(B') \tag{6.3}$$

For the case at hand, $P(B') = .45$, the probability that Mr. Poor is elected, and $P(B) = .55$, the probability that Mr. Rich is elected. Also, $P(A \mid B) = .3$, the probability that there is a graduated tax increase if Mr. Rich is elected, and $P(A \mid B') = .9$, the probability for this type of tax increase when Mr. Poor is elected. Thus, the probability that there will be a graduated tax increase is determined by Eq. (6.3) to be

$$P(A) = (.3)(.55) + (.9)(.45) = .570$$

Formula (6.3) is quite general and can be used for many situations. For example, suppose a test has been found which correctly diagnoses cancer in 90% of the cases. Unfortunately, however, it incorrectly diagnoses cancer in 1% of the cases. The probability of having cancer is .05.

EXAMPLE. What percent of the cases in which the test has been given are diagnosed as having cancer?

Solution: Let A be the event that a patient selected at random has been diagnosed by the test to have cancer and let B be the event that a patient does not have cancer. Then $P(B') = .05$ so $P(B) = .95$. Moreover, $P(A \mid B) = .01$ and $P(A \mid B') = .9$. By the formula

$$P(A) = P(A \mid B)P(B) + P(A \mid B')P(B') = (.01)(.95) + (.9)(.05) = .0545$$

is the probability that the test diagnoses cancer. Thus, it does this in 5.45% of the cases.

There is a psychological problem, therefore, since this percent is higher than the cancer rate in the population. Unfortunately, somebody is being diagnosed as having cancer who does not have the disease.

EXAMPLE. What is the probability that this occurs?

Solution: In our notation the question is the determination of $P(B \mid A)$, that is, the probability that a person diagnosed as having cancer actually does not have the disease. We know that

$$P(B \mid A)P(A) = P(A \cap B)$$

by the results in Section 6.7. Also,

$$P(A \cap B) = P(A \mid B)P(B) = (.01)(.95) = .0095$$

by the same results. It follows that

$$P(B \mid A) = \frac{P(A \cap B)}{P(A)} = \frac{.0095}{.0545} \approx .174$$

since $P(A) = .0545$ by the previous result.

What was accomplished in this example is the determination of $P(B \mid A)$ when $P(A \mid B)$ and some other facts are given. In general, from the results in Section 6.7,

$$P(B \mid A)P(A) = P(B \cap A) \qquad \text{and} \qquad P(A \mid B)P(B) = P(A \cap B)$$

Now $A \cap B = B \cap A$ so the above are equal to each other. Written in an algebraically equivalent form, the above is

$$P(B \mid A) = \frac{P(A \mid B)P(B)}{P(A)}, \qquad P(A) \neq 0 \tag{6.4}$$

which is known as Bayes rule. It computes the probability of *B given A* which is quite different from the probability of A given B. For example, consider the problem of drawing two cards from a deck without replacement. Let A be the event of drawing an ace and let B be the event of selecting a numbered card from the deck. Since there are four sets of numbered cards (2, 3, 4, . . . , 10) of nine each, $P(B) = (4 \cdot 9)/52 = 9/13$. Now $P(A \mid B) = 4/51$ since there are four aces in the deck and 51 cards if a numbered card was removed. On the other hand, $P(B \mid A) = 36/51 = 12/17$ since there are $4 \cdot 9 = 36$ numbered cards and 51 cards remaining after an ace has been removed. Note also that by Eq. (6.4) we obtain the same result since

$$P(B \mid A) = (4/51)(9/13) \div (4/52) = 36/51 = 12/17$$

where $P(A) = 4/52$.

Exercises

1. The probability that a person selected at random from the population is smart is .05. The probability that a smart person is a mathematician is .1, whereas the probability that a person who is not smart is a mathematician is .01. What is the probability that (a) a person is a mathematician and (b) a mathematician is smart?
2. The probability of having an auto accident is .001. The probability that you are drinking if you have an auto accident is .68, and the probability that you are drinking if you do not have an auto accident is .009. (a) What is the probability that you are drinking? (b) What is the probability, if you are drinking, that you have an auto accident?
3. Three candidates are running for the office of Governor of Ohio, namely, Mr. G, Mr. B, and Mr. A. The probabilities that each is elected are, respectively, .55, .40, and .05. If Mr. G is elected the probability that there will be an increase in the income tax is .1, whereas this probability is .9 and .6 for the candidates A and B, respectively.
   a. What is the probability that there will be an increase in the income tax?
   b. If someone in California discovers after the election that Ohio has increased its income tax, what is the probability that Mr. B was elected?
   c. What is the probability that Mr. A was elected under the conditions of b?
4. A man tosses a coin repeatedly and obtains heads each time. He decides that the probability that the coin is biased is .95. If the coin is biased, the probability that he gets a head on the next toss is one.
   a. What is the probability that he gets a head on the next toss of the coin if we accept his analysis?
   b. If he gets a head on the next toss, what is the probability that the coin is biased?
5. There is a .3 probability that Firm A will bid on an ecology contract. If A does not bid, Firm B has a probability of .8 of landing the contract, whereas Firm B's chances are .2 if A does bid.
   a. What is the probability that Firm B will get the contract?
   b. If B lands the contract, what is the probability that Firm A did not bid?
6. The chance that you will pass a course if you study is .99. The chance that you will pass if you do not study is .11. Suppose that there is a fifty-fifty chance that you will study for the course. What is the probability that you will pass?
7. On a local audience participation show, a contestant is to select one of two identically appearing boxes in which money has been placed. In one box

there are four one dollar bills and in the other box there are two one dollar bills and a hundred dollar bill. The contestant is to draw out one bill which he can keep. What is his probability of winning $100? Suppose the first contestant selected a box and drew a dollar. If the second contestant also drew a dollar from this box, what is the probability that the third contestant will draw $100 from that box?

8. A game show sponsor wants to place three red balls and two green balls into indistinguishable boxes so as to maximize the probability of a contestant drawing a red ball after randomly selecting a box. How many boxes should the sponsor use and what is the assignment of the balls into the boxes?
9. In Philadelphia it was reported (*New Republic*. May 26, 1973. pp. 19–21) that individuals under 30 who plead guilty or are found guilty of a violent crime are sent to jail in 65 percent of the cases. Persons 30 or over who are convicted or plead guilty to such crimes are imprisoned 48 percent of the time. If 40 percent of the adult population in Philadelphia is under 30, what is the probability that a person who pleads guilty or is found guilty of a violent crime will be sent to jail? What is the probability that a person who went to jail for such a crime is under 30?

6.11. Infinite Games

Many games, such as the game of craps, have the potential of never ending. Consider, for instance, a game where a player repeatedly rolls a die until he obtains a six or a one. If he rolls a six (before he rolls a one) he wins, and if he rolls a one (before a six) he loses. What is his chance of winning the game? Since one and six are equally likely, his chance of winning is clearly 1/2. Yet, this game has the potential of never ending since he could roll only 2, 3, 4, or 5 in every roll.

More generally, let p and q, respectively, be the probabilities of winning and of losing in a *single* turn in a game that has the potential of never ending. The odds in favor of winning the game are p to q. Indeed, if the probability of winning each turn of such a game is 2/5 and the probability of losing the turn is 1/5, then clearly the odds in favor are 2 to 1 (or 2/5 to 1/5). This is reasonable since there is twice the chance to win as to lose. Whereas a few examples can make the odds in favor of winning the potentially infinite game plausible, a proof cannot be supplied based on the axioms at hand. For such a proof it is necessary to extend Axiom 3 to include infinitely many events that are mutually exclusive, taken two at a time. Rather than get carried off into this esoteric mathematical adventure, we believe it sufficient to convince you that the odds quoted are correct.

Let us now turn to a special version of the game of craps. A player rolls a pair of dice. If he initially rolls a total of 7 or 11, he wins. Should he roll a 2 or 12, he loses. When some other total is rolled, it is termed his point and he must repeatedly roll the dice until he obtains the point or a seven. If he obtains a seven he loses, and if he first makes his point, he wins. It is proved in the exercises that the player with the dice has better than a one half chance of winning. The analysis is by consideration of particular cases.

EXAMPLE. Suppose he rolls a four. What is the probability that he makes his point (four)?

Solution: The probability of a four with a pair of dice is 1/12. (There are three ways to obtain a total of four with a pair of dice and 36 ways for the dice to fall.) The probability of a seven is 1/6. Hence, the odds in favor of a four before a seven are 1/12 to 1/6 or, equivalently, 1 to 2. The probability of making the point is $1/(1 + 2) = 1/3$.

A similar analysis provides the probability of making the various points:

TOTAL	Prob. of the total	Prob. of making point
3	2/36	1/4
4 or 10	3/36	1/3
5 or 9	4/36	4/10
6 or 8	5/36	5/11
7	6/36	----
11	2/36	----
2 or 12	1/36	----

Exercises

1. Verify that the third column of the above chart is correct according to the principles of this section. Each entry in the third column is a conditional probability, namely, the probability that you win if your point is as given.
2. Suppose three coins are repeatedly tossed. What is the probability of obtaining three heads *before* obtaining two tails?
3. An urn contains 20 balls, three are red, five are green, eight are yellow, and yellow, and four are blue. You draw a ball. If it is red you lose, if it is green, you win, and if it is any other color the ball is replaced for a new turn. What are the odds in favor of winning?
4. What is the probability you will win on the first roll of a crap game?
5. What is the probability of winning when you have the dice in a crap game?
6. In the usual game of craps, the player loses if he rolls a three on the first roll. What now is the probability that he wins?
7. Change the game of craps so that you lose with a 2 or 3 on the first roll, win with a 7 or 11, and have to make your point for a 4, 5, 6, 8, 9, 10, or 12. What now is the probability that you will win if you have the dice?

Chapter 7 Average, Expectation, and Games

7.1. Is the Average Average?

During the course of a day the mayor of a small town in Nevada has three occasions to speak about the average income of the families in his town. First, when speaking to a wealthy prospective migrant to the town, the mayor claims that the average income of the families in the town is $34,000. Shortly thereafter, the mayor dictates a letter to the director of HEW in which this average income is reported to be $4,000. Finally, in an argument presented to the town council late that day, the mayor states that his $10,000 salary is only equal to the average income of the families in the town.

Is it possible for the mayor to have told the truth in each case? An affirmative answer could be obtained if different interpretations are placed on the terms used in each case. In particular, the word "average" has a loose meaning in general conversation. There are a number of quantative measures that are quite distinct which qualify as averages. For instance, assume that there are nine families in the town, four have incomes of $4,000, one has an income of $10,000, three have incomes of $20,000 and the last has an income of $220,000. (The latter family is tenth owners of a casino.) One average, called the *mean*, is the sum of the incomes of the families divided by the number of families, namely, \$306,000 ÷ 9 = \$34,000. Another average, called the *median*, is a middle value. In this case, the family with the \$10,000 income is in the "middle"; four families have larger incomes and four families have smaller incomes. A third average that is occasionally used is the *mode*, defined as the value that occurs with highest frequency. For the nine families in the town, the mode is \$4,000 since there are more families with a \$4,000 income than with any other income.

For our hypothetical town, the mean income conveys no useful information since it is strongly influenced by the single family whose income is \$220,000. The mode and the median are meaningful here. However, the peculiar distribution of income in the town still makes each of the averages somewhat misleading. The question of which average should be used is not always easily answered and, perhaps, is dictated by the information that one intends to transmit.

Generally speaking, the most widely used average is the mean. It is precisely the fact that the mean, unlike the mode or the median, is influenced by each piece of data that makes it more useful as a mathematical predictor.

Exercises

1. What reasons would you give to refute each of the following arguments?
   a. The average income of the readers of the *Wall Street Journal* is, according to a survey, \$20,000 per year. There is thus little chance that you, a college student, read this publication.
   b. A survey indicated that nine out of 10 drivers who had accidents during sunny days were without sunglasses. Hence, your chance of an accident is less on a sunny day if you wear sunglasses.
   c. Three of each four doctors surveyed recommend product x for a certain physical problem. Hence take product x for this condition.
   d. This lake is perfectly safe. It averages only two feet in depth so your children cannot drown.
2. In each of the following examples which average—median, mean, or mode—do you suspect is being cited?
   a. The Fly-by-Night Travel Agency states the average cost of a hotel room in Miami Beach, Florida, is \$15 a night.
   b. The average letter grade in this course is a C.
   c. The average wage for a company reported by the union is \$12,000 while the management reported \$18,000. (The same group of employees was used when computing each average.)
   d. The average height of the American male is 5 ft 8 in.
   e. The Motor Vehicle Bureau reports that each car on the road has a 2.3 occupancy during 1970.
3. When there are an odd number of items, there is always a middle item whose value (median) is half way between the values of the other items. Define a median for the case when there are an even number of items.

4. Find the median for each of the following sets of data:
   a. 1,2,3; b. 1,2,100; c. 2,4,6,8; d. 2,4,60,61; e. 1,2,2,2,3,3.
5. Roll six dice repeatedly and record the numbers appearing on each die. Compute the mean and the median for each roll. Does the mean or the median seem to be less variable? (Example: If the dice fall on 1,1,2,4,4,6, the mean is 3 and the median is also 3.)

7.2. What To Expect!

There is an average, more appropriately called the *expectation*, associated with problems in probability. Suppose, for example, that you will receive either \$12, with probability 1/4, or \$8, with probability 3/4, when you play a certain game. The gain for each game is either \$12 or \$8. However, if you play the game a hundred times, how much would you *expect* to win?

The conditions in the problem state that you would *expect* to gain \$12 for 25 percent (= 1/4) of the games and \$8 for the other 75 percent. In 100 games, you should expect a gain of \$12 X 25 + \$8 X 75 = \$900, or a \$9 average per game. This does not mean that you necessarily average exactly \$9 per game after 100 games; it is a *prediction* about your average gain based on the probable outcomes.

The example motivates the general definition of mathematical expectation. If A and B are mutually exclusive events in a sample space S and if $A \cup B = S$, then the expected value of obtaining the amounts of S_1 or S_2 respectively when A or B occurs is

$$E = S_1 P(A) + S_2 P(B)$$

For instance $E = 12(1/4) + 8(3/4) = 9$ in the example of the preceding paragraph. The expected value E is the sum of the amounts gained multiplied by the probability of gaining these amounts.

EXAMPLE. Suppose you hold a ticket in a raffle for which the prize is \$1000. If 20,000 tickets are sold, your expected value is E = \$1000 · (1/20,000) = 1/20 = .05, since your gain is zero when you lose. What this means is that if you repeatedly buy tickets in numerous raffles of this type you will win an average of 5¢ per raffle. Another meaning is that 5¢ would be the cost of each ticket provided the raffle is assumed equitable, that is, the runners of the raffles do not make a profit. If the raffle is to make a profit the cost per ticket must be greater than 5¢.

EXAMPLE. Each player on the winning team of the world series gets \$15,000, whereas each player on the losing team receives \$8,000. What is the expectation of the players on team A if the odds are 3 to 2 in favor of team A winning the series?

Solution: The probability that team A wins the series is $P(A) = 3/5$ in view of the odds. Thus, $P(A') = 2/5$ is the probability that team A loses the series. Now

$$E = (3/5) \cdot \$15{,}000 + \$8{,}000(2/5) = \$9{,}000 + \$3{,}200 = \$12{,}200$$

No ball player actually gains $12,200 in any one year—clearly they receive either $15,000 or $8,000. But, should the team A enter the series for a number of years and should the odds in their favor always be 3 to 2, then the average gain of the continuing players on team A would be around $12,200. (The New York Yankees went through such a cycle some years ago.)

The definition of expectations was stated for two mutually exclusive events. It extends in a natural way to any finite number of events that are mutually exclusive in pairs. Perhaps it is also worth noting that the amount gained can be negative; most of us, of course, find gaining a negative amount costly.

EXAMPLE. Suppose that you roll a die and win an amount equal to the number showing on the die, provided that number is even. If the number on the die is odd, you lose $3. What is your expectation?

Solution: There are four mutually exclusive events that are important in the example. They are the events of obtaining an odd number, a two, a four, or a six. The first has probability 1/2 of occurrence and the other each have probability 1/6. Therefore, the expectation is

$$E = \$2\left(\frac{1}{6}\right) + \$4\left(\frac{1}{6}\right) + \$6\left(\frac{1}{6}\right) + (-\$3)\frac{1}{2} = +\$.50$$

You should expect to win an average of 50¢ per game after a large number of games.

EXAMPLE. The runners of the following game charge $1 for the privilege of playing it. An urn containing 97 red balls, 2 yellow balls, and 1 green ball is placed before the player. He draws one ball. If it is green, the player receives $50, if it is yellow, the player receives $20, and if the ball drawn is red, the player has no gain. What is the average profit (per game) for the runners of the game?

Solution: $P(R) = 97/100$, the probability of drawing a red ball; $P(Y) = 2/100$, the probability of a yellow; and $P(G) = 1/100$, the probability of a green. The expected value is, therefore, $E = 50(1/100) + 20(2/100) + 0(97/100) = \$.90$. Since the runners of the game charge $1 per game, their average profit per game is 10¢.

Another approach is to note that you win $49 when a green ball is drawn since $1 of the $50 received was the charge to play the game. Similarly you *win* only $19 when a yellow ball is drawn and, finally, you lose a dollar if the ball is red.

Your expectation is, therefore, $E = 49(1/100) + 19(2/100) - 1(97/100) = -.10$. Your average loss, which is the runner's gain, is 10¢. It should not be a surprise that the answer obtained from each method was the same.

Exercises

1. A player rolls a die and wins the number of dollars equal to the number showing on the die. What should the player pay per game if there are to be no profits for the house?
2. A coin is tossed four times. If you receive $1 for every head, what should you pay for the privilege of playing the game? The answer is also the expected number of heads.
3. The runners of a dice game plan to give $50 to a player if he rolls a total of 11 with a pair of dice and $5 to the player should he roll a 7. They wish to average 50¢ profit per game. How much should the charge be for each game?
4. Since the owners of the game in Exercise 3 find that they have a few takers, they decide to charge $2.75 per game and change the rules to $20 for an 11 and $50 for a 12. What is their average gain per game?
5. A coin is tossed four times and you receive $2 if there are equal numbers of heads and tails. Also, if either all coins are heads or all coins are tails you receive $4. In other cases you must pay $2. What is your expected gain?
6. A builder is bidding on a construction contract which promises a profit of $100,000 with probability of .7 if all goes well. However, the profit will be cut to $10,000 due to a time clause in the contract should certain materials not arrive on time. The latter has by experience a .2 probability of occurrence. Finally, should there be a strike, which has probability .1, the builder would lose $20,000 due to higher wages and delays. What is his expectation for this contract?
7. In roulette, the wheel is numbered 0 to 36 with the numbers 1 to 36 alternately colored red and black. A player who bets on the red or black receives an amount equal to the bet should the wheel stop on that color. If the wheel stops at 0, the player loses.
   a. What is the expectation of a player who bets $10 on black?
   b. Suppose each number pays 30 to 1. What is the expectation of a player who bets $10 on red and $1 on zero?
8. A bakery has found that there is almost never a demand for more than 4 birthday cakes per day. The unit cost of a birthday cake is $1 and their selling price is $3. If the cakes are not sold, they are given to the employees free of charge. It is known from experience that the demands for exactly 0, 1, 2, 3, and 4 cakes have probabilities .09, .3, .4, .2, and .01 respectively.
   a. What is the expected profit of the bakery if they stock 1 birthday cake?

b. What is the answer when they stock 2, 3, or 4 cakes?
c. How many cakes should they stock in order to maximize their expected profit?

7.3. Decisions, Decisions

Assume that Mr. X is running against only Ms. Y for the office of president. A certain company sells political buttons which they buy from a supplier in either of two batches: Batch I contains more Mr. X buttons than Ms. Y buttons and Batch II contains just the reverse. The company's previous experience with presidential elections indicates their profits follow the pattern in the following chart.

	Y = Ms. Y wins	X = Mr. X wins
Batch I	\$10,000	\$15,000
Batch II	\$12,000	\$ 8,000

Which batch should they order from the supplier?

Suppose we know that $P(Y) = .4$ and, hence, that $P(X) = .6 = 1 - .4$. Then the expected value from Batch I is $E_1 = 10{,}000(.4) + 15{,}000(.6) = 13{,}000$ whereas it is $E_2 = 12{,}000(.4) + 8{,}000(.6) = 9{,}600$ from Batch II. Clearly Batch I is the best choice under the assumed conditions.

Suppose something happens that changes $P(Y)$ to .8 before the company places its order. What then should they do? Now $P(X) = 1 - P(Y) = .2$. Therefore, we have $E_1 = 10{,}000(.8) + 15{,}000(.2) = 11{,}000$ and $E_2 = 12{,}000(.8) + 8000(.2) = 11{,}200$. In this case Batch II should be ordered.

The probability that an individual wins an election is, of course, not as definite as stated in the two instances above. Even polls will give you only estimates of the actual probability. If the company is to make a decision, such as the one mentioned, it would be better for them to find first the probability that produces the same expected value for each decision. If such a probability exists, then it will be clear which batch to order based on whether or not the probability $P(X)$ is less than or greater than the probability that produces the common expected value.

In order to illustrate this, let us replace $P(Y)$ by p so $P(X) = 1 - p$ and set $E_1 = E_2$. This yields $10{,}000p + 15{,}000(1 - p) = 12{,}000p + 8{,}000(1 - p)$, or $15{,}000 - 5{,}000p = 8{,}000 + 4{,}000p$. Transpose to obtain $7{,}000 = 9{,}000p$ or $p = 7{,}000/9{,}000 = 7/9$. We have already seen that Batch I is better when $P(Y) = .4 < 7/9 \approx .778$. It should, in fact, be a better choice as long as $P(Y) < 7/9$ as seen in the graph (Figure 7.1). On the other hand, should $P(Y) > 7/9$ then Batch II is the better choice for the company. If $P(Y) = 7/9$, it would not make any difference which decision on batches was made; they both produce the expected value of $10{,}000(7/9) + 15{,}000(2/9) \approx 11{,}111.11$.

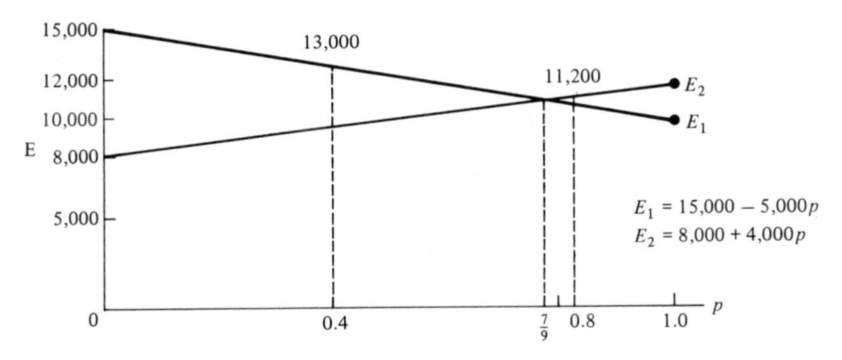

Figure 7.1

(In Figure 7.1 $E_1 = 15{,}000 - 5{,}000p$, $E_2 = 8{,}000 + 4{,}000p$. Since the company attempts to maximize its profits, it would select at any point the largest (highest on the graph) expected value.)

This example illustrates how a decision can be based on mathematical expectations. When a business or industrial problem can be suitably quantified, the decision is often removed from the hands of the pessimists or the optimists and placed into the hands of the statistician. The demand for individuals trained in statistics proves that decision making through mathematical analysis is an important consideration for management in modern companies.

Exercises

1. There are two games which are played with a single die at a casino. The first awards you \$13 if you roll a six and costs \$3 to play. The second game costs \$2 to play and yields \$3 should you roll a one or a two. Which game is the better to play if you wish to minimize your losses?

2. A catering company prepares box lunches to sell on the streets of the campus. They pay \$.50 for each box lunch and sell them for \$.90. Past experience indicates that they will sell 2,000 lunches if it rains and 3,000 lunches if it does not rain. Furthermore, unsold lunches must be destroyed by order of the Health Commission.
   a. If the chance of rain for the next day is reported as .4, should the company prepare 2,000 or 3,000 box lunches?
   b. Suppose that the chance of rain is .5. How many lunches should they prepare?
   c. For what probability of rain p would the expected value from preparing 2,000 lunches equal that of 3,000 lunches?
   d. Graph E_1 and E_2 as functions of p.

3. An investor intends to purchase shares in one of two mutual funds. The Bull

fund offers a $2000 gain if the market is on the upswing and a $1000 loss if the market tends downward. The Bare fund offers no gain if the market tends upward and $3000 if the market drops. If the probability is .6 that the market increases, what fund should the investor select?

4. A couple plans to spend one day of their honeymoon in Des Moines, Iowa. From previous visits to the city, they decide to reserve a room in one of two hotels. The first hotel cost $38 a day and the second cost $34. The couple recalls that one of these two hotels houses an outstanding restaurant but they cannot remember which one. Since they plan to dine at the restaurant, they will have to pay a cab fare of $6 should the restaurant not be in the hotel they select.
   a. Where should they make their reservation if they wish to minimize expected expenses and if they believe that there are two-to-one odds that the restaurant is in the first hotel?
   b. Suppose the odds are reversed. Where then should they stay?
   c. Suppose the couple finds the expected expenses are the same with either choice of hotel. What are the odds in favor of them rooming in the first hotel?

7.4. Game Theory

The theory of games, brought to widespread notice by Von Neumann and Morgenstern in 1944 in their book entitled *Theory of Games and Economic Behavior*, attempts to present a mathematical theory for conflict among rational beings. Conflict takes many forms, not necessarily antagonistic, such as attempting to win at a card game or a sport. Problems in competitive economics and in social behavior have more than a superficial connection with games of conflict.

When dealing with a rational opponent in a "game," decisions are based on the strategy which provides your best average gain knowing that your competition is striving for the same objective for himself. For example, in Section 7.3, Exercise 2, the following gains are obtained according to the occurrence or nonoccurrence of rain:

		RAIN	NO RAIN
Catering	PLAN I	$800	$800
Company	PLAN II	300	1200

Suppose Mother Nature suddenly becomes rational and attempts to keep the Catering Company's gain to a minimum. Clearly, she would make it rain each day. For then she could confine the profit of the company to either $800 or $300 rather than $800 or $1200. On the other hand, the company, knowing Mother Nature is against it, would pick strategy I to maximize its profits. The game (conflict) ends with the Catering Company making $800 and it raining each day. This answer is quite different from the result in Exercise 2 since we have changed the game to one with rational opposition. This change makes it a problem in what is currently classified as the theory of games.

EXAMPLE. Suppose players A and B have two cards each. A has a black 5 and a red 10 whereas B has a black 10 and a red 10. Each person selects a card and plays it. If they match in color, then A wins the difference between the numbers on the cards played; otherwise, B wins the difference. What are the best strategies for A and for B? We begin the analysis by listing what is known as A's payoff matrix, namely, a listing of the amounts player A *has to pay* to B for the various possibilities:

	B \ A	Black 5	Red 10
Black	10	-5	0
Red	10	5	0

When A pays B a negative amount, this means that A actually wins the numerical value of the listed amount. Thus, A wins \$5 when B plays his black 10 and A plays his black 5. Suppose now that you are Mr. A. Playing a black 5 may offer the possibility that you will win \$5, but it requires that B play his black 10. Knowing that your opponent is intelligent, you realize that he would not do this. In fact, B would play his red 10 since, with this choice, he wins either \$5 of \$10. Hence, you must play your red 10 or lose \$5. This proves that a rational A plays his red 10. In the same way, B protects himself by playing his red 10. The end result is that nobody wins any money although A has the satisfaction of winning all the games (the colors match). The value of the game, that is, the amount won by B, is zero in this case.

This illustration is an example of a 2 by 2 zero-sum two-person game. It is 2 by 2 since each player has two strategies from which to choose. The game is zero sum if the amount won by one player is lost by the other player (no house cut). Practical games usually involve more than two strategies as well as more than two players!

Let us abstract things and turn to a 3 by 3 zero-sum two-person game given by the following matrix.

B \ A	I	II	III	Row minima
1	3	-3	-2	-3
2	2	1	0	0
3	1	0	-1	-1
Col. maxima	3	1	0	

One strategy which can be used is the so-called "minimax" strategy. In this, A looks at the maximum loss that each strategy produces. For instance, strategy I causes A to lose 3, 2, or 1 according to whether B plays his strategy 1, 2, or 3. The *maximum loss* is, therefore, 3. For strategy II, A wins 3, loses 1, or loses 0 and, hence, 1 is his maximum loss. With strategy III, his maximum loss (column maximum) is 0. If A plays the minimax strategy, he would play that strategy which *minimizes* his maximum loss, that is, strategy III. On the other hand, B looks at his minimum gain with each strategy. His smallest gain with his strategy I is -3, the minimum of the numbers 3, -3, and -2. With his strategy 2 the minimum gain is 0, and with strategy 3 it is -1, since -1 is the smallest of the numbers 1, 0, or -1. Now B would like to maximize his minimum gain if he is to play the minimax strategy. Hence, he selects strategy 2 since this provides the maximum 0 of the row minima $-3, 0, -1$.

Note that when B plays his minimax strategy in this game, A will lose more than 0, much to B's pleasure, should A play any strategy other than III. Similarly, if B varies from strategy 2 while A holds fast on his III, B can do nothing other than lose. This always happens if the row minimum *equals* the column maximum. The common value is called a saddle point (since it is both a minimum and a maximum like the base of a saddle). See Figure 7.2. Only in problems where there is a saddle point is the minimax strategy the best possible one.

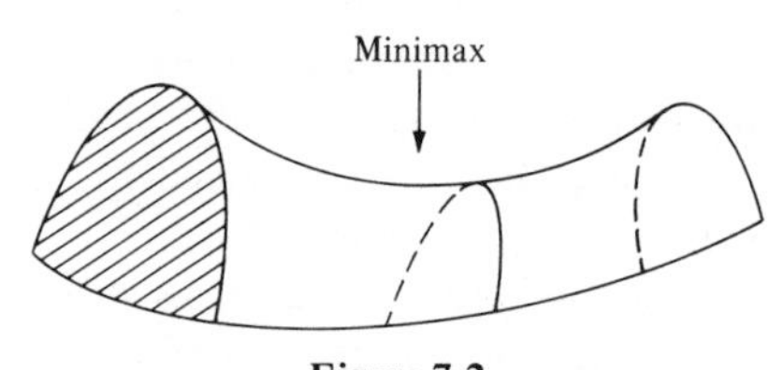

Figure 7.2

Exercises

1. What is the minimax strategy for each of the following games? In each case show whether the minimax strategy is the best strategy that can be invented.

a.

B \ A	I	II
1	2	3
2	-10	5

b.

B \ A	I	II
1	2	3
2	-2	1

c.

B \ A	I	II	III
1	2	-1	-2
2	-1	2	3
3	4	2	2

d.

B \ A	I	II	III	IV
1	1	2	3	4
2	3	2	4	4
3	-1	0	1	2
4	7	-5	2	-4

2. In each of the previous problems, the matrix is A's payoff matrix, that is, the numbers are quoted in terms of A's payment to B. Determine in each case B's payoff matrix for the game. (The payoff matrix of B should list B's strategies on the top.)
3. Compare the minimax strategy for each matrix in Exercise 1 with the minimax strategy for the corresponding matrix in Exercise 2. Does it matter if the payoff matrix is that of A or of B as far as the determination of strategies is concerned?
4. Two people match pennies according to the following rules: Player A wins the coins if both pennies are heads and player B wins the coins if both are tails. When the coins do not match, neither player wins. Construct the payoff matrix for A. Is there a minimax strategy? Why?
5. Build a payoff matrix for the coin matching game in which player A wins the coins provided both are heads or both are tails. In all other cases, player B wins the coins. What strategy should A use if the players intend to match a large number of coins?

7.5. Sometimes This, Sometimes That

Suppose that you and a friend engage in a coin matching contest. The rules are that you win when the revealed coins match and you lose if one coin is a head while the other is a tail. If your opponent is aware of the selection (head or tail) you intend to make, then clearly he can win the turn. In repeated turns, moreover, if you play according to a predictable pattern, an intelligent friend can take advantage of the situation and force you to lose more frequently than necessary. For instance, if you select heads on every other turn, your opponent would counter by selecting tails in these turns, forcing you to lose. In some way you must make the selection in a random manner so that no pattern can be discovered by the other player. Furthermore, your selection should not favor either heads or tails. Indeed, if your strategy selects heads 75 percent of the time, your opponent can win 75 percent of the time by always playing tails! For these reasons, it is better to flip the coin in a coin matching game rather than making the choice of heads or tails with each play. However, if you detect a pattern to your opponent's strategy, then you can vary your selection to take advantage of the situation.

A randomized strategy for the coin matching game is the strategy where both players flip coins to select the head or the tail. This strategy, as we shall prove, produces even odds for each player to win. If either player strays from this randomized strategy, the other player can vary his strategy and increase his chances of winning the game. Justification of this fact is left as an exercise. The methods of proof for this problem parallel the discussion in our next example.

EXAMPLE. The following is the payoff matrix of a two-person zero-sum game.

B \ A	I	II	row min
1	5	-3	-3
2	-2	3	-2
col max	5	3	

What randomized strategy should A use? What about B? What is the (expected) value of the game?

Solution: The game does not have a saddle point since the minimum of the maximum loss for A is 3 (strategy II), whereas the maximum of the minimum gain for B is -2 (strategy 2). Should A and B each play their second strategy all the time, then A would have to pay B three units (dollars?) after each game. Hence, a smart Mr. A would not stick with II if he discovers that B always is using strategy 2. He would rather switch than fight so he would play I, thereby making B pay him \$2 after each game. But Mr. B is not stupid so he starts to play strategy 1 and A has then to pay him \$5. When A discovers this, he goes back to strategy II making B pay him \$3 a time until B switches to strategy 2, and so on. Is there a better way for A and for B to play the game?

To answer the question, let us assume that B is going to play his strategy 1 with probability p and his strategy 2 with probability q. Clearly $p + q = 1$ or $q = 1 - p$ since he has to play one or the other strategy if he plays the game at all. For example, if $p = 1/2$ and $q = 1/2$, then B plays strategy 1 half the time and the other half of the time he plays strategy 2. If $p = 2/3$ and $q = 1/3$, then he plays strategy 1 two-thirds of the time in the long run. This does not mean that he plays 1 two times and strategy 2 once in every three games since, if A knows what strategy B plans to use, he can use a strategy of his own which gives him a win. Indeed, if A knows that B is going to play strategy 1, he should play II and, therefore, win \$3.

What $p = 2/3$ and $q = 1/3$ is intended to mean is that in the long run (say 99 games or 999 games) he would have played strategy 1 two-thirds of the time (66 or 666 times) approximately. The way he should do this is to find some method to randomize his playing of strategies so that, on the average, this method would tell him to play strategy 1 two-thirds of the time. He can do this, for example, by rolling a die. Should the die land on 3, 4, 5, or 6, he plays strategy 1 while he plays 2 if the die lands on 1 or 2. (Notice that the probability that the die lands on a number from 3 to 6 is $p = 4/6 = 2/3$ whereas the probability it lands on 1 or 2 is $q = 2/6 = 1/3$.)

Without yet stating the value of p, let us continue to assume that B plays strategy 1 with probability p and strategy 2 with probability $q = 1 - p$. Then, should A play his strategy I, his expected value (that is, average long term payoff) is

$$EI = 5p + (-2)(1 - p) = 7p - 2 \tag{7.1}$$

If, instead, A plays his strategy II, his expected value is

$$EII = (-3)p + 3(1 - p) = 3 - 6p \tag{7.2}$$

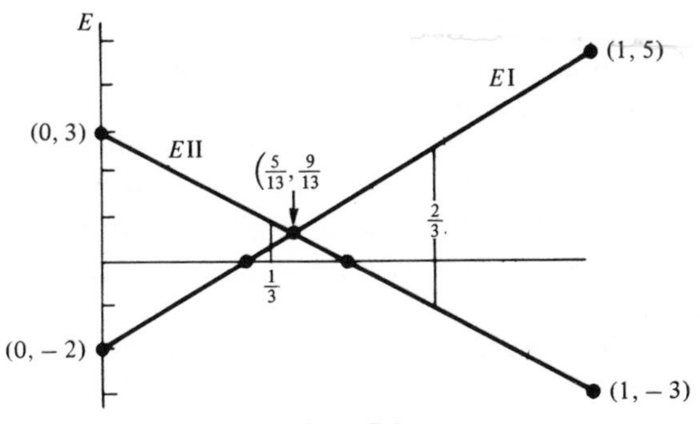

Figure 7.3

Now suppose $p = 2/3$ (B plays strategy 1 two-thirds of the time). Since A is smart, he will soon discover this and, hence, he will start to play his strategy II all the time. A's payoff from that time onward will be $EII = (-3)(2/3) + 3(1/3) = -1$. In other words, A will start to *win* \$1 per game on the average. B would be silly to do this since he can actually win money from A by other strategies.

For example, if $p = 1/3$, then, when A discovers it, he will play strategy I since it is clear from the graph in Figure 7.3 that this will lower his expected value and, decrease his loss. Once he does this, A's expected value will be $EI = 5(1/3) + (-2)(2/3) = 1/3$.

Now if B changes his p value to the point where $EI = EII$, then A is trapped. No matter what A does he cannot change his expected value even though he knows B's strategy. The choice of p for which $E\text{I} = E\text{II}$ is found by equating Eq. (7.1) to Eq. (7.2): $7p - 2 = 3 - 6p$ or $p = 5/13$. Hence, B should play strategy 1 with probability 5/13 and strategy 2 with probability $q = 1 - 5/13 = 8/13$. A's expected value is then $E\text{I} = E\text{II} = 7(5/13) - 2 = 35/13 - 26/13 = 9/13$. Thus, A will have an average loss of 9/13 dollars even though he knows B's strategy probabilities. The probability $p = 5/13$ for B to play strategy 1 is the optimum strategy for B. The value of the game is 9/13. This represents the average loss for A after a large number of games are played.

If A plays strategy I with probability p and plays II with probability $q = 1 - p$, the same analysis leads to the expected values

$$E1 = 5p + (-3)(1 - p) = 8p - 3, \qquad E2 = (-2)p + 3(1 - p) = 3 - 5p$$

These are the average gains for B should B play strategy 1 and strategy 2 respectively. Setting them equal, we obtain $E1 = 8p - 3 = 3 - 5p = E2$ or $p = 6/13$. This is the probability with which A should play strategy I. $q = 1 - p = 1 - 6/13 = 7/13$ is the probability he should play strategy II. The expected value as before. The latter is no surprise since it is a zero-sum game, that is, one's loss is the other's gain.

You may wonder why A should bother to play his optimum strategy since, once B plays his optimum strategy, A is going to lose on the average 9/13 per game. Suppose A feels defeated and starts to play his strategy II all the time. Soon our cunning Mr. B discovers this, changes his strategy to 1 and, therefore, makes A lose \$3 per game. What this illustrates is that if either player deviates from the optimum strategy, the other can take advantage of this and increase his average gain. Hence, both players, being cunning and intelligent, would stick to the optimum strategy.

Finally, how does B play strategy 1 five out of thirteen times on the average? One way is to put five black balls and eight white balls into a bag. Before he plays he draws a ball. If it is black, he plays strategy 1 and if it is white he plays strategy 2. (Of course, he had better not let smart Mr. A see what type of ball has been drawn!) In this way he will "randomize" his choice of strategy 1 and 2 so that the probabilities of selecting them are, respectively, 5/13 and 8/13.

Exercises

1. What is the optimum strategy for A and for B in the coin matching game?
2. What are the optimum strategies for A and B and what is the expected value of the game for each of the following payoff matrices for A?

a. (A across the top, B down the side)

	A	
B	5	3
	-4	2

b. (A across the top, B down the side)

	A	
B	5	-2
	-1	3

c.

	A	
B	5	-5
	2	4

d.

	A	
B	2	-2
	-2	2

3. How would you randomize the optimum strategy for A in each of the cases in Exercise 2?

4. Consider the following payoff matrix:

	I	II
1	4	-1
2	6	8

Show that there is a saddle point at (I, 2). Apply the graphical ideas in the solved problem and determine the value of p such that $E\text{I} = E\text{II}$. What is the optimum strategy? Why?

5. Can you solve three equations in three unknowns? If so, find the optimum strategy for B in the 3×3 matrix game:

		I	II	III
	1	2	-4	4
B	2	-6	3	-3
	3	6	6	-6

(Hint: Let strategies 1, 2, and 3 be played with probabilities p, q, r, respectively. Then $p + q + r = 1$ since 1, 2, 3 are the only strategies. Set $E\text{I} = E\text{II}$ $E\text{I} = E\text{III}$ to get two other equations in p, q, and r). Is it a fair game?

6. Two people play a game in which the first writes one of the integers 2 or 5 on a slip of paper whereas the second writes one of the integers 3 or 4. If the sum is even the first player wins the sum in dollars and if the sum is odd the second player wins the sum. Write the payoff matrix for the first player, find the optimum strategy for each player, and find the value of the game.

7. On a Delta Queen trip, a stranger comes to Jack and offers to play the following game. They each show one or two fingers. The stranger states that he will give Jack \$3 when he displays one finger while Jack shows two and \$1 if he shows two while Jack displays one. To make the game fair, the stranger states, Jack must give him \$2 before each matching of fingers.
   a. What is Jack's optimum strategy?
   b. What is the value of the game?
   c. How should the stranger play the game?

8. In the Civil War, the Gray Army has two positions to defend and it is capable of successfully defending one position but not both. On the other hand, the Blue Army can attack only one of the two positions. One of the two positions contains supplies and is worth four times as much to capture as is the other. If Blue attacks a defended position, it will lose a large

number of troops. Assign a loss of -3 to Blue in this case, a 4 to the event of taking the position with the supplies, and a 1 to winning the other position. What should each army do?

7.6. Linear Programming

The graphical analysis in the two-person games of the last section can be applied to other practical situations. One of these is the subject called "linear programming" which is ultimately closely allied with game theory although on the surface the connection may appear superficial. For our purposes it suffices to observe the similarity of the graphical approach to these two subjects.

To illustrate the type of problem considered in linear programming, assume that you are resistant to inflation in food prices and yet your family requires at least 3 lbs of meat per week. You know from past experience that family problems surface should you serve a larger quantity of pork than beef in a week. Furthermore, the family expects to be served each week at least 1 lb of pork. When you check the meat at the local grocery store, there is only pork and beef that even appears edible. The pork cost is $1.50 per lb and the cost of beef is $1.80 per lb. You decide not to buy more than 5 lbs of meat at these prices. How should you purchase your week's supply of meat to minimize your cost?

The answer depends on psychological factors. (a) Do you strictly want to minimize the cost to fight inflation or (b) do you want to minimize the cost and obtain a full quota of meat for your family? We will consider both issues, but first, let us translate the problem into equations.

Let x be the amount, in pounds, of beef purchased and let y be the amount of pork. Since your family requires at least 3 lbs of meat per week, we must have $x + y \geqslant 3$. Also, you do not wish to exceed 5 lbs of meat so $x + y \leqslant 5$. The amount of pork must be at least 1 lb which translates as $y \geqslant 1$.

We next graph the inequality $x + y \geqslant 3$. This is accomplished by graphing the line $x + y = 3$. The set of points for which $x + y \geqslant 3$ all lie on the same side of this line. For example, $x = 2, y = 2$ has $x + y = 4 > 3$ and, hence, all the points on the same side of $x + y = 3$ as $(2,2)$ satisfy the inequality $x + y \geqslant 3$ (see Figure 7.4).

On the same graph, we also plot points that satisfy the inequalities $x + y \leqslant 5$ and $y \geqslant 1$. Furthermore, the condition that the quantity of pork served is not to exceed the quantity of beef becomes $y \leqslant x$ with our symbolism. The set of points that satisfy *all* these inequalities is indicated by the shaded region in Figure 7.5.

The cost C of the meat is $C = 1.80x + 1.50y$ and is to be minimized in this situation. The cost C goes up no matter if pork or beef is purchased. Hence, we must select as little meat as possible in order to minimize the cost. This implies that the point for which C is as small as possible must be on the line

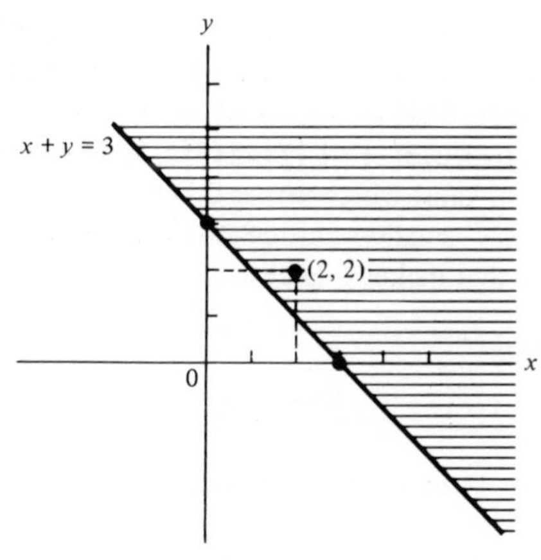

Figure 7.4

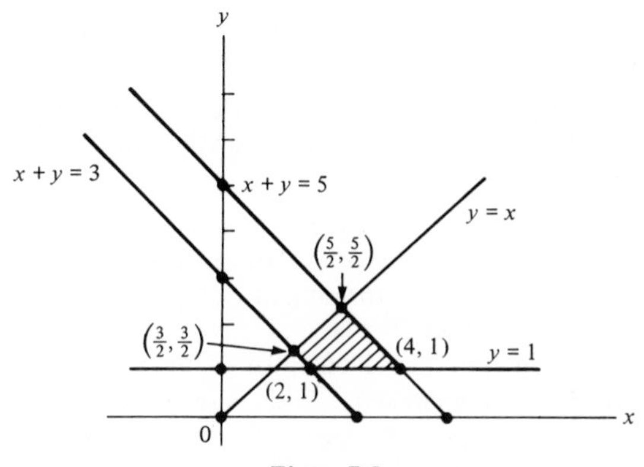

Figure 7.5

$x + y = 3$. Indeed, the problem requires that at least 3 lbs of meat be served each week. Next, since beef is higher priced than pork, we should purchase as much pork as possible for our 3 lbs of meat per week. This implies $x = y$ because the family insists on at least as much beef as pork. Together $x + y = 3$ and $x = y$ imply $x = y = 3/2$. The minimum cost C is $1.80 \cdot (3/2) + 1.50 \cdot (3/2) = \$4.95$ per week. To minimize cost, you must purchase a pound and a half of beef and the same quantity of pork.

Linear programming generally follows the patterns in this example. The problem in two variables is to minimize or maximize a linear expression $ax + by + c$, where a, b, and c are known values (like our price per pound in the example) and (x, y) lies in a region bounded by a finite number of line segments. There is a general theorem in such problems which states that the *minimum* (or the maximum) *always occurs at a corner of the region.* When applied to the example at hand, it means we need only check the cost C at each of the four corners $(2,1), (3/2, 3/2), (4,1)$ and $(5/2, 5/2)$ to determine the extremal values of C. That corner value which makes C the smallest of these four values is the value which makes C smallest throughout the region. A similar statement holds about the corner that makes C largest.

Exercises

1. Graph together the straight lines that bound the regions defined each of the following inequalities:

   a. $x \geqslant y$ b. $2x + 3y \geqslant 0$ c. $y \leqslant 10$ d. $3x - 2y \leqslant 13$

2. Shade the region of points (x, y) in the plane that satisfy all the above inequalities.

3. Find the maximum of $3x - 2y + 5$ in the region of Exercise 2. What is the minimum value?

4. Suppose that the price of pork is \$2.00 per lb and the price of beef is \$1.90 per lb. What is the minimum cost to your family under the conditions in the example of this section?

5. Assume that two bluegrass groups are to be invited to campus, the "Country Gents" and the local "Crabgrasses." The program is to be three hours in length with the Crabgrasses performing at least $\frac{1}{2}$ hour. On the other hand if the Country Gents refuse (for lack of music) to perform more than two hours, the remaining time could be filled in by a member of their group who considers himself a comedian. If the Crabgrasses charge \$1 per minute to perform, the Country Gents charge \$3 per minute, and the comedian charges \$.50 a minute, what is the minimum cost and the schedule of time for the University?

   (Hint: Let x and y be the hours that the Crabgrasses and Country Gents perform, respectively. Then $x + y \leqslant 3, y \geqslant 0, x \geqslant 1/2$, and $y \leqslant 2$. The time the comedian performs is $3 - x - y$ and, hence, the cost $C = \$60x + \$180y + \$30(3 - x - y)$ in dollars.)

6. Suppose, now, that the profit from this show is $p = \$30x + \$10y + \$60$. What is the distribution of time between performers that maximizes the profit?

7. Suppose the profit is $p = \$20y - \$2x + \$100$. What is the distribution of time that maximizes the profit?

8. Give a reason why a linear expression $ax + by + c$ defined for (x, y) on a line segment is either smallest at one end of the segment or else always has the same value on the segment.

Chapter 8 Who Makes Decisions by Statistics?

8.1. Hypothesis Testing

H. G. Wells is quoted as saying, "Statistical thinking will one day be as necessary for efficient citizenship as the ability to read and write." This day is approaching rapidly and, in the transition, people are sometimes being misled by improper applications of statistical principles. The most common error is the belief that with statistics we can "prove" cause. That smoking, for example, "causes" lung cancer cannot be proved statistically. The fact of the matter is that statistical arguments indicate that the incidence of lung cancer among smokers has a higher probability than among nonsmokers. It may, however, be that something in the make-up of a person who allows himself to be caught in the smoking habit also makes him more susceptible to lung cancer. The fact that there is a high probability for lung cancer among smokers means that scientists should, and do, spend efforts seeking a *cause* for lung cancer from cigarettes, but certainly statistics do not establish this cause.

How does a statistician infer certain conclusions, and how reliable are these conclusions? In order to illustrate the techniques instead of answering the questions directly, consider the following example. A man declares he can tell the difference between good beer and Cincinnati beer. To test his claim, an experiment is designed calling for eight glasses of beer, half of which are Cincinnati beers and the others good beers. The man knows that there are four good beers and states his opinion on each glass. We are to judge if he is depending solely on chance for his conclusions.

CASE 1: Suppose that the man correctly identifies all the beers. What is the probability that he did this by chance? There are $C_4^8 = (8 \cdot 7 \cdot 6 \cdot 5)/(1 \cdot 2 \cdot 3 \cdot 4) = 70$ ways to select four beers, which he will call good, from the eight. Now only one of these 70 ways properly identifies all the beers, so the probability that he makes a correct identification by chance is $1/70 \approx .0143$.

CASE 2: Suppose that the man correctly identifies three of the four good beers. What is the probability that he has done this by chance? There are still 70 ways to pick the four beers to call "good" from the eight. On the other hand, there are $C_3^4 = (4 \cdot 3 \cdot 2)/(1 \cdot 2 \cdot 3) = 4$ ways to select three of the four good beers. Also, there are C_1^4 ways to pick one Cincinnati beer from the four and classify it as good. Hence, there are $4 \cdot 4 = 16$ ways to pick three good beers and one Cincinnati beer. Thus the probability of three good and one Cincinnati beers being selected is $16/70 \approx .229$.

Now the statistician would initially make a hypothesis like "the man cannot identify good beer from Cincinnati beer." He then would decide that the hypothesis is probably false if the experimental result is a sufficiently rare event. For example, the probability that the man correctly identifies all the beers strictly by guessing is .0143 (Case 1) and, thus, observing this outcome is indeed rare when guessing is the only criterion for the distinction. In this case, we would naturally assume that our man can tell the difference in the beers, rejecting the initial hypothesis. However, if the man correctly identifies three of the four good beers (Case 2), an event that has probability .229 of occurrence when only guessing is used for the judgment, then the outcome is not so rare. You might, and then again, you might not, reject the hypothesis that the man cannot distinguish between the beers! The statistician would have set up criteria for rejection of the hypothesis before the experiment was conducted. Our prejudice against, or in favor of, the beer tester might otherwise determine how to judge the hypothesis when the man selects three of the four good beers.

Before the experiment is performed, therefore, it is necessary to select a real number α between 0 and 1, called the significance level, for our judgment about the hypothesis. The initial hypothesis, often called the null hypothesis, is to be rejected if the outcome falls among those rare events that have collectively a probability not exceeding α of occurrence. For example, at a .05 ($\alpha = .05$) significance level in the beer testing experiment, we would reject the hypothesis that our man cannot distinguish the good beer from the Cincinnati beer only if he correctly identifies all the beers (Case 1). If the significance level is raised to .3, then we would reject the hypothesis if the man identifies at least three beers correctly. Indeed, the chance that he picks at least three good beers is $16/70 + 1/70 \approx .243$, the probability of selecting exactly three good beers plus the probability of selecting all the good beers. This is less than our .3 significance level. (The most popular significance levels are .05 and .01. Not many individuals with some knowledge of statistics would trust a result based on a .3 level.)

There is something "one-sided" about the beer tester experiment. We were interested in validation of the man's claim that he could identify beers. It is possible that he might be able to distinguish between the beers and, yet, not properly label them. He could, for instance, be wrong and misclassify all the beers. While it is likely that the man can distinguish between the beers in this case, it is in what is classified as good beer that the man has erred.

Let us introduce a new null hypothesis, "The man cannot distinguish between beers." We no longer will be interested in whether or not he properly classifies the beers. With this new hypothesis at the .05 significance level, we reject the hypothesis if either he identifies the four good beers or he classifies the four Cincinnati beers as the good beers. The probability of the first event is 1/70 as is the probability of the second of these events. Thus the probability that one or the other of these events occurs by guess work is $2/70 \approx .0285$, which is less than the prior 5% significance level. If you add the probability that the three beers have been correctly identified or that three beers have been incorrectly classified, you obtain a number (.486) that well exceeds the significance level. This means you reject the hypothesis only if the man is always right or always wrong!

EXAMPLE. A coin is tossed 10 times and we obtain 9 heads. Should we accept the hypothesis that the coin is fair?

Solution: The probability that we get all heads (or all tails), under the assumption that a head or a tail is equally likely, is $1/2^{10} = 1/1024$. The probability that we obtain exactly 9 heads is 10/1024. For, we could obtain the single tail in the first, the second, the third, . . . , the tenth toss–a total of 10 ways. This, of course, means that the nine other coins are heads in each case. Therefore, the probability of at least 9 heads is $1/1024 + 10/1024 = 11/1024 \approx .0107$ which is less than even an a priori 2.5% significance level for rejection. We would reject the hypothesis in this case and decide the coin is biased.

One should note in each example that the rejection of the hypothesis could be wrong. We might reject the hypothesis that the man cannot tell the difference in beers when, in fact, he really cannot. On the other hand, we might decide he cannot tell the difference when, in fact, he can. We have set the a priori probability so that the chance that we made either of these types of errors is small, but we have not eliminated the chance that they could occur. Additional experimentation would make the conclusions more reliable.

The two types of errors mentioned in the last paragraph and the selection of an a priori significance level for rejection of a hypothesis are related. If it is better to accept a false hypothesis then to reject a true one, you would take a very small value for this a priori probability of rejection. If the reverse is true, the a priori significance level could be somewhat larger. For example, if our beer tester is to be shot when he cannot tell the difference in beers, a humanitarian would make the a priori probability for rejection of the hypothesis that he cannot tell the difference as large as he possibly can, whereas a sadist would set this level at a small number. The a priori significance level is, therefore, a judgment based on which type of error has the more impact.

Exercises

1. What is the probability that the beer tester will correctly identify at least two beers? At least one?
2. What is the probability that we obtain exactly eight heads in the toss of a coin 10 times? At least eight?
3. The probability that a person smokes is .3. The probability that a person dies from lung cancer is .01 and the probability that, if a person dies from lung cancer, he is a smoker is .99. What is the probability that a smoker will die from lung cancer?
4. Suppose that the beer tester must identify the one good beer in a collection of four beers? What is the probability that he will do this by chance? Change the problem and have the tester identify the two good beers in a collection of four. What is the probability that he will identify exactly one of the two good beers in a collection of four?
5. Set up the hypothesis that a dime is not biased. Decide on significance limits for rejection of this hypothesis when the coin is to be tossed 10 times in the test. (You should set limits so that if there are too many tails as well as too many heads, the hypothesis is rejected.) Now check the hypothesis by tossing a dime 10 times.
6. A type I error is a rejection of the hypothesis when in fact it is true. A type II error is acceptance of a false hypothesis. Suppose, in exercise 5, that we set the significance level at .01, that is, that we reject the hypothesis if the experimental result (number of heads or of tails) has probability less than or equal to .01. What is the minimum and maximum number of heads one can obtain to accept the hypothesis? Set the level at .20. What now is the answer to the last question? Find the type I and type II errors for each case.

8.2. What Is Normal?

One of the most common questions asked around examination times of a teacher is, "Are you going to assume the class is normally distributed in the determination of grades?" The reason for the question, to which the questioner hopes he will receive an affirmative answer, is that the student generally feels more secure if his performance on a test is judged against other students' instead of in relation to some a priori standard set by the instructor. The student likely knows that he has not mastered all the material in the course and doubts that more than a very small number of his fellow students have approximate mastery of the subject. The teacher, moreover, is most often believed to require that his students walk away from a course knowing as much as the teacher knows about the material.

What is interesting about the question on grading is that many who ask it have little understanding of the meaning of a term used. What indeed is a "normal distribution?" To provide a partial answer, define first a standard normal function by

$$y = \frac{e^{-x^2/2}}{\sqrt{2\pi}} \tag{8.1}$$

where π is the celebrated number 3.14159 . . . and e is another important constant, 2.71828 . . . . When x values are substituted into the formula to compute corresponding y values and when the points (x,y) are plotted, the standard normal function is found to have a graph that is bell-shaped, is always above the x-axis, is small for large values of x, and is symmetric (same shape) on the right and left of the y-axis (Figure 8.1).

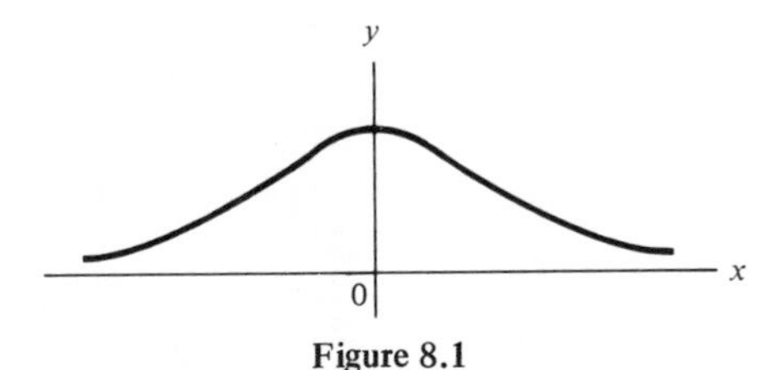

Figure 8.1

One fact that is not easily ascertained from the figure is that the *total area* under the curve and above the x-axis *is one*. This fact, which is the reason for the $\sqrt{2\pi}$ in the formula, makes the area under the curve correspond to the probability fact that $P(S) = 1$ for a sample space S. Now a "*normal distribution*" is a function that tells us how much of the area under the normal curve lies to the left of each point x on the horizontal axis. It is, in fact, the probability that an experiment has resulted in a value of x or less. For example, if t has a standard normal distribution, then the probability that t does not exceed, say, 1 is .84134, a number found from a table. This means that .8413 is the total area above the x-axis and under the normal curve to the left of the point $x = 1$. (See the shaded area in the Figure 8.2(a).) We write this as $P(t \leqslant 1) = .8413$.

Tables for the standard normal distribution, such as Table 8.1, quote the probability of the standard normal variable t being between zero and a positive number. For example, Table 8.1 cites for 1 the value .3413. By the symmetry of the normal curve, half the total area under the curve is to the left of 0. Hence, the area to the left of 1 is $.5 + .3413 = .8413$, half the area plus the area from 0 to 1.

What is the probability that t does not exceed -1? The table contains only positive values of x, where $t \leqslant x$; the symmetry of the normal curve, however, can be utilized to obtain results for negative values of x. Indeed

$$P(t \leqslant -1) = P(t \geqslant 1)$$

as indicated in Figure 8.2(b). But, $P(0 \leqslant t \leqslant 1) = .3413$ and, hence, $P(t \geqslant 1) = .5 - .3413 = .1587$.

Finally, we can combine answers to discover the probability for t to fall within a given interval. For instance,

$$P(-1 \leqslant t \leqslant 1) = P(t \leqslant 1) - P(t \leqslant -1) = .8143 - .1587 = .6826$$

The symmetry also suggests that $P(-1 \leqslant t \leqslant 1) = 2P(0 \leqslant t \leqslant 1) = .6826$.

TABLE 8.1

x	$P(0 \leqslant t \leqslant x)$	x	$P(0 \leqslant t \leqslant x)$	x	$P(0 \leqslant t \leqslant x)$
.05	.0199	.55	.2088	1.5	.4332
.10	.0398	.60	.2257	1.9	.4713
.15	.0596	.65	.2422	1.96	.4750
.20	.0793	.70	.2580	2.0	.4772
.25	.0987	.75	.2734	2.1	.4821
.30	.1179	.80	.2881	2.5	.4938
.35	.1368	.85	.3023	3.0	.4987
.40	.1554	.90	.3159	3.5	.4998
.45	.1736	.95	.3289	4.0	.5000
.50	.1915	1.00	.3413		

Once one learns to use tables of the normal distribution, it is natural to ask as to the origin of the table itself. Perhaps it suffices to state that building the tables involves numerical methods and calculus.

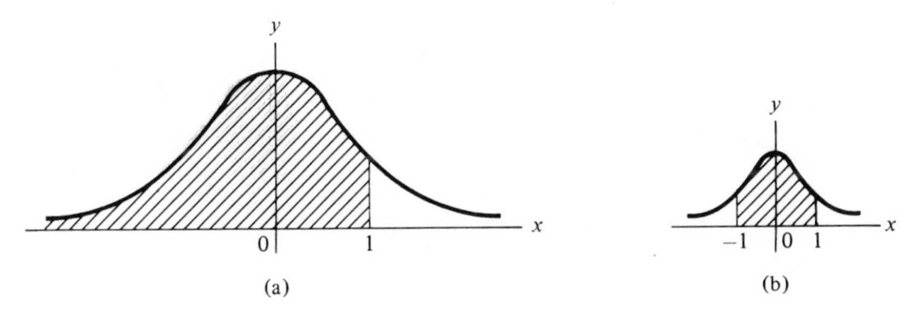

Figure 8.2

Exercises

1. Find $P(t \leqslant .75)$, $P(t \leqslant -.75)$, $P(-.75 \leqslant t \leqslant .75)$, $P(-1.5 \leqslant t \leqslant 2)$, $P(0 \leqslant t \leqslant .5)$, $P(-.5 \leqslant t \leqslant 0)$, $P(0 \leqslant t \leqslant 1.5)$, and $P(-1.5 \leqslant t \leqslant 0)$. What are these values in terms of areas under the standard normal curve?
2. Show that $P(t \leqslant x) + P(t \leqslant -x) = 1$ for $x = .5$, 1, 1.5, and 2 using Table 8.1.
3. Use symmetry to make the results in problem 2 reasonable.
4. Tables can be utilized in a backward manner. Use Table 8.1 to determine an approximate x such that $P(t \leqslant x) = .80$. Find an x such that $P(-x \leqslant t \leqslant x) = .95$ and an x such that $P(-x \leqslant t \leqslant x) = .605$.
5. Plot the points (x, y) where $y = 3^{-x^2}$ and $x = 0, 1, -1, 2, -2, 3$, and -3. The curve has a similar shape to that of the normal curve.

8.3 Applications

Thus far, we have considered only the standard normal distribution, although most applications do not initially have data that follows this distribution. For instance, the scores on the College Entrance Examination Board (CEEB) tests have a distribution that can be transformed into a standard normal distribution. For this purpose, it is necessary to know the mean of the distribution of scores and another measure of central tendency, called the *standard deviation*.

The standard deviation of a distribution is another type of average which measures how much the scores deviate from the mean. A large standard deviation indicates that the distribution is somewhat flat, as in Figure 8.3(a), while a small standard deviation indicates that many scores are close to the mean (Figure 8.3(b)). Without any attempt at this point formally to define the standard deviation, let us assume it to be a given number, denoted by σ. In the case of the CEEB scores, σ is known to be 100. The mean of these scores is $\mu = 500$. If x denotes a score, then the variable

$$t = \frac{x - 500}{100} \tag{8.2}$$

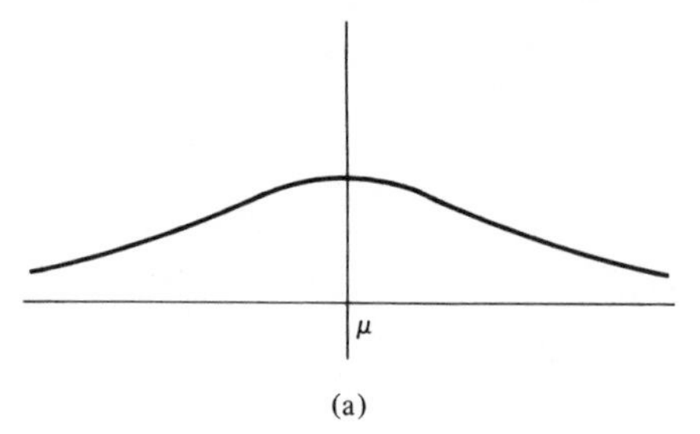

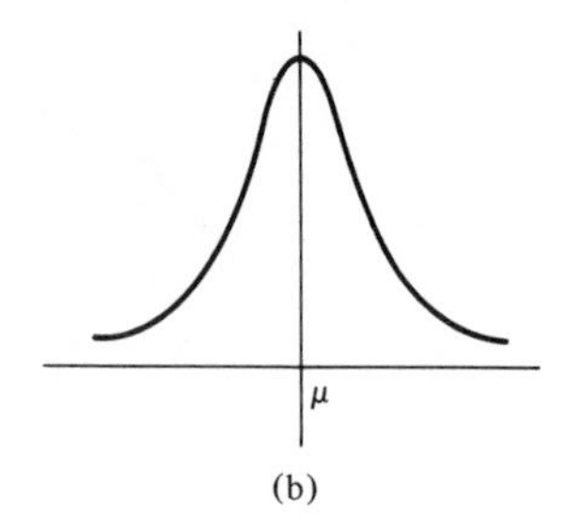

Figure 8.3

is known to be a standard normal variable. In particular $P(0 \leqslant t \leqslant 1) = .3413$. Now, by (8.2), $0 \leqslant t \leqslant 1$ is equivalent to $500 \leqslant x \leqslant 500 + 100 = 600$ since $0 \leqslant \dfrac{x - 500}{100} \leqslant 1$ or $0 \leqslant x - 500 \leqslant 100$.

This means that $P(500 \leqslant x \leqslant 600) = .3413$; in other words the probability that a test score x falls between 500 and 600 is approximately .3413.

In a similar way, we can find an interval for the scores such that 95% of the scores fall in this interval. Indeed, $P(-1.96 \leqslant t \leqslant 1.96) = .95$ and from this we obtain

$$-1.96 \leqslant \frac{x - 500}{100} \leqslant 1.96 \qquad \text{or} \qquad -196 \leqslant x - 500 \leqslant 196$$

It follows that 95% of the scores are between $500 - 196 = 304$ and $550 + 196 = 696$. By the same argument, 60.5% of the scores fall in the range 415 and 585. For $P(-.85 \leqslant t \leqslant .85) = .6046$ by Table 8.1 and

$$-.85 \leqslant \frac{x - 500}{100} \leqslant .85 \qquad \text{or} \qquad -85 \leqslant x - 500 \leqslant 85$$

Now if we believe that certain data can be transformed into a standard normal distribution, how do we determine the mean and the standard deviation? Without going deeply into the subject of statistics, we can only estimate these parameters from the data. (One should worry about the possible errors in our approximation techniques although the estimates are considered adequate when the sample contains at least 40 entries.) We simply replace the mean μ of the distribution $t = (x - \mu)/\sigma$ by the mean $\bar{x}$, obtained by averaging the data. The standard deviation σ is, furthermore, estimated by calculating the quantity

$$S = \sqrt{A - \bar{x}^2}$$

where A is the mean of the *squares* of the entries that comprise the data. To illustrate the computation of S, let the data be 1, 3, 4, 5, 7. The mean is the sum divided by the number of points, that is, $(1 + 3 + 4 + 5 + 7)/5 = 4$ and the mean of the squares is $(1 + 9 + 16 + 25 + 49)/5 = 20$. It follows that $S^2 = 20 - 4^2 = 4$, or $S = 2$.

Suppose next that a teacher of a large class discovers the mean score on a test given the class to be $\overline{x} = .75$ and the standard deviation to be .10. Letting x be the grade, the teacher then assumes that

$$t = \frac{x - .75}{.15}$$

is a standard normal variable and assigns grades according to the following probabilities:

A: $P(t > 2) = .0228$,	$x > .95$;
B: $P(1 < t \leqslant 2) = .1359$,	$.85 < x \leqslant .95$;
C: $P(-1 < t \leqslant 1) = .6826$,	$.65 \leqslant x \leqslant .85$;
D: $P(-2 \leqslant t < -1) = .1359$,	$.55 \leqslant x < .65$;
F: $P(t < -2) = .0228$,	$x < .55$

The intervals for the various grades are determined as before. For instance, $1 < t \leqslant 2$ implies $1 \leqslant (x - .75)/.10 \leqslant 2$ or $.10 \leqslant x - .75 \leqslant .20$. The probabilities represent the approximate fractions of the class that receive the corresponding grades. Should the teacher notice a serious discrepancy between these fractions and the proportion of students in the class who received each grade, then the assumption that t is a standard normal variable is likely to be fallacious.

Things would be much simpler if we would progress to a pass-fail system of grades. We would need only the interval $t < -2$ in this case provided the teacher believed that the level for passing is the same as when grades were assigned.

Exercises

1. Suppose that you have a score of (a) 600, (b) 550, or (c) 700 on the CEEB test. What is the probability that an individual has a score greater than yours in each case?
2. What is the probability that a person obtains a CEEB score between 400 and 650?
3. On one type of I.Q. test the mean is 100 and the standard deviation is 15. What percent of the population should have IQ scores (a) above 130, (b) below 90 or (c) between 94 and 112?
4. Compute the standard deviation for each of the following sets of data:

   a. 1, 2, 3, 4, 5 b. 2, 2.5, 3, 3.5, 4 c. 2, 3, 3, 3, 4
5. The usual manner to assign grades to a class based on the normal distribution is to determine the intervals for each grade from the standard normal

variable t where

$$A\colon\ t>1.5;\quad B\colon\ .5<t\leqslant 1.5;\quad C\colon\ -.5\leqslant t\leqslant .5;$$
$$D\colon\ -1.5\leqslant t<-.5;\quad F\colon\ t<-1.5.$$

a. What percent of the class receives each grade?
b. If the mean of the class is .75 and the standard deviation is .15, what intervals of scores determine each letter grade?

6. Let F be the distance in inches between fingertips of a person when the person's arms are extended parallel to the ground. Let T be the person's height in inches. The *ape ratio* is defined as F/T. (The name is based on the fact that this ratio is generally larger for apes than for man.) Estimate, by computing the ape ratio of a few friends, the mean and standard deviation for these scores.

7. Leonardo Da Vinci, with his sketch entitled *Study in Proportion*, indicates that the ape ratio should be one for his ideal man. Assume that the mean ape ratio for the human population is 1 and that the standard deviation is .03. Find the probability that a person has an ape ratio greater than yours.

8.4. The Binomial Distribution

Let p be the probability that an event will occur in a single trial and let $q = 1 - p$ be the probability that the event will fail to occur. In n trials let r denote the number of times the event occurs. What is the probability of r successes in n trials? For example, if we roll a die, what is the probability of 2 sixes in 10 trials?

In order to answer questions like this, we consider all possible "words" made from the $n = 10$ letters SSFFFFFFFF, where S is a success (a six) and F is a failure (other than a six). Each "word" represents a particular way in which we obtain exactly two sixes in ten trials. For instance, SSFFFFFFFF means the sixes were on the first and second trials, whereas the other trials yielded numbers other than six. FSFFSFFFFF means sixes on the second and fifth trials. Now, from our methods on "words" with duplication of letters, there are

$$\frac{10!}{8!2!} = C_2^{10} = \frac{10 \cdot 9}{1 \cdot 2} = 45$$

"words" of 10 letters each that can be formed from 8 letters F and 2 letters S. Next, observe that the probability of the SSFFFFFF event is $(1/6)^2(5/6)^8$ since 1/6 is the probability of a six and 5/6 is the probability of a number other than six (1, 2, 3, 4, or 5). Each of the 45 ways to obtain exactly two sixes has probability $(1/6)^2 \cdot (5/6)^8$. Hence, the probability of two sixes is $45 \cdot (1/6)^2(5/6)^8$.

Similarly, the probability of four sixes is $C_4^{10}(1/6)^4(5/6)^6$, the probability of seven sixes is $C_7^{10}(1/6)^7(5/6)^3$, etc. The exponent on the 1/6 is the number of sixes; the exponent on the 5/6 is the number of trials that do not result in a six. In general, we have that the probability $p(r)$ of r sixes in 10 trials is

$$p(r) = C_r^{10}(1/6)^r(5/6)^{10-r}$$

where $r = 0, 1, 2, \ldots, 10$. This is an example of the *binomial "curve"* (actually a set of points).

When we pass to a general problem, instead of the particular one with a die, we have an event with probability p of occurrence in a single trial, probability $q = 1 - p$ of failure, and n trials. The probability of r occurrences in n trials is

$$p(r) = C_r^n p^r q^{n-r}$$

For example, the probability of 5 heads when a coin is tossed 7 times is $p(r) = C_5^7(1/2)^5(1/2)^2 = 21/128$. The *binomial distribution* is a function or rule $D(r)$ which gives the probability of *at most r* occurrences in n trials. It is nothing more than the sum of the probabilities of $0, 1, 2, 3, \ldots, r$ successes, that is $p(0) + p(1) + p(2) + \cdots + p(r)$.

Working with large values of n, one finds the computations with the binomial distribution tedious, to say the least. There is, however, an elegant theorem in statistics that enables us to use the normal distribution in place of the binomial.

THEOREM. If r is the number of successes in n trials of an event for which p is the probability of a single success, then $t = (r - np)/\sqrt{np(1-p)}$ is approximately a standard normal variable for sufficiently large n.

Experience indicates that the approximation is good if $np > 5$ and $p \leqslant \frac{1}{2}$.

Suppose that you are taking a multiple choice (5 options for each question) examination with 40 questions. You guess at each answer. What is the probability that you will obtain a grade of at least 8? Here $p = 1/5$ (the probability of a correct guess on each question) and $n = 40$ (the number of questions). Hence $t = (r - np)/\sqrt{np(1-p)} = (7.5 - 8)/\sqrt{32/5} \approx -.198$ and t is approximately a standard normal variable. Thus, $P(t \geqslant -.198) = .422$ from a table of the normal distribution, that is, the probability that you obtain a score of at least 8 by guessing is nearly a half!

You should question why the value $r = 7.5$ was used in the illustration rather than the value $r = 8$. The reason is that the standard distribution is defined for all real values of t whereas the binomial distribution that is being approximated is defined only for nonnegative integer r. When we wish to find the value of the binomial from 8 onward, it is better to begin with a point half way between 8 and its closest predecessor 7. Then, for example, we can answer questions like "what is the probability that you obtain a score of *exactly* 8 by guessing?" Indeed, you need only consider the area under the normal curve from $(7.5 - 8)/\sqrt{32/5}$ to $(8.5 - 8)/\sqrt{32/5}$. From a table of the normal distribution, the answer is found to be approximately .145.

What is the probability that you get a score of the least 10 by guessing? Again $t = (r - np)/\sqrt{np(1-p)} = (9.5 - 8)/\sqrt{32/5} \approx .593$. Now $P(0 \leqslant t \leqslant .593) = .223$ so $P(t \leqslant .593) = .723$ and $P(t > .593 = 1 - P(t \leqslant .593) = .277$. You have around a 28% chance of getting at least 10 correct! Some other probabilities for at least r correct in 40 questions is given in the following table:

0	1	2	3	4	5	6	7	8	9	10	15
1	.998	.995	.985	.962	.917	.838	.723	.578	.422	.277	.005

Exercises

1. Construct a table for the probability of 0 to 5 heads in tossing a coin 5 times.
2. What is the probability that you will guess two correct answers in a three question multiple choice (five possible answers) test? At least two? At least one? None correct?
3. With a 100 question true-false test, what is the probability that you can obtain a score of 40 by strict guessing? At least a score of 60? A score of 80, at least? A score of exactly 80? (Hint: $79.5 < r < 80.5$)
4. Suppose that when a coin is tossed 100 times we obtain only 40 heads. Would you, statistically, judge the coin to be biased? What would your answer be with 36 heads?
5. Plot the number of heads (x-axis) against the probabilities of the heads (y-axis) in problem 1. Draw a smooth curve through the plotted points. (The result should be a curve similar to the normal curve. This observation leads to the conjecture and eventually to the proof of approximation of the binomial by the normal.)
6. According to Mendel's inheritance law the odds in favor of a yellow pea over a green pea are 3 to 1. In an experiment with 400 peas, 120 came out green. Does this run contrary to Mendel's law?
7. What is the probability that you get a score of exactly 9 in a 40 question multiple choice (5 options) test? What is the probability that you obtain a score of 8, 9, or 10?

8.5 Testing a Relation between Two Conditions

There are numerous problems in which we must decide if the difference between two sample percentages or proportions is due to chance or to other factors. A manufacturer may note that 2% of the items are defective in one sample of his product and that 2.3% are defective in a later sample. If this cannot reasonably be attributed to chance, the manufacturer should examine the production process to discover other possible causes.

Karl Pearson (1857–1936), one of the world's foremost statisticians, discovered a method to judge the difference between two relations.

Before discussing Pearson's method in detail, it is necessary to speak about some limitations connected with statistical tests in general. Most statistical tests require that a "random sample" be drawn from the population being tested. A sample is random if the probability that any one entry in the population is selected for this sample equals the probability that any other entry is chosen. Two samples are independent if both are random samples that were selected without any dependence on one another. For example, if you select one random sample and then select a second sample from the entries in the first, the samples are not independent. The outcome of the second sampling is influenced by the results in the first sample.

The tests in this book are large sample tests meaning that one should have a sample of size 40 at least before the test is applied. Many statistical problems are of a nature that large sampling is impossible. As a consequence, statisticians have developed other methods to deal with decisions based on small samples. These can be found in most the introductory books on statistics and are not included here.

Pearson's test is based on a selection of two large independent samples from the population. Let $S1$ and $S2$ be the number of "favorable" cases observed in the two random samples of sizes $N1$ and $N2$, respectively. The variable

$$t = \frac{\frac{S1}{N1} - \frac{S2}{N2}}{\sqrt{p(1-p)\left(\frac{1}{N1} + \frac{1}{N2}\right)}} \qquad p = \frac{S1 + S2}{N1 + N2}$$

is approximately a standard normal variable provided there is no difference between the two relations.

Consider, by way of illustration, the conjecture that "women are more eager to marry than are men." We supply 204 students in a class with a questionnaire asking for the following information: Are you male? Do you desire to get married within the next six years?

Suppose the answers are as indicated in the chart below:

		Male: Yes	Male: No
MARRY	YES	48	59
	NO	50	47
		98	106

Would this justify the conjecture? Here $N1 = 48 + 50 = 98$, the number of males in the sample and $N2 = 59 + 47 = 106$, the number of females sampled, $S1 = 48$ and $S2 = 59$. Thus,

$$t = \frac{(48/98) - (59/106)}{\sqrt{\frac{107}{204} \cdot \frac{97}{204} \cdot \left(\frac{1}{98} + \frac{1}{106}\right)}}$$

$$\approx \frac{.4898 - .5566}{\sqrt{(.525)(.476)(.0196)}} \approx -\frac{.068}{.070} \approx -.97$$

Now the probability that $t \leqslant -.97$ is from a table of the standard normal distribution, .1660, so the conjecture is not "proved" using any reasonable significance level (5 or even 10 percent). In fact, the result indicates that the conjecture is false.

We have tested in this example the hypothesis that there is no difference in the male-female reaction to the question on marriage. The experiment has led us to the conclusion that the hypothesis is true.

Applications of statistical tests tend to suffer from one or more weaknesses. For example, are the samples consisting of 98 males and 106 females drawn from a particular class really random samples of the population consisting of all adults? Clearly it is not; indeed the average age of the group sampled is less than that of the population as a whole and all are college students. Hence, the conclusion is questionable.

One solution to the problem is to change the hypothesis to read "female college students are more eager to marry than male college students." Then we should be prepared to answer the objection that the students in the class do not represent a random sample of the population of all college students. Perhaps the type of student that registers for the particular class introduces a bias in the sample and perhaps the students at your college are not representative of college students elsewhere. These objectives, while less serious than the initial one, should be considered. You should attempt, at least, to feel confident that your sample is reasonably representative of the entire population. Otherwise who would believe your conclusions? Certainly not me!

Exercises

Use a table of the standard normal distribution and the value of t to determine the validity of the following:

1. Theory: Age and swinging go together. Questionnaire: Are you over 20? Do you date at least three times a week? Answers: Yes, Yes: 25; Yes, No: 45; No, Yes: 50; No, No: 120.

2. Theory: Beer drinkers are fat. Questionnaire: Are you a regular consumer of beer? Are you at least 5-lbs. overweight? Answers: Yes, Yes: 120; Yes, No: 45; No, Yes: 35; No, No: 92.
3. Theory: Good students find college dull. Questionnaire: Have you at least a 3.0 grade point average? Do you find most of you courses a challenge? Answers: Yes, Yes: 20; Yes, No: 60; No, Yes: 220; No, No: 140.
4. Theory: Males favor women's liberation more than females. Questionnaire: Are you a male? Do you approve of the women's liberation movement? Answers: Yes, Yes: 80; Yes, No: 25; No, Yes: 40; No, No: 12.
5. Theory: Freshmen, more than other college students, favor the president. Questionnaire. Are you a freshman? Do you generally approve of the president's policies? Answers: Yes, Yes: 123; Yes, No: 87; No, Yes: 95; No, No, 56.
6. Professor A flunked 10 out of 50 calculus students. Professor B flunked 5 out of 50. Dean C said to Professor A: "You're too tough." Was the dean justified?
7. Theory: More men than women have an ape ratio (See Exercise 6 of Section 8.3) greater than one. Set up a survey to prove or disprove this.

Chapter 9 Naive Fortran Programming

9.1. You and the Computer

We are being infiltrated by machines. There are now more than 50,000 large-scale computers in operation in the U.S.A. Only a decade ago there were 1,700 such machines and 40 years ago there were none. The computer population explosion has already affected your existence and it promises to have a greater influence on you in the future. You may desire to resist by, perhaps, folding all your computer-card bills, but it would be more profitable to learn about these "thinking" machines. Through knowledge we can control those monsters and even solicit their aid in providing us with a better society.

The computer is simply a tool that significantly shortens the time it would take man to manipulate symbols and to perform calculations. While it is true that such machines do nothing we cannot do ourselves, this statement ignores the precious time factor. In fact, a job of one-hour duration on a modern computer would take a man, working continuously 24 hours a day, more than a hundred years to complete. This staggering new capacity of mankind enables us to reconsider problems that, because of their length and the drudgery involved, were virtually impossible in the past, as well as to treat many of the new problems that face us today.

Each technological development of man has had the potential for evil, sometimes in the name of progress. A computerized police state, such as the one painted by Orwell in *1984*, is not an impossibility in this computer age. The citizenry can, however, protect society by becoming aware of the startling capabilities of the computers and, thereby, assure through law that these capabilities are directed toward constructive goals.

A more likely result of the computer revolution is the virtually complete automation of industry. While the benefits of increased productivity, lower costs, and the elimination of many routine tasks are apparent, there is a difficulty in that automation may lead to unemployment. As unemployment increases, demand for goods lessens, which in turn adds to unemployment. A vicious downward spiral is created. The negative economic implication must be offset by positive planning on a national scale. Negative income taxes, reduced work weeks, increased demand for services, and retraining programs are some methods of overcoming the underemployment problem.

Although the danger of malevolence by computers exists, there is a much greater potential for beneficial progress and improved understanding of many areas of endeavor through their use. We already have computerized reservations on airlines, computerized billing and payrolls in companies, computerized catalogs of regional libraries and computerized methods to aid in the resolution of urban problems, such as traffic, crime, housing, and schools.

Attempts to produce computer programs to translate one language into another have stimulated a great deal of research and a better understanding of the structure of language. A new field called "artificial intelligence" has appeared in which at least some of the researchers are attempting to understand how humans reason by simulating various models on the computer. Computers can assist a student in learning a subject by providing him with "individualized" questions and permitting him to pace himself in the learning process. Music and the arts have been affected by the computer. Perhaps the future in many areas of endeavor is in the hands of those who understand and can utilize the computer's capacity.

9.2. Fiction or Fact, Who Knows the Future?

What are the implications of the computer revolution for the future? Projecting from the current progress, we can forecast with high probability some major changes in our country during the next 25 years.

First of all, the banking system will be considerably altered. No longer will an employer pay you by check; instead your pay will instantly be credited to your account on payday. The employer will enter the data from a terminal in his office which will transmit the data to the bank of your choice. Billing, for such things as time payments, will be accomplished by a computer transfer of funds from your account. Writing checks and sending them through the mail will be a thing of the past. Checks as we know them today will cease to exist.

Even the use of cash will decrease. In general, you will carry a bank card and the computerized cash register will draw directly and immediately from your account when you make a purchase. Money, in the form of government issued coins and bills, could even vanish unless traditionalists insist on its continuation as a form of barter.

Charge accounts will remain, but the periodic or the delayed payment will be automated. The bank statement will be a computer printout telling the sources of the funds in your account and the destination of funds dispersed by charges, purchases, and periodic payments from your account.

In the field of education, students at all levels, from early grade school through college, will have access to computer terminals to supplement other methods of learning. The terminal can be programmed to ask a question about a subject and to adjust each additional question according to the answer it receives on the previous question. For example, an incorrectly answered question might be followed by a terminal printout of some review material, followed by a new, perhaps simpler, question. In this manner the terminal can provide each individual with instruction based on his ability to learn. No longer would all students be required to keep the pace set by the teacher. Furthermore, miniature inexpensive computers will be developed which, among other things, will enable students to learn via the computer in the comfort of their homes.

The degree of progress of the student, while at home or at school, will be recorded by the computer and available at any time to the teacher for evaluation. The computer will maintain records of the time needed to complete each step in the learning process as well as the attainment level for each student. Thus, the student with a lower achievement level who is a slow learner can be distinguished from the one who has really not attempted to learn. Similarly, the bright student can be distinguished from the average student who reaches a high achievement level by extraordinary effort.

This section discusses only two of a multitude of potential consequences of a computer-oriented society. Other applications are left to your imagination!

Exercises

1. What are some of the employment implications in the development of the checkless checking account cited in this section? For example, would tellers be necessary? What would be its impact on postal service?
2. Discuss the question, "Will the computer replace the teacher?"

3. Write an essay on consequences of a computer-oriented society other than those mentioned in this section.

9.3. The Past

The idea of using mechanical means for performing arithmetical operations is very old and the origins of the early devices, such as the abacus, are lost in antiquity. Many of these calculating devices are very much in evidence today. The slide rule, invented by Robert Bissaker in 1654, is still found in the pockets of budding engineers although an inexpensive mini-calculator could soon replace the slide rule. The mechanical adding machine was invented by Blaise Pascal (1623–1662), a mathematician, who found routine addition and subtraction a chore. Gottfried Leibniz (1646–1716), a philosopher and mathematician, designed a machine that would multiply as well as add and this invention is the forefather of the modern office calculator.

All of these calculating devices required constant attention and information from their human operators. The first general purpose computer, with its memory units and some ability to follow a flexible pattern of instruction, was envisioned by an Englishman, Charles Babbage (1792–1871), around 1820. Babbage spent his life and fortune attempting, without success, to build what he called an analytical engine using the mechanical means of his day. The engineering difficulties encountered in attempting to build his complicated machine with its multitude of gears were enough to doom his project. His theories, however, were sound, although they were virtually unnoticed for the next hundred years.

Around 1937, Howard Aiken, an American applied mathematician, shaped a plan for an electrically powered computer that used many of Babbage's ideas. Aiken was initially unaware of the work of Babbage and duplicated much of the efforts of his predecessor. With the support of IBM, Aiken and some coworkers built the first working computer, called the Mark I, in 1944.

The first electronic computer, named the "ENIAC" which stands for "Electronic Numerical Integrator and Computer," was built at the University of Pennsylvania in 1946 by J. Eckert, an electrical engineer and J. Mauchy, a physicist. The replacement of mechanical devices of the Mark I by electronic ones in the ENIAC increased the speed of a single computation more than a thousand fold. The ENIAC was huge; it contained 18,000 vacuum tubes generating heat, and had a high probability of failure. The introduction of "solid state" electronics, with its transistors and printed circuits, helped resolve these problems of size and failure of the early electronic computer.

Furthermore, the arithmetical speed and the memory capacity have significantly increased with each new generation of computers. The next generation of computers will be smaller by a factor of 1000 than those of 25 years ago and considerably faster than today's machines.

A major development for the advancement of computers was the stored program concept of John von Neumann, an outstanding mathematician of this century. Whereas the early calculating machines could store data for a problem in their memory units, special wiring was needed for each problem to control the sequence of operations on the data at hand. Many laborous hours were spent rewiring boards so that the computer would be capable of proper action whenever a new problem was encountered. In 1945 von Neumann suggested that instructions be coded and stored electronically, thereby eliminating the special wiring problem. A machine could be designed to interpret these coded instructions and to establish the necessary connections between electrical circuits. No longer was any external physical change needed to adapt the computer for each problem to be resolved. The coded instructions, or *program*, was sufficient.

There has been a rapid sequence of improvements in technology and simplification of coding instructions in the years since 1945. Today's computer is much smaller, faster, and considerably less likely to malfunction than its forefathers. At the same time, moreoever, it has greater capacity, additional memory units, and is easier to instruct and to operate.

9.4. What Is a Program?

A program is simply a set of instructions used to tell a computer how to provide an answer to a specific problem. The day when the computer can directly accept sophisticated kinds of instructions in ordinary English is yet to come. Consequently, we must write our program in some special language which is translated by the compiler (assembler) into binary code for the computation.

There are numerous computer languages, some of which are more suitable than others for certain missions. For example, FORTRAN, ALGOL, and PL/1 are commonly used for scientific problems, COBOL is used for data processing, and JOVIAL for military applications. It has been estimated that 90% of all scientific programs for the computer are written in FORTRAN so we shall study a subset of this language in this text.

In addition to being understood by the computer, all programs must be completely and precisely stated. The machine is both an electronic brain and an electronic moron. While it can quickly carry out well-stated instructions, it has no ability whatsoever to infer what you mean. When an answer of zero appears while one of 1000 is expected, it is not the machine that has erred; rather, it is the instructor (Programmer). You should expect to have some errors (bugs) in your programs.

Although programming techniques have been immensely improved and simplified since the early days, the process of finding and correcting errors in a program (debugging) consumes considerable time and effort for even the experienced programmer. There are usually special compilers at each computing center that assist programmers in the debugging process. Most errors in the program prevent the computer from completing the problem solving task. The special compilers are designed to print a message describing the types of programming errors that cause the premature termination of the computation. Since there is some variation from one computer installation to another, it will be necessary to acquaint yourself with any special instructions and compilers at your computing center.

Computers contain two basic kinds of components: logic elements (called switching elements) and memory elements. In almost all computers these elements are binary in that they are two stage devices. The logical elements process (perform arithmetic, etc.) the data that is stored as sets of binary memory units.

Each number in the data is stored as a binary number in a collection of memory elements. Now the count of the memory elements reserved for a number is finite–there are only a finite number of these physical memory devices that can ever be in existence in the real world. Hence, an infinite decimal such as π or .333 . . . must be stored by only a finite number of binary memory elements. Such numbers are rounded before storage leading to an unavoidable error, called the round-off error. Each set of memory elements containing a stored number is given a unique address, selected by the user, and can be recalled to be processed only by citing the particular address. How this is accomplished will be discussed in connection with general programming.

9.5. Flow Charts

Before communicating with a computer, one must be certain that he has an unambiguous plan of how to proceed in the computation and that his plan can be completed in a finite number of steps. A device, called a *flow chart*, is commonly used which graphically describes the steps in the computation and helps us avoid the much too common human error of asking for the impossible or of asking for something we really do not want.

By way of illustration, suppose an individual can answer each of the three questions in a test. The steps the person uses are to (1) accept the test, (2) answer a question, (3) answer another question, (4) answer the last question, and (5) turn in the answers. The flow chart for the completion of the test is a diagram indicating the sequence of steps. Either of the two flow charts in Figure 9.1 satisfies this requirement.

The arrows in the chart tell us where to go after completion of the step. There are two arrows for a decision; the next step is dependent on the answer to the question. The chart on the left of Figure 9.1 lists the steps directly. Should the test contain twenty questions instead of three it would be necessary, in the first diagram to have 17 additional rectangles containing the statement "ANSWER ANOTHER QUESTION." The chart on the right, however, does not change if there

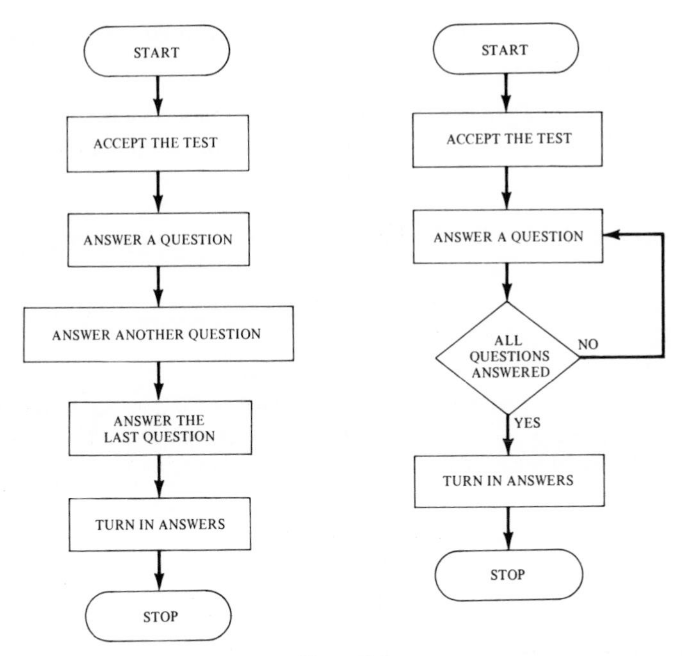

Figure 9.1

are three, 20, or any finite number of questions. Indeed, as long as the instruction "ANSWER A QUESTION" is interpreted as answering a question not previously answered, the chart indicates that the individual passes from one question to another until all are answered. Once all questions are answered; the test is turned in.

Each geometric figure in the flow charts has a meaning. Although flow charts are in common use, there is considerable variation as to the meaning of the geometric figures used in these charts. For our examples, nevertheless, certain conventions will be followed.

Terminal. The beginning, the end, or a place of interruption of the flow chart.

Instruction. One or more instructions can be placed in a rectangle in the flow chart.

Decisions. This figure is used when you must decide which of two directions (yes or no) to proceed, that is, a reply to a question with a "yes" or "no" answer.

Another figure in common use, which was not necessary in the previous example is the trapezoid:

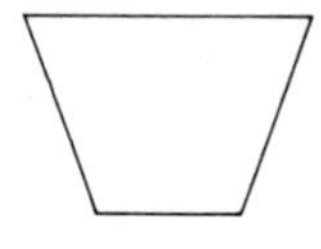

Data. This represents data that is entered into the problem or printed from the computation in the problem.

This figure use will be illustrated later in connection with numerical problems.

Exercises

1. Construct a flow chart for a *1-hr* test containing 20 questions. Adjust your chart, if necessary, so that the resulting flow chart applies to any test containing a finite number of questions and a fixed time limit.

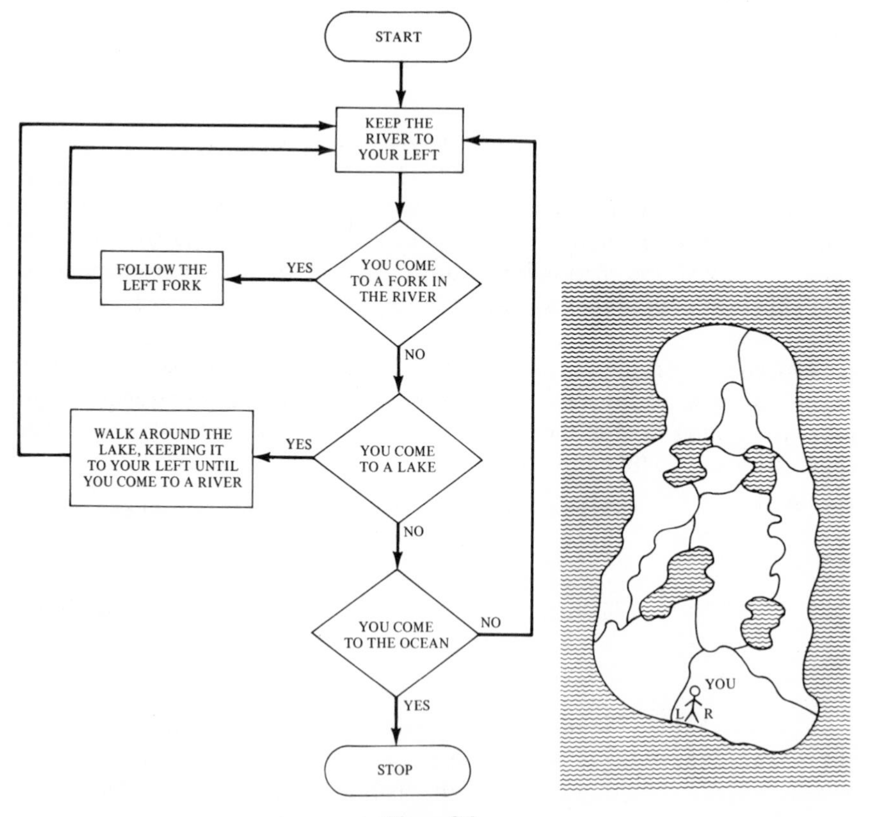

Figure 9.2

2. Build a flow chart for a timed test in which each answer is checked provided time remains.
3. Write a flow chart for changing a tire. The steps should include checking the spare to determine if it is flat, jacking up the car, removing the lugs one by one, replacing the flat by the spare, replacing the lugs, and jacking down the car.
4. Construct a flow chart for placing a telephone call.
5. Draw the path you follow if the flow chart in Figure 9.2 guides your travel on the illustrated island in the Pacific ocean.
6. Change the problem so that all instructions direct you to the right instead of the left. Then find the path when you begin the journey at point R. Do you ever reach the ocean?

9.6. Input Media

Oral communication with a computer has not been sufficiently developed for practical purposes as yet. Meanwhile, other means are utilized for communication between the human and the computer, the most common of which is the punch card. Information is transmitted by a code that depends on the location of rectangular holes in the columns of each card. A keypunch machine, which has a keyboard that is like those on an ordinary typewriter with some special symbols replacing the usual nonnumerical and nonalphabetical characters, is used for the preparation of the coded cards. The cards are read into the computer by a punch card reader and the program is executed.

In recent years time-sharing computer systems have come into widespread usage. These systems enable a number of individuals to communicate directly and simultaneously with the computer from remote access points. Basic components of such a system are a *terminal* for input/output and, frequently, an acoustic coupler that enables the user to transmit messages to the computer by telephone. The terminal, like the keypunch, is similar to a typewriter.

In computing centers that have time-sharing systems, advanced programmers usually store a number of games that enable an individual to play against the computer. For example, the card game of blackjack is easily programmed with the computer acting as the dealer. Some random process is utilized to arrange the cards and the computer types the content of each hand. A few lines of the game as played on a computer are given on p. 204. If it exists at your center, the program might be stored under an informative name selected by the programmer.

This text emphasizes the use of a terminal rather than a key punch. There are some minor differences and they are listed in later sections. It is highly recommended that a student attempt both methods of communication with the computer whenever this is possible.

```
THIS PRØGRAM SIMULATES THE GAME ØF BLACKJACK.
IT ALLØWS YØU TØ PLAY FIVE HANDS ØF BLACKJACK WITH THE CØMPUTER.
DØ YØU NEED MØRE INFØRMATIØN AND INSTRUCTIØNS
?YES

RANDØM NUMBERS WILL BE GENERATED AND CØNVERTED
TØ RANDØM CARDS.  THE DECK WILL BE RESHUFFLED
PERIØDICALLY.

WHEN ANSWERING QUESTIØNS,
    0 = NØ
    1 = YES
    2 = DØUBLE DØWN
    3 = SPLIT

Ø.K., LET'S PLAY!

THE CØMPUTER IS THE DEALER, AND GETS A 15 MIN. BREAK
AT 1945 HØURS. WHAT TIME IS IT NØW?200


WAGER?300

I SHØW              5     ØF DIAMØNDS
FIRST CARD IS       7     ØF DIAMØNDS
NEXT CARD IS        ACE ØF DIAMØNDS
HIT?1
NEXT CARD IS        8     ØF DIAMØNDS
HIT?1
NEXT CARD IS        6     ØF CLUBS
YØU BUSTED, YØUR TØTAL IS  22
MY HØLE CARD IS     9     ØF HEARTS
YØU'RE BEHIND $ 300

WAGER?500

I SHØW              8     ØF SPADES
FIRST CARD IS       KING ØF SPADES
NEXT CARD IS        8     ØF CLUBS
HIT?0
YØUR TØTAL IS  18
MY HØLE CARD IS     4     ØF CLUBS
I DRAW              10    ØF DIAMØNDS
I BUSTED***MY TØTAL IS  22
YØU'RE AHEAD $ 200
```

If you wish to contact the computer from a terminal that is not directly connected to it, you dial a phone number assigned to the time-sharing system at your computing center. When (and if) you obtain a steady whistle from the other end of the line, insert the phone into the cradle at the side of the terminal. Turn the terminal on and suppress the return button. Sooner or later the terminal types (among other things) a message such as:

USER NUMBER, PASSWORD–

The password is a special sequence of letters or numbers obtained from the center. If you know the password type it in after the dashes and press the return key to transmit your response to the computer. (The computer does not stand idle while you type a message letter by letter. It receives the message on each line you type only when the return key is pressed.) The terminal answers an acceptable response to the password request by typing a message such as READY.

You are now ready to load a program from the library or to initiate a new program. In the former case, type LOAD followed by a space and then the name of the program.† For example, if someone has written a blackjack program stored under, say, the title BLJK, then type LOAD BLJK and return the carriage. The computer should reply READY. You can execute the program by typing RUN and depressing the return key. Thereafter, follow the instructions that the terminal types until you have had enough. Press the ATTENTION or BREAK button whenever this occurs and the computer will stop. Type OFF and press the return when you are finished with the terminal. The computer will respond with a time-used message. After this, turn off the terminal (and cradle if necessary) and hang up the phone.

If terminals are not available, you are going to miss the fun and games on the computer. You will, nevertheless, have the pleasant task of punching computer

Figure 9.3

cards like the one in Figure 9.3. The card has 80 columns and rectangular holes may be punched in any of the 12 different positions in each column. The keypunch uses these holes to code the FORTRAN message.

If you wish to enter your own program, you must first become familiar with some basic ideas in FORTRAN or some other computer language.

†Terminal words such as LOAD, READY, RUN, OFF are not standardized. Check your installation for the equivalent commands.

9.7. Rudiments of FORTRAN

In order to have the computer consider your problem in this language, you must first tell the computer you intend to use FORTRAN. This is frequently done by typing ENTER FORTRAN and then pressing the return button. (In fact, you must always press the return after each statement in order to have it *register* with the computer. Rather than repeatedly reminding you of this fact, we will assume in the remaining sections that you have carried out this instruction.)

FORTRAN statements are mixtures of letters, numbers, and symbols, many of which are common usage with mathematics. Some basic departures from normal mathematical symbols or meanings are mentioned here.

a. Multiplication: "*", for example, 4*3 = 12.

b. Division: "/", for example, 8/4 = 2.

c. Exponents: "**", for example, X^2 is X**2 and $X^{1/2}$ is X**.5

d. Numbers: Distinguish between integers and real numbers by always using decimal points for the latter. For example, 2 is the integer 2 whereas 2. or 2.0 is the real number two. One example where this distinction is important is in the computation of X^2. If you write X**2., the computer converts to logarithms to complete the computation and this could introduce a round-off error into the problem.

e. Equals: "=" appears like it does in mathematics but in FORTRAN it is often an assignment statement. For example, X = Y reads "REPLACE X BY Y" and is not the same as Y = X (goodbye symmetry). The latter takes whatever is stored in position X of the machine and places it in position Y whereas X = Y stores whatever is located in Y into position X. (There are exceptions to this rule which we presently ignore.)

   To illustrate the new interpretation for equals consider the two programs (read downward):

```
Y = 3.49      Y = 3.49
X = 2.35      X = 2.35
Y = X         X = Y
```

   The first will store 2.35 into position Y as well as retain 2.35 in position X. The second, however, assigns what it finds in position Y into position X and the number 2.35 is displaced. Thus, both X and Y are 3.49 in this case.

Owing to this definition of the equal symbol, some commands do not make reasonable mathematical equations. For example, S = S + .1 stores in position S whatever was formerly in S and adds one tenth. Mathematically, however, it is a nonsense equation (no solution for S).

f. *Inequalities.* Although some terminals have the symbols $>$, $\geqslant$, $<$, $\leqslant$, which mean respectively "greater than," "greater than or equal," "less than," and "less than or equal," these symbols are not universally accepted by computing centers for teleprocessing.

It is better, therefore, to become familiar with the FORTRAN equivalent for each of the inequality symbols given in the accompanying chart.

Mathematical Symbol	FORTRAN Equivalent
$>$	.GT.
$\geqslant$	.GE.
$<$	.LT.
$\leqslant$	.LE.
$\neq$	.NE.
$=$	.EQ.

For example 3.GT.2, 7.LE.14/2, 5.NE.6. The FORTRAN equivalents are easily remembered since each is a natural two letter abbreviation of the verbal description of the symbol. Thus, "less than" is .LT. in FORTRAN, "not equal" is .NE. in FORTRAN, etc. The periods before and after the abbreviation are part of the FORTRAN code; no period can appear between the two letters in the code.

Multiplication, division, addition, and subtraction are binary operations which means that only two numbers at a time are combined by each of these operations. It is ambiguous, therefore, to write $7 + 5 + 3$ since the operations could be performed in two different ways, namely, $(7 + 5) + 3 = 12 + 3 = 15$ or $7 + (5 + 3) = 7 + 8 = 15$. The associative law applies to addition so, ambiguous as it was, we obtain the same total sum in either case. Should the operation be subtraction, which is not associative, we have a problem: $7 - 5 - 3$?, $(7 - 5) - 3 = 2 - 3 = -1$ whereas $7 - (5 - 3) = 7 - 2 = 5$. In order to avoid complications of this sort parentheses can be used to define which operations are to be performed first.

The computer understands the meaning of parentheses and does perform operations in the inner-most parentheses first. It follows the standard mathematical rules on parentheses. However, to avoid the necessity of a large number of parentheses, there is another convention that is also permissible on the computer. In a line without parentheses, the computer computes first the exponentiations in the formula, then it computes the multiplications and divisions, and finally it computes the additions and subtractions. Furthermore, the multiplications and divisions are performed in order of their appearance from left to right. Then the additions and subtractions are compiled from left to right. For example, consider the problem (4./2.*2.**2). First, the machine computes the exponentiation to obtain 4./2.*4. Then it computes the multiplications and division *from left to right:* 4./2.*4. = 2.*4. = 8. If, instead, we desired to have the expression 2.*2.**2 divide 4., it is necessary to use parentheses:

4./(2.*2.**2)=4./(2.*4.)=4./8.=.5

The parentheses omission rule is relatively simple and does make the formulas easier to read. By way of illustration, all the parentheses in the following computation are unnecessary:

((5./2.)*3.+7.*(3.)**2)-4

Indeed,

5./2.*3+7*3.**2-4.=

5./2.*3+7.*9.-4.=

2.5*3+7.*9.-4.=7.5+63.-4.=

70.5-4.=66.5

by the hierarchy of performance of the arithmetic operations.

A word of caution about parentheses. FORTRAN demands every "(" have a closing ")". An error message is received if some parenthesis is not closed. In addition, negative numbers should be enclosed in parentheses except at the beginning of a line. For example, 4.*-.3 is in error; it should be 4.*(-.3)=-1.2. However, -.3*4.=(-.3)*4.=-1.2. Note -3**2=-9 while N**2=9 when N=-3 and (-3)**2=9.

Exercises

1. Compute the value of each of the following FORTRAN problems:
   a. 3*4**3/3
   b. (2*3)**(4/2)
   c. 2*3**(4/2)
   d. 2*3**4/2
   e. (2*3)**4/2
   f. 4+3*2/3-3
   g. 3*4+2**4-3*8/6+4*2**2/8
   h. -2**4
2. Write each of the following formulas in FORTRAN:
   a. $y = 2x^2 + 3x - 35$
   b. $y = 3x + \dfrac{2x^2 + 1}{3x - 1}$
   c. $y = 5x^x$
   d. $y = 1 + \dfrac{x}{1 + \dfrac{x}{1 + x}}$
3. Exponentiation is, according to the parentheses omission rule, performed from *right to left*. Using this fact, express 2**2**2**3 as a power of 2.

4. Consider the following programs:

a.	b.
X = 5.1	X = 5.1
Y = 3.2	Y = 3.2
X = Y	Y = X

What is stored in locations X and Y in each case?

5. Find a FORTRAN error in each of the following:

a. 2*-3	b. 4+3***2	c. 3**-2
d. 4*(3+2-(7+1)	e. 6/-3	f. 2(3+2)*2

9.8. Some Rules for Simple Computations

If you wish to write your own program, the type of input device and minor local variations play a role in what is acceptable. Therefore, we divide the discussion into two parts, A and B. Section A is devised for terminal (remote entry) users while B treats the analogous problem when the program is entered via cards at a computing center that has a WATFOR or WATFIV compiler. If neither is available, Appendix X indicates methods that can be used at any installation which has a FORTRAN compiler.

I. ENTER FORTRAN

A. After you give the password, and return the carriage, it is often necessary to inform the computer that you intend to program in FORTRAN. A typed command such as ENTER FORTRAN calls the FORTRAN compiler into service for your program.

B. The first card is $JOB card and this code is typed from a keypunch in the first four columns of the card. Columns 5 through 15 are left blank while columns 16 onward contain user identification information required by your local computing center. Your instructor will describe the exact format for the $JOB card. The $JOB card brings the program to the attention of the appropriate compiler so your program will be recognized as a FORTRAN program.

II. ARRANGEMENT OF THE PROGRAM

A. Every FORTRAN statement in your program must carry a line number. The computer follows the instructions in your program in the order of the line numbers. Usually you should number the lines by tens so that there is room to insert additional statements that you detect are necessary when you RUN your program. If you add a line 15 after typing the other lines in the program, then line 15 will be considered *before* all other FORTRAN statements with higher line numbers even though it was the last line entered from the terminal. Terminal commands such as ENTER FORTRAN and RUN are not part of the program and, hence, are not numbered.

B. The program cards are physically arranged in sequence by the user. Some people number their cards in columns 73 through 80 since it is quite "painful" to drop a deck of unnumbered cards.

III. NAMES

A name is an address (storage location) for the current value of a variable in your program. A name consists of one to six letters or digits. The first character, however, must be a letter from the alphabet. HEIGHT, AREA, B007, X123, Q, AA are grammatically correct identifiers in FORTRAN whereas UNDERRATE, COMPUTE, 2MUCH are not, since the first two contain more than six alphanumeric characters and the last does not begin with a letter.

IV. END

Finish off each program with the statement END. This statement informs the compiler that your program is over and permits it to go on to other tasks.

V. SIMPLE WRITE STATEMENTS

These statements print results. Without them, the computer keeps the answers a secret.

A. The simple WRITE statement "WRITE(6,*)X,Y,Z" causes the terminal to print in sequence on a line whatever is stored in the addresses X, Y, and Z. Remember here that FORTRAN is a code and *all* the symbols in the WRITE message are part of this code. (It is common, but not universal, to use the cited WRITE on a terminal. Check for variations at your computing center. You can substitute whatever is accepted as a free-form write at your institution for the WRITE(6,*) throughout these notes.)

B. In WATFOR or WATFIV the simple WRITE is "PRINT, X,Y,Z." This prints side-by-side the numbers stored at the addresses X, Y, Z. Notice the comma after the word PRINT; this is a necessary part of the code.

With a simple WRITE statement you can print numbers stored in one or more of the locations in the computer memory, not simply three as in the above illustration. All that is necessary is that each address to be printed be listed after the WRITE command. A computation is not permitted in a WRITE statement. To ask the computer to print X**2, even though it knows the number in X, is an error. If you wish to compute X^2 and print the answer, preceed the WRITE by a FORTRAN statement such as "Y=X**2" and then print the contents of the address Y.

VI. SIMPLE READ STATEMENTS

A. The command "READ(5,*)A" tells the computer to assign to the address A whatever number you type during the running of the program. To illustrate, consider the following program:

```
10   READ(5,*)A
20   B=A**2
30   WRITE(6,*)B
40   END
```

When this program is RUN, the computer will type a question mark on the left margin of the paper after it has compiled the program. You then type a number after the question mark and this will be the number assigned to position A. The printout is the square of A. The number 5 in READ(5,*) is part of the code and is necessary. (Again, this code is commonly used for READ statements, but there are variations at some installations.)

B. In WATFOR or WATFIV the simple READ statement is "READ, A." Should there be a READ statement in your program, there must be a corresponding data card which contains the numerical value for the address in the statement. For example, consider the program above (without the line numbers 10, 20, 30, and 40). A data card is necessary when this program is RUN and the numerical value on this card is assigned to location A. The card is then so-to-speak set aside. Should you have a longer program containing a second READ command, the computer would look at the next data card for the required information; the first card is used only for the first READ statement. There must be at least as much data as your program requires via the READ statements. If there is a READ, A, B statement in your program, you can put two numbers on a single card. A third number on the card would be lost unless the READ command contains a third variable. Each READ uses as many cards as necessary to assign data to each variable in the READ. If only part of the data on a card is used, the remaining data is *not* available for future READ statements in the program.

The numerical value for a READ can be typed anywhere on a card from space 1 through 80. No blanks may appear between the digits of the number punch on a card.

VII. SPACING

The spacing of FORTRAN statements is often critical and is necessary at least for programs that are entered from cards. For the simple programs discussed in this section, there should be *six* empty spaces on each card and the FORTRAN statement can start in *any* space from the seventh onward as long as it ends on or before space 72. Words such as READ can be typed with spaces in unnatural places without misinterpretation by the computer. For example, any of the following are equivalent to the simple READ in the form READ(5,*)A:

```
R  EAD   (5   ,*)   A ;   READ  (5,*)  A;
RE        AD(5,     *)A  ; R    E   A  D  ( 5, *) A
```

To avoid errors, however, numbers should not be split with empty spaces. Otherwise blank spaces are ignored by the computer.

If teleprocessing is your mode, it may be necessary to place seven empty spaces between the last digit of the line number and the first letter of the FORTRAN statement for the simple programs in this section. You should check to determine if these empty spaces are necessary at your computing center since it is easier to type a program without them if this is an option.

VIII. ERRORS

If you made a typing error, you can retype the line (with the same line number) on a terminal or repunch the card. There are easier ways to correct mistakes, but these vary according to the type of terminal or keypunch being used.

IX. CONTROL CARDS

We have mentioned that the first card of a deck is a $JOB card for the WATFOR of WATFIV compiler. The next sequence of cards contains the FORTRAN program with one FORTRAN statement on each card. The cards for the program are arranged in the order of execution by the computer. After the FORTRAN program cards comes the $ENTRY card with $ENTRY punched in columns 1–6, the

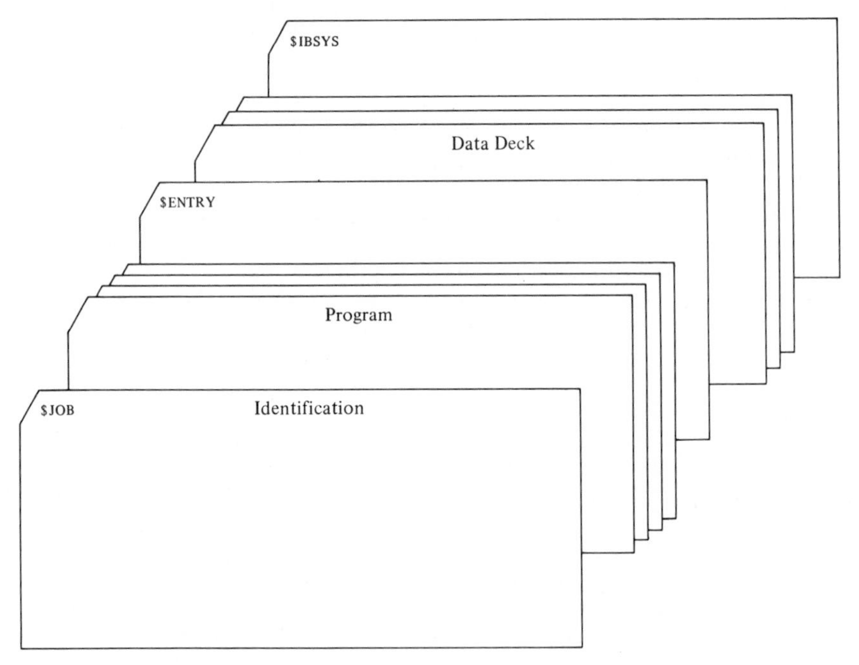

Figure 9.4

remaining columns are left blank. Next comes the data deck for the READ statements (no READ, no cards). For the final card punch $IBSYS in columns 1–6, leaving the other columns blank. After the deck is completed and arranged in this order, the program is submitted to your computing center for execution. (See Figure 9.4.)

The exercises throughout these notes are based on remote entry input (terminal). To convert any exercise to batch processing, simply ignore each line number and one blank space following the line number. Except for the differences in the READ and the WRITE statements noted above, this conversion is sufficient for those who are required to use cards.

Exercises

1. Which of the following are grammatically correct FORTRAN names?

   a. WRITE b. RADUIS c. N012 d. FORTRAN
   e. 1AREA f. VARIABLE g. G h. I984
   i. AT BAT

2. What is printed by each of the following programs:

a.
```
10  A=2.2
20  B=A+A
30  WRITE(6,*)B,A
40  END
```

b.
```
10  ONE=13.5
20  TWO=2.7
30  THREE=ONE/TWO
40  WRITE(6,*)THREE
50  END
```

3. Suppose the following program is RUN and the data entered for X is (a) 1 (b) - 1 (c) 5. What is printed in each case?

```
10  READ(5,*)X
20  Y=-X**2+X
30  WRITE(6,*)X
40  END
```

4. Find all the errors in each of the following programs.

a.
```
10  READ X
20  X**2=Y
30  WRITE Y
40  END
```

b.
```
10  READ(6,*)A
20  B=A+A
30  WRITE(6,*)B
40  FINISH
```

c.
```
10  A=B
20  B=.5
30  WRITE(6,*)A
40  WRITE(6,*)B
```

9.9. FORTRAN Number Systems

FORTRAN distinguishes between integers and real numbers and uses different methods to perform the arithmetic operations for each of these number systems. The integers in FORTRAN are the numbers $0, -1, 1, -2, 2, -3, 3, \cdots$ whereas the real numbers are all possible decimals within certain machine dependent limits for each set. The decimal point serves to distinguish a whole real number from its corresponding integer. Thus, the integer 2 is different from the real number 2., since the latter contains a decimal point.

Real number arithmetic is the standard arithmetic of grade school. For example, 7.2 * 2.1 = 15.12, 3.6 - 9.21 = -5.61, 9./4. = 2.25. The operations +, -, and * provide no surprise answers in integer arithmetic. Indeed, 2 + 2 = 4, 3 * 7 = 21, 9 - 11 = -2. With division, however, the integer quotient is sometimes different from what you might expect. While 4/2 = 2 in integer arithmetic, we have 9/4 = 2 as well. The answer you expect for a problem 9/4 is 2.25 but the latter is *not* an integer. What the computer does is suppress the decimal point and the fractional part of the answer in a division problem with integers. Thus, the quotient of two integers is itself an integer. For example, 11/3 = 3, -34/7 = -4, 1/2 = 0.

You can mix integers and real numbers in a computation but this activity can produce errors if you are not careful. Consider, for example, the calculation 2/4 * 3.2 which mixes the integer arithmetic and the real arithmetic. Performing the operations from left to right we have 2/4 = 0 and 0 * 3.2 = 0.0. On the other hand, 3.2 * 2/4 = 1.6! The reason for the different answers for this calculation is that the computer performs an operation involving an integer and a real number in real arithmetic. Thus, in the problem, we have 3.2 * 2 = 6.4 and 6.4/4 = 1.6. In the first case 2/4 was performed in integer arithmetic and then the answer was multiplied by 3.2 in real number arithmetic to obtain 0.0. The point of this discussion is simple; avoid mixing real numbers and integers (called a mix-mode) in any calculation.

The letters I, J, K, L, M, N stand for integers in FORTRAN while the other letters in the alphabet are reserved for real numbers. A variable (address) that *begins* with one of the letters I-N is interpreted as an integer variable regardless of the choice of alphanumeric characters in the other positions of the address. Similarly, a variable that begins with A-H, O-Z is a real variable. For example, IT, INT, J, LOT, K123, NREAL, MUST are integer variables and A, BIG, STORE, XI52, ZIP are real variables.

When an integer is assigned to a real address, the decimal point is added. For example X=7, ATTIC=9 assigns 7. and 9. to memory locations X and ATTIC respectively. If a real number is assigned to an integer address, only the part preceding the decimal point in the number is stored. Thus, I = 2.99 assigns 2 to the memory location I; there is no rounding before the assignment.

The real arithmetic is called "floating point" arithmetic in computing circles. The manner in which the machine uses exponents to shift the decimal point of a real number suggests this title. Because of this name, programmers frequently add decimal points to integers through an abbreviation of the word "float." For example, if I = 6, we add the decimal by using FLI = I and FLI starts with a real letter and, hence, stores 6. in the location called FLI.

Exercises

1. What is IT in each of the following cases:

    a. IT = 7/2
    b. IT = 3/9
    c. IT = (−7)/2
    d. IT=3.999
    e. IT=4+3+5/2
    f. IT=(−3)*(−4)
    g. IT=1/2
    h. IT=1/2+1/2
    i. IT=2*1/2
    j. IT=2*(1/2)

2. a. X=7/2. What is stored in location X?
   b. X=4.3 and YES=2.6. What is stored in location KISS when KISS=X+YES?

3. What is wrong with the statement NUMBER3=1+Y+MUST*WORK?

4. Suppose you RUN each of the following FORTRAN programs. What is printed?

a.

```
10    X=7.
20    Y=2.
30    I=X/Y
40    WRITE(6,*)I
50    END
```

b.

```
10    I=7
20    J=2
30    X=I/J
40    WRITE(6*)X
50    END
```

c.

```
10    X=7.
20    Y=2.
30    Z=X/Y
40    WRITE(6,*)Z
50    END
```

d.

```
10    I=7
20    J=2
30    K=I/J
40    WRITE(6,*)K
50    END
```

   Try each of the above programs on a terminal.

5. What is stored in location I in each of the following cases:

    a. I=3.2*5/2
    b. I=5/2*3.2
    c. I=2/5+3/5
    d. I=9−3.7
    e. I=7.9+11.9
    f. I=3+2/4

6. What is stored in location D in each of the following cases?

    a. D=17/11
    b. D=2*3.2/4
    c. D=2/4*3.2
    d. D=3.2*2/4
    e. D=9/10+19/10
    f. D=3+2/4

7. What errors can you detect in the following programs?

```
10        READ; NATURE
20        NATURE=X
Y=2.*X
40        WRITE, Y, ZNATURE
50        END
```

```
10    Y=5.5
20    Y=I
30    J=2I+Y
40    WRITE(5,*)J
50    END
```

9.10. Telling a Computer Where To GO

Unless instructed to do otherwise, a program is performed according to ascending line numbers or, in batch, according to the order (arrangement) of the cards. It is frequently important, however, to transfer attention to parts of the program that are out of sequence. The GO TO command is one way to accomplish this task.

For illustration, suppose you wish to print the squares of the numbers 1, 2, 3, 4, · · · . A partial program for this purpose is the following:

```
10       N=0
20   12  N=N+1
30       M=N**2
40       WRITE(6,*)N,M
50       GO TO 12
```

When this partial program is executed, line 20 replaces the 0 in location N by 1 and then computes the square of 1, which is stored in location M by line 30. The values stored in N and in M are printed side-by-side in line 40. When line 50 is reached, control is transferred back to the FORTRAN statement *numbered* 12, that is, line 20. A new N ($1 + 1 = 2$) is then calculated and squared in line 30. The printout is 2 4 and control is returned by line 50 to statement 12. Once a computer transfers control to a statement, it follows the command of that statement and then each subsequent statement in order until it hits, if one exists, a GO TO statement. Thus, the partial program as written continues the process of returning to statement 12 for ever and ever, amen. Unless one has an infinite amount of time this program should not be RUN. We have built an *infinite loop* into the program. Some controls, one of which is introduced in the next section, are necessary to terminate the looping after a finite number of steps.

The number 12 in the illustration can be replaced by any positive integer less than a hundred thousand†; the choice of 12 is no better than any other choice within the indicated range. The statement number, whatever it is, should begin with at least one space between it and the line number and it should end on or *before* the sixth space after the line number. On cards, the statement number can appear anywhere within the first five spaces on the card. The FORTRAN statements appear anywhere *after* the sixth space on the card. The sixth space is left blank.

There are limitations as to the type of FORTRAN statement control that can be transferred by a GO TO command. For our purposes, we permit transfer to (a) a WRITE statement, (b) an assignment statement such as N=N+1 or X=3.2

†Check this limit at your center if you intend to use large statement numbers.

and (c) a STOP statement. The command STOP terminates further consideration of your program by the computer.

Line numbers are *terminal* instructions to tell the computer in what order the statements are to proceed. When we transfer control to a statement via a GO TO in a program, line numbers are ignored. If we say "GO TO 35," there must be a *statement* beginning with 35 and *only one such* statement. Otherwise the computer would not know to which statement control is to be transferred. The computer never uses the line numbers for control purposes so it makes no difference if there is a line 35 or not.

9.11. IF Statements

In logic "If A, then B" has a meaning that is slightly different from the FORTRAN meaning. The machine carries out the command B whenever A is true. When A is false, the machine skips to the next line.

EXAMPLE. "IF (X.GE.2.) GO TO 22." Such a statement tells the computer to go to step 22 whenever what is stored in position X equals or exceeds the real number 2. There must be a step numbered 22 and this number must appear in addition to the line number. Should $X < 2.$ in the problem, the computer goes on to the next step.

EXAMPLE.

```
10       Y=1.1
20       X=Y**6
30       IF(X.GE.2.) GO TO 22
40       Z=(2.-X)**.5
50       WRITE(6,*)Z
60    22 STOP
70       END
```

Note the dual numbering in line 60. The first number, used on terminals only, the line number 60, tells the compiler in what order to carry out the step. The second number is part of the FORTRAN program itself. When X is greater than or equal to 2., control passes directly from line 30 to line 60 (Statement 22). Lines 40 and 50 are skipped in this case.

The IF statement in the above program avoids the impossible task of grinding out a square root of a negative number.

Observe that we have used parentheses and have not used a comma or a "then" in the IF statement. You will confuse the machine should you not follow our example. IF (A)B is the proper way to tell the machine to follow command B whenever statement A is true.

There are limitations in the antecedent A and the consequent B in the FORTRAN IF(A)B statement. The antecedent must be a comparison that can be true or false–it is not an assignment statement. For example, X.GT.Y, 2*N.EQ.14, 2.–X.LT.O† are permissible antecedents in the IF statement provided the addresses X,Y,N have previously been defined. *We* limit the consequent B to be (a) a GO TO statement, (b) a WRITE statement, or (c) a STOP statement.

An IF statement can be used to terminate loops. For example, the following program computes and prints the squares of the integers 1, 2, 3, · · · , 100.

```
10        N=0
20     12 N=N+1
30        M=N**2
40        WRITE(6,*)N,M
50        IF(N.LT.100) GO TO 12
60        END
```

The infinite loop cited in the last section has now been replaced, via the IF statement in line 50, by a finite loop.

In addition to the above use of the IF, there is what is called "the arithmetic IF" statement. It has the form IF (A) n_1, n_2, n_3 where n_1, n_2, n_3 are statement numbers in the program. Control is transferred to statement n_1 if and only if A is negative, to n_2 if and only if A is zero, to n_3 if and only if A is positive.

EXAMPLE:

```
150       IF(A-B)10,35,71         175       GO TO 17
160    10 Y=(B-A)**.5             180    71 Y=(A-B)**.5
165       GO TO 17                185    17 WRITE(6,*)Y
170    35 Y=0.                    190       END
```

This assures us that we are taking the square root Y of a positive number. When $A - B < 0$, then $B - A > 0$ in line 160.

Exercises

1. What is the printout of each of the following programs?

```
          a.                                b.
10        X=1.1                   10        X=1.
20        GO TO 30                20        X=X+1.
30        X=X**2                  30        IF(X.GT.10.) GO TO 35
40     30 WRITE(6,*)X             40     35 WRITE(6,*)X
50        END                     50        END
```

†Arithmetic operations are performed before comparisons such as .LT., .GT., or .EQ.

c.

```
10       X=1.
20    35 X=X+1.
30       IF(X.LE.10.) GO TO 35
40       WRITE(6,*)X
50       END
```

d.

```
10       X=1.1
20    25 X=X+1.
30       IF(X.EQ.10.) GO TO 36
40       GO TO 25
50       END
```

Remark: Equality within parentheses of an IF statement is standard mathematical equality and is *not* an assignment statement.

2. What is wrong with each of the following programs?

a.

```
10       X=1.1
20       X=X+1.
30       IF(X.LT.10.) GO TO 20
40       WRITE(6,*)X
50       END
```

b.

```
10       I=1
20    27 I=I+2
30       IF(I.GT.10) GO TO 27
40    27 WRITE(6,*)I
50       END
```

c.

```
10       I-1
20    49 I=I+1
30       GO TO 49
40       WRITE(6,*)I
50       END
```

d.

```
10       X=1.1
20       Y=X**2
30       J=Y
40       IF(J.EQ.1.21) GO TO 34
50    34 STOP
```

3. Consider the following program.

```
10       READ(5,*) A,B
20       IF(A*B) 15,30,45
30    15 Y=A*B
35    35 GO TO 18
40    30 Y=A+B
45       GO TO 18
50    45 Y=A/B
60    18 WRITE(6,*)Y
70       END
```

a. What is printed when A = 2.4 and B = .6?
b. What is printed when A = 5.0 and B = –.2?
c. What is printed when A = B = –7.0?
d. What is printed when A = 0. and B = 11.75?

4. Construct a flow chart for each of the examples in Exercise 1.

5. Study the following program.

```
10       READ(5,*)N
20       I=1
30    19 I=I+1
40       J=N/I
```

```
 50        K=I*J
 60        IF(K.EQ.N) GO TO 22
 70        GO TO 19
 80     22 WRITE(6,*)I
 90        GO TO 19
100     30 STOP
110        END
```

a. Add a statement that terminates computation once I has reached 25.
b. What is printed when $N = 7$, $N = 8$, and $N = 21$?
c. In general, what is printed by this program when the line (35 IF(I.GT.N) GO TO 30) is added.

9.12. DO Statements

EXAMPLE. DO 13 I = 1,10

The command in the example iterates for ten steps. Statement 13 should be numbered 13 independent of the line number it has in your program. It indicates the end of what is called a "DO LOOP" and is 13 CONTINUE. Counting must be performed by integers and not real numbers. Thus, in a DO LOOP make certain the variable name after the initial number in the statement is the integer type, that is, it must start with one of the letters I, J, K, L, M, N. The command "DO 13 A = 1,10" would be in error since the letter A indicates a real variable.

There is nothing sacred about the number 13 in this command. Any other positive integer less than a hundred thousand can be substituted provided only that the number is not used elsewhere in the program. Similarly, the command is not limited to 10 steps; "DO 29 J=1,87" means the iteration is performed 87 times.

EXAMPLE. What is the sum of the first 20 counting numbers? Computer solution:

```
10        J=0
20        DO 19 I=1,20
30        J=J+I
40     19 CONTINUE
50        WRITE(6,*)J
60        END
```

The computer performs this program as follows: It begins with 0 in location J. Then it adds 1 to 0 and stores the answer 1 in the location J, replacing the former number 0. Next it adds 2 to J(=1) to obtain 3 which is now stored in J, replacing the former integer 2. This process is continued until the loop is completed. The latter occurs when I=20 is added to the former J. At that time, the next line after

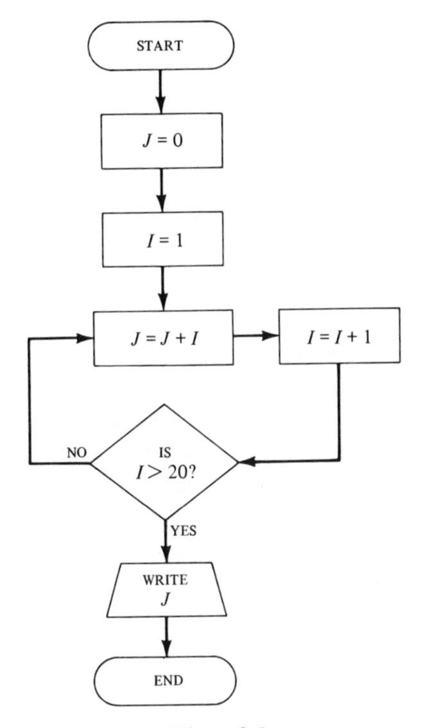

Figure 9.5

the loop, namely, line 50 is performed and the answer 210 is printed. Figure 9.5 shows the flow chart for this process.

Some final precautions are in order. We used the variable I in DO LOOP. If I had been used before the loop, the value stored in I would be lost once the loop variable I begins with its initial value. If I is used later in the program, a new value should be assigned to I; it is not 20 as the end of the loop indicates.

The loop variable I should *not* be redefined within the loop. Addition of a line 25 I=I+1 to the DO LOOP in Figure 9.5 above would be improper since it is an attempt to change the value of the loop address I.

The counting in a DO LOOP does not have to begin with 1, although it must start with some positive *integer*. The terminal integer, indicating the end of the counting for the loop variable, need not be larger than the initial integer. However, unless the terminal integer is at least two larger than the initial integer of the loop variable, the steps in the loop will be carried out only once.

We cannot transfer control from outside the loop, that is, to any statement between the DO through the CONTINUE statements. You can, however, *exit* from a loop before the loop is completed.

EXAMPLES:

a. *Improper*

```
10        DO 15 J=1,20
20     12 N=J+1
30     15 CONTINUE
40        GO TO 12
50        END
```

b. *Proper*

```
10        DO 15 J=1,20
20        N=J+1
30        GO TO 12
40     15 CONTINUE
50     12 N=N**2
60        END
```

The program (a) attempts to enter the loop from line 40 which is outside the loop. In the second program (b), the value $N = 2$ is transferred to statement 12 even though the loop variable J has not completed its sequence.

The range of the loop variable in a DO LOOP can be a previously defined address. For example, "DO 17 I=K,KK" is acceptable as long as K and KK have previously been defined as positive integers. No computation can be used directly in the definition of the range of the loop variable. For example, "DO 17 J=2,2*N" is in error; the computation 2*N has to be performed and stored under a name, say, M. Then "DO 17 J=2,M" is a suitable FORTRAN statement to begin the loop.

Exercises

1. Add the line "35 WRITE(6,*)J" to the first program in this section. What is printed?
2. What is printed by each of the following programs?

a.

```
10        J=1
20        I=1
30     15 I=I+1
40        J=J+I
50        IF(I.LT.21) GO TO 15
60        WRITE(6,*)J
70        END
```

b.

```
10        X=0.
20        DO 33 K=2,5
30        FLK=K
40        X=X+1./FLK
50     33 CONTINUE
60        WRITE(6,*)X
70        END
```

c.

```
10        J=2
20        DO 31 I=1,3
30        J=J**2
40     31 CONTINUE
50        WRITE(6,*)J
60        END
```

d.

```
10        JSUM=0
20        DO 95 M=1,100
30        JSUM=JSUM+M
40        IF(JSUM.GT.10) GO TO 16
50     95 CONTINUE
60     16 WRITE(6,*)JSUM,M
70        END
```

3. What is wrong with each of the following programs?

a.

```
10   N=0
20   DO M=0,5
30   N=N+M
40   CONTINUE
50   WRITE(6,*)N
60   END
```

b.

```
10      X=1.
20      DO 40 A=.5,9.5
30      X=A*X
40   40 CONTINUE
50      WRITE(6,*)X
60      END
```

c.

```
10   DO 40 N=10,-5
20   J=N**2
30   WRITE(6,*)J
40   CONTINUE
50   END
```

d.

```
10      DO 92 N=1,10
20      N=N+1
30      WRITE(6,*)N
40   92 CONTINUE
50      END
```

4. Construct a flow chart for the problem of obtaining the sum of the squares of the first 100 counting numbers. Write a program for this problem.

9.13. Unnumbered Commands

We have seen the commands ENTER FORTRAN, LOAD, and RUN. There are a number of others which could prove useful. Again, there are variations from one computing center to another so it is necessary for you to discover their equivalent forms at your center.

a. LIST This lists your program by line numbers in ascending order. All that matters is the number you assign to a line and not the order that you entered the step into the terminal.

   If you load a program from the library, the command LIST will write out the program unless the writer of the program wanted to protect it from others' eyes. Some programs are protected this way at various centers since they are lessons for programmed courses.

b. SAVE The computer saves your program for future use or corrections. You will be asked to name the program when you use this command. Names should be one "word" with a limited number of characters.

c. CAT This lists the names of the programs that are in the library of users, somebody's saved program. CAT is an abbreviation of the word "Catalogue."

d. DELETE The command DELETE 15 eliminates line 15 from the program. This command is unnecessary if you plan to retype the line since the computer *keeps only the last entered line with the same line number.*

e. CLEAR After you are finished with the programs from the library and you desire to write your own program, type CLEAR before you ENTER FORTRAN.

This command removes the program which you have loaded into the machine. It is unnecessary if you desire to load a different program from the library.

f. PURGE To remove your program from the library sometime after it has been saved, type PURGE followed by the name you have selected for your program.

Supplementary Exercises

Exercises 1–5 assume use of the WATFOR or WATFIV compiler.

1. Correct the following statements:
   a. IF A.GT.B GO TO 152
   b. IF (A EQ B) GO TO 10
   c. IF (PROGRAM.EQ.RUBBISH GO TO 2000
   d. IF (X>2,300) GO TO 9
   e. IF A≠B GO TO 12
   f. PRINT A,B,C.
   g. DO 32 I = 1,15
   h. K = O

2. Convert the programs in Exercise 2 of Section 9.12 into programs for cards instead of a terminal.

3. Suppose there is a single READ statement in a program and it calls for three variables A, B, and C. If the data cards each contain seven numbers, how many times can the variables A, B, and C be read from 27 data cards?

4. What numbers would be printed by the following programs?

a.

```
      X=0
      DO 50 I=1,4
      X=X+1.0
      IF(X.GE.2.5) GO TO 60
50    CONTINUE
60    PRINT,X
      END
```

b.

```
      X=2.
      Y=3.
39    X=Y+X
      Y=Y+2
      IF(Y.LT.7.) GO TO 39
      PRINT, X,Y
      END
```

5. What is the FORTRAN value of each of the following?

   a. 2**2 + 1/3
   b. 2.**2+1./3.
   c. (2.+3**2)**2/11.0
   d. (3.**2+4.**2)**.5
   e. 3.*(2.**2+8./4.)/9.
   f. 1/2+1/4+1/8+ · · ·
   g. ((2**2)**2)**2

6. What is *wrong* with the following programs?

a.

```
10           A=1.0
20     1000  IF(B.EQ.10.0) STOP
30           A=A+1.
40           B=A**2
50           GO TO 1000
60           END
```

b.

```
10           A=1.
20           B=2.
30           C=B**2-4.*A
40           WRITE(6,*)A,B,C
```

c.

```
10           DO 1000 I = 1,10
20           A = A + I**2
30     1000  CONTINUE
40           WRITE(6,*)A
50           END
```

d.

```
10           J = 0
20           DO 198 I = 1,20
30           J = J + 1/I
40     198   CONTINUE
50           END
```

7. Correct each of the following FORTRAN statements:

a. `2*J=I`
b. `X=2.(3.**2)`
c. `DO I = 2,212`
d. `11 CONTINUED`
e. `WRITE, A,B,C`
f. `IF I.NE.J, GO TO 12`
g. `IF(I.EQ.3.) STOP`
h. `WRITE(6,*)X,X**2`
i. `DO 29 I=1,2*N`
j. `READ(5,*)`
k. `GO TO N`
l. `DO 30 I=0,9`
m. `DO 30 I=9,99.`
n. `IF(I.LT.3,400) GO TO 20`

8. What is wrong with the following programs:

a.

```
10           S=0
20           DO 31 K=1,100
30     15    FLK = K
40           S = S + 1./FLK
50     31    CONTINUE
60           IF(S.LT.6.) GO TO 15
70           END
```

b.

```
10           DO 17 I = 1,10
20           J = I**2
30           I = I + 1
40           K = I**3
50           WRITE(6,*)J,K
60     17    CONTINUE
70           END
```

c.

```
10           KSUM=0
20           DO 97 L=1,7
30           KSUM=KSUM + L**2
40     97    CONTINUE
50           LL = L*KSUM
60           WRITE(6,*)L,LL,KSUM
70           END
```

d.

```
10           L=0
20     15    IF(L.GT.20) GO TO 35
30           L=L+1
40           DO 19 L=1,10
50     19    CONTINUE
60           GO TO 15
70     35    WRITE(6,*)L
80           END
```

Chapter 10 How Much Does It Cost?

10.1. Compound Interest

A few years ago banks compounded interest on a semiannual or quarterly basis. Now the trend appears to be to compound the interest more frequently—daily or perhaps hourly. Is there really much difference?

We shall write a computer program for this problem. First of all, let us assume the yearly interest rate is $4\frac{1}{2}\%$. By annual compound interest it is meant that the interest at the end of each year is added to the principal (amount invested) to form a new amount. This new amount is used in the computation of the next year's interest. The interest each year is .045 times the amount invested that year. A program for N years of investing \$1000 at $4\frac{1}{2}\%$ annual compound interest can be written by performing the calculation of the new amount each year with a DO LOOP. The output is J, the number of years, and P, the amount accumulated after those years, for the original \$1000 investment.

```
10    READ(5,*)N
20    P=1000.00
30    R=.045
40    DO 23 J=1,N
```

```
50        P=P+R*P
60        WRITE(6,*)J,P
70     23 CONTINUE
80        END
```

Suppose next the investment of $P = \$1000$ is compounded semiannually (twice a year) and the rate remains at $4\frac{1}{2}\%$. This means that half of the .045 annual rate is the rate for each six month period, so .0225 is multiplied by the amount invested in that period and added to the old amount to form a new amount. Quarterly compound interest divides the year into four periods (3 months each) and, hence, the interest over each period is $.045 \cdot P/4$. Programs for semiannual and quarterly compound interest on \$1000 at $4\frac{1}{2}\%$ for N years are given below.

Semiannual

```
10        READ(5,*)N
20        P=1000.00
30        R=.045
40        L=2*N
50        DO 75 J=1,L
60        P=P+P*R/2.
70        WRITE(6,*)J,P
80     75 CONTINUE
90        END
```

Quarterly

```
10        READ(5,*)N
20        P=1000.00
30        R=.045
40        L=4*N
50        DO 75 J=1,L
60        P=P+P*R/4.
70        WRITE(6,*)J,P
80     75 CONTINUE
90        END
```

The output is J, the *number of payment periods* (two or four per year according to whether the interest is compounded semiannually or quarterly) and P, the amount accumulated at the *end of each period*. The L in the problem is the total number of payment periods in N years. It was necessary to have the L in a memory location before we introduced the DO LOOP. (See Section 9.10.)

We could use the method to obtain answers for the same problem with the interest compounded daily, which for many banks means 360 times per year. However, some difficulties would be encountered. First of all, there would be 360 lines for each year in the printout since the WRITE statement is inside the loop. This would be wasteful of paper and printer time. This problem, however, is eliminated by deleting the WRITE statement from within the loop and replacing it by a WRITE(6,*)L,P outside the range of the loop. Another, more serious, problem might then be noticed. The printed answers could have significant errors! This difficulty occurs when there is a large error obtained by accumulating the errors due to rounding off the numbers in each step of the calculation.

In order to illustrate the round-off problem, consider the computation of $P + P * R/360$ when $P = 1000.$ and $R = .045$. Initially,

$$P + P * R/360. = 1000.+ 1000.*.045/360.$$
$$= 1000.+.125 = 1000.125$$

The computer retains only a limited number of digits, usually seven digits on IBM machines, of each real number (unless given special instructions). Hence, it would store the answer 1000.125 if the above calculation were part of a program. The next step in an interest problem would be to use this answer in the computation of a new amount: $P = P + P * R/360$. Since only seven digits are retained, the computer would store only 1000.250 in location P whereas the exact answer to the calculation is 1000.250015625. You may believe this is close enough; surely it is, provided the problem ends with just a few more computations. Remember, however, that there are to be 358 more calculations before we even obtain the interest in a one year daily compound interest problem. These small errors could accumulate to affect the result with such a long computation. The round-off error cannot be ignored in large scale computations. In the next section, a method is given that reduces the round-off error for the compound interest problem.

Exercises

1. Write a program for simple interest (the interest is paid to the investor after each period and the principal remains fixed) for N years at rate R on P dollars.
2. Write a program for P dollars at a rate R per year, compounded monthly, for N years. Try your program for one year on $P = 1000.$ at $R = .045$ and at $R = .05$ on the computer.
3. Rewrite the first program in this section without a DO LOOP.
4. Evaluate 1.6**4 in the following ways:
   a. Carry all digits.
   b. Round off *each* step in the calculation to two places, for example, 2.52→2.5; 2.58→2.6; 2.55→2.6.
   c. Truncate each calculation to two digits, for example, 2.52→2.5, 2.58→2.5; 2.55→2.5 (all digits after the second are dropped).
5. Assume you have a computer which rounds off each number to 4 digits. What would it print for the solution to the problem of \$1000 invested at $4\frac{1}{2}\%$, compounded daily, after one year? (The program should be an analogue of the second and third programs in the text.)

10.2 Compound Interest Revised

Suppose the initial investment of P dollars at a fractional annual rate of R (this is the percent rate divided by 100) is compounded K times per year. For example, $P = \$1000.00$ at an $R = .045$ ($4\frac{1}{2}\% = 4.5/100$) annual fractional rate might be compounded quarterly ($K = 4$). We seek a formula for the value of the investment after N years.

Set $L = K * N$ where L represents the number of payment periods in N years.

At the end of the first period the value of the investment is

$$P1 = P + P \cdot R/K = P(1 + R/K) \tag{10.1}$$

using the distributive law to factor out the common factor P. Now $P1$ plays the role of P in the next period. Thus, the value of the investment at the end of the second period is

$$P2 = P1(1 + R/K) \tag{10.2}$$

where P has been replaced by $P1$ on the right side of formula (10.1). However, we already have a formula (10.1) for computing $P1$ in terms of P. If this is substituted for $P1$ in (10.2), we obtain

$$P2 = [P(1 + R/K)](1 + R/K) = P(1 + R/K)^2 \tag{10.3}$$

The amount after three periods is $P3 = P2(1 + R/K)$ and substituting the value of $P2$ in terms of P given by (10.3) yields

$$P3 = [P(1 + R/K)^2](1 + R/K) = P(1 + R/K)^3 \tag{10.4}$$

Continue this process. Evidently, we find that the value of the investment after L payment periods is

$$PL = P(1 + R/K)^L \tag{10.5}$$

Replacing L by $K \cdot N$, we have

$$PL = P(1 + R/K)^{K \cdot N} \tag{10.6}$$

These formulas can be used to avoid the DO LOOP in the programs of Section 10.1. For example, the following programs print out the value of the original investment P after L periods or after N years. The investment is at a rate R compounded K times per year.

```
10  READ(5,*)N,K,P,R
20  L=N*K
30  FLK=K
40  P=P*(1.+R/FLK)**L
50  WRITE(6,*)P,L
60  END
```

```
10  READ(5,*)N,K,P,R
20  FLK=K
30  P=P*(1.+R/FLK)**(N*K)
40  WRITE(6,*)P,N,K
50  END
```

Suppose instead of daily, a bank decides to compound interest every minute or every second. Would the return on the investment of \$1000 be much different? Actually, the answer is no. As the number of payment periods K increases, the value of PL in (10.6) gets closer to the number $Pe^{N \cdot R}$, where $e = 2.7183\ldots$ is a certain constant. (This is proved in calculus courses.) This fact is helpful in that for a large number of payment periods per year it is a good approximation and, furthermore, it requires less computer time to calculate. (Computer time is costly.) The table below illustrates the relatively insignificant changes that occur when $P = \$1000$ is invested at $R = 4\frac{1}{2}\%$ annual interest compounded daily ($K = 360$) and "instantaneously," that is, by the formula $PL = Pe^{N \cdot R}$. K = number of payment periods per year. PL = amount after one year.

K	1	2	4	360	1000	Instantaneous
PL	1045.00	1045.51	1045.76	1045.98	1045.99	1046.00

Whereas there is a 76¢ increase in yield from a quarterly ($K = 4$) compounding over an annual ($K = 1$) compounding, there is only a 2¢ increase from daily compounding to an instantaneous compounding formula.

Exercises

1. Write a program for computing the value of an investment of $1000 at an annual rate of 5%, compounded daily (360 times per year), after one year. Use (a) the iteration implied in Section 10.1, and then use (b) the formula (10.5) RUN these programs on a computer and compare the results.
2. Which is the better (maximum return) way to invest $1000? (a) At 5% interest, compounded annually, or at 4.8% interest, compounded quarterly, for one year? (b) for 10 years? (c) 20 years?
3. Prove that at 5% interest, compounded annually, and at 4.8% interest, compounded quarterly, the first investment will *always* have a higher yield regardless of the number of years that pass.
4. The value of e^x, where $e = 2.7183 \ldots$, can be obtained on a computer for each real x. The command is Y=EXP(X) where X is a real number. Use this command to write a program for instantaneous compound interest at a rate R on an investment of P dollars for N years.
5. Use the program in Exercise 4 to determine which is the better investment, 5% compounded annually or 4.8% compounded instantaneously.
6. What is the value of an investment of one dollar at 8% instantaneous compound interest after 50 and after 100 years?
7. A bank advertises that one can earn interest today on money that was added to an account yesterday. At a 6% instantaneous compound interest rate, how much would you have to invest in order to earn $1 today on yesterday's interest?

10.3. Time Payments

Recall that for an original investment P (principal) at an annual rate R, compounded K times per year, for L payment periods the amount in the account is

$$PL = P(1 + R/K)^L \tag{10.5}$$

At K payment periods per year, we have $N = L/K$ is the number of years of the investment. We shall use formula (10.5) to compute annuities.

An annuity is a sequence of periodic payments (rents) usually of equal amounts. Suppose, for example, you put \$100 per year into a fund that pays 3% annually compounded, for three years. What is the value of your fund after the three years? Notice that the first \$100 is in the fund for three full years so by the formula (10.5) it is valued at $\$100(1 + .03)^3 = \$109.27$ ($R = .03$, $K = 1$, $N = 3$, $P = 100$). On the other hand, the second \$100 is invested, after the first year, for a total of two years. Thus, by (10.5) it is valued at $100(1.03)^2 = \$106.09$. The last \$100 is in the account for only one year and, hence, is valued at \$103 after that year. In total, therefore, we have \$109.27 + \$106.09 + \$103 = \$318.36 in the account after three years.

Most annuities extend for longer periods than that of the previous problem. For example, suppose you join a Christmas Club in a bank that pays 5% interest, compounded monthly. You agree to pay \$50 into the account on the 10th day of each month. If you start the account (pay the first \$50) on January 10, 1972, and withdraw the money (plus interest) on December 10, 1972, how much will you have at the withdrawal date?

The first \$50 (January 10) remains in the account for a total of 11 months (January 10 until December 10). Hence, by (10.5), its value is $\$50 \cdot (1 + .05/12)^{11}$ since $L = 11$, $K = 12$, $R = .05$, and $P = 50$. The second \$50 (February 10) remains in the account for only 10 months (February 10 until December 10) and its value by (10.5) is $\$50 \cdot (1 + .05/12)^{10}$. Continue in this way and total the results for each \$50 invested. We obtain

$$50 \cdot (1 + .05/12)^{11} + 50 \cdot (1 + .05/12)^{10} + \cdots + 50 \cdot (1 + .05/12) \tag{10.7}$$

The last entry in this sum is the November 10th investment of \$50 which the bank holds only for one month. To obtain a numerical answer we write a FORTRAN program:

```
 10       P=50
 20       R=.05
 30       K=12
 40       FLK=K
 50       L=K-1
 60       A=0
 70       DO 36 I = 1,L
 80       A=P*(1.+R/FLK)**I+A
 90    36 CONTINUE
100       WRITE(6,*)A
110       END
```

When you RUN this program you will obtain the sum (10.7). (Why?) We have written the program so that lines 10, 20, 30 can be replaced by 10 READ(5,*)P,R,K and, hence, the program can be run for other values of the principal P, the annual rate R, and the number of payment periods K.

Suppose, next, you purchase an auto on time payments. The dealer states that he charges "bank rates" so there is no need to obtain a bank loan. After the usual so-called reductions in list price, you owe $2142.86 and you wish to pay this off in two years (24 months). The dealer states "Our annual rate is 6%, just like the various banks, so your interest will be (.06)*(2142.86) = $128.57 per year or $257.14 in two years. Adding the interest to your loan we get $2142.86 + $257.14 = $2400.00. Hence, you will pay us $2400.00/24 = $100 per month." Are you really paying 6% on the loan?

The only fiction in this example are the numbers. This is, in fact, the way auto agencies, and banks as well, compute time payments for the purchase of an automobile. Before we directly answer the question above, notice that the first $100 you paid to the dealer has been debt by you for only one month. Yet you paid interest at 6% per year on this amount as if you had held it for the full time (at simple interest). A similar observation applies to each time payment.

Now suppose you got the dealer to wait (some chance!) a full two years and accept a lump sum payment of $2400 at that time. Suppose also you find a "bank" which pays you 6%, compounded monthly, and you enter $100 into an account at this bank at the end of each month from the time of purchase. Then the bank would hold the first $100 for 23 months, at which time you intend to withdraw the money to pay the dealer. Computing the interest by formula (10.5), we have $100(1+.06/12)^{23} = 100(1.005)^{23}$. The second $100 deposited is held by the bank for 22 months so the value of this investment at the end of the period is $100(1.005)^{22}$. Continue in this way and add all the results (this is an annuity). We obtain a sum of $2543.20. Thus, you pay the dealer $2400 and have $143.20 left.

This example illustrates the fact that there is a difference between the quoted (simple) interest rate and the actual or effective rate of a smaller loan. The Truth in Lending Law requires the dealer or bank to quote the effective rate for each loan. The effective rate takes into consideration the fact that each payment of the loan has not been held for the full period of the loan.

When banks prepare mortgage loans they subtract some principal as well as the interest with each payment, and they charge you on the new principal. They, in effect, only charge you interest on the outstanding principal at each stage unlike the auto loan situation. This method may be a little less simple but it is not difficult to program on a computer.

Consider, for example, a loan of $20,000 on a home at an annual rate of 9% for 20 years. The monthly payment is determined so that the loan is retired after 240 = 20 * 12 payments of equal size. Furthermore, the first payment must include the interest on \$20,000 held for one month at a rate of 9% a year. That is, the first payment must be larger than \$20,000(.09)/12 = \$150 with the excess being subtracted from the principal. Indeed, if the monthly payments are \$180, then \$30 = 180 - 150 is taken from the principal and the outstanding principal is \$19,970 after the first month. The interest on this amount for the second month is 19,970(.09)/12 = \$149.78. Hence, \$30.22 = \$180 - 149.78 is subtracted from the outstanding principal after the second month. This means \$19,939.78 of the original loan is outstanding; \$60.22 of principal has been returned to the lender. The next month's interest is based on the outstanding principal of \$19,939.78. This process continues until the principal has been paid back to the lender.

How much should the monthly payment be if a mortgage of \$20,000 is obtained at an annual rate of 9% for a 20 year period? The selection of this monthly payment must be such that after 240 = 20 · 12 equal payments the loan has been entirely repaid. A formula for the monthly payment can be derived after some effort. The formula is

$$M = BR/[1-(1+R)^{-L}] \tag{10.8}$$

where B is the amount borrowed, R is the rate *per month*, L is the number of months of the loan, and M is the monthly payment. M is usually adjusted to the closest dollar, which with round-off errors, might force the final monthly payment to be a slightly different amount than all other monthly payments.

In computing effective rates for short term loans, such as auto loans, the same formula (10.8) is used. For this type of loan, however, the monthly payment M is known and also known is the number of months L that the loan is to continue and the amount of the loan B. The rate R (per month) is the unknown and it is computed from the given information by formula (10.8). This is a complicated equation. In the next chapter we illustrate how to use the computer to resolve this problem.

Exercises

1. Suppose you invest \$100 each month in a company that pays monthly compound interest at a rate of 12% a year. What is the value of the investment in three months, four months, and two years? (Use the computer for the last result.)

2. Write a program for an investment of \$200 per month at $4\frac{1}{2}$% compounded daily and for a two year period. (Use 30 days in a month; 360 days in a year.) DO NOT RUN THIS PROGRAM.

3. Suppose a bank pays 5% continuously compounded interest. What is the value of an account in which \$100 is entered each month for 24 months? (Use the computer.)

4. Construct a program to compute the monthly payments of a mortgage loan using formula (10.8). The program should permit the user to read into the problem the amount of the loan, the yearly interest rate and the time period in years of the loan.

5. Suppose you borrow \$31,000 from a loan shark at 100% interest per year. If the lender agrees to receive one payment of \$32,000 a year, for how many years will the loan continue? (The figures are absurd but the computation can be completed by hand.)

6. Construct a program that prints the amount charged to principal and to interest for each month when a mortgage loan of B dollars is obtained at an annual rate of R for a period of N years.

10.4 More Annuities

There are formulas and tables that can be used to compute annuities. The mathematical derivation of a simple formula for an annuity is based on summation of a geometric progression and is deferred to later in this book. For the moment, we intend to assume the summation formula.

We will change the perspective on the annuity problem from the point of view of the borrower to that of the lender. Suppose you purchase an item on time and you agree to pay \$100 per month for 24 months. What is the value of this to the seller? Clearly you will spend \$2,400 for the item. If the dealer has a large-scale business, he should be able to reinvest and to obtain a return on each payment at the rate, say of 6%, that he quotes you. Hence, he holds the first payment for 23 months so at monthly compound interest it has the value of $100(1 + .06/12)^{23}$ to him. The second payment has value $100(1 + .06/12)^{22}$ to him since it is held for 22 months, and so on. The sum of these monthly payments with interest is

$$S = 100(1 + .06/12)^{23} + 100(1 + .06/12)^{22} + \cdots + 100(1 + .06/12) + 100$$

which is the value of your investment of the dealer. Now if r is the interest rate *per period* ($r = .06/12 = .005$ here) and if n is the number of payment periods ($n = 24$ here), the formula for this sum on $P(= 100$ here) dollars paid per period is equal to

$$S = P\frac{(1+r)^n - 1}{r} = P(1+r)^{n-1} + P(1+r)^{n-2} + \cdots + P(1+r) + P \qquad (10.9)$$

and it is the value of sale to the dealer. In the special case,

$$S = 100\frac{(1 + .005)^{24} - 1}{.005} \approx \$2543.19 \qquad (10.10)$$

Formula (10.9) can be programmed on the computer and has the advantage that it avoids the DO LOOP of the previous section. This reduces the round-off error for problems involving large n.

Exercises

1. Write a program for an annuity at a rate of r per period for n periods using formula (10.9) with $P = 100$. Use this program to check (10.10).
2. Suppose you place yearly payments (periodic rents) of $6000 at an interest rate of 5% into an account. How long would it take to accumulate $120,000? (Annually compounded.)
3. Suppose you invest $1000 per year into an account for 10 years. What compound annual interest rate does the account have if your return is $12,577.89?
4. Compare the answers obtained on a computer from the program in Exercise 1 with the program in Section 10.3 where line 10 is replaced by 10 P=100.

10.5. Present Value

The present value at compound interest is the worth now of future dollars invested at compound interest. For example, if you have $100 due to you at 5% compound annual interest in two years, how much is the investment worth now? The compound interest formula (10.6) provides the necessary means to compute this quantity. Indeed, $PL = 100, K = 1, N = 2$, and $R = .05$, so

$$100 = P(1+.05)^2 \quad \text{or} \quad P = 100/(1.05)^2 \approx 90.70$$

Thus, investing $90.70 at 5% annual compound interest for two years will yield $100 after two years.

It is clear from this discussion that the present value formula is

$$P = S/(1 + R)^N \tag{10.11}$$

where R is the rate per period and N is the number of periods. S is the future dollars of the investment. Formula (10.11) is easily programmed.

There is also a *present value* connected with an annuity. The amount S (future dollars) associated with the annuity is given by formula (10.9). The present value of this amount at a compound interest rate of r per period for the n periods of the annuity is defined as the *present value* of the annuity. Suppose, for example, that an expert estimates that a particular mine would produce a net income of $20,000 per year for the next six years. At a 6% annual compound interest rate, what should be the most one pays for the mine? The answer is the present value of the mine. First, however, we observe that we hope to receive at the end of the six year period

$$S = 20{,}000(1.06)^5 + 20{,}000(1.06)^4 + \cdots + 20{,}000$$
$$= 20{,}000[(1.06)^6 - 1]/.06 \approx 139{,}500$$

since each year's return can be invested at 6% annual compound interest. The present value of the projected income from the mine is

$$P = 139{,}500/(1.06)^6 \approx 98{,}342$$

This means you should pay less (because mining is not a sure thing) than \$98,342 for the mine. Indeed, if you invest the \$98,342 at 6% compound interest per year, the investment will be worth \$139,500 in 6 years.

The formula for the present value of an annuity is by (10.9) and (10.11)

$$A = S/(1+r)^n = P\frac{(1+r)^n - 1}{r(1+r)^n} = P\frac{1 - 1/(1+r)^n}{r} \qquad (10.12)$$

where r is the rate per period, n is the number of periods and P is the payment per period. Again, the formula (10.12) is easy to program.

The similarity between formula (10.12) and formula (10.8) of Section 3 is more than coincidental. Indeed, the present value A of the annuity in (10.12) is the amount B of the loan and the periodic payment P equals the monthly payment M in formula (10.8).

This survey of time payments, compound interest, and annuities is quite brief and serves mainly to illustrate some programming ideas. A more detailed discussion is found in most texts on accounting. Every consumer should learn certain of these concepts for his own protection.

Exercises

1. Program the formula for the present value of an annuity. Apply this program to the mining example in this section.
2. What is the maximum that you should pay for an oil claim if a geologist assures you that you could pump \$1000 worth of oil a day for two years from the source? (Compare with bank rates of 6% compounded daily, that is, 360 times a year.) Use the computer for this problem.
3. What is the present value of an annuity of \$15,000 per year for 10 years at a 5% rate?
4. What is the present value of \$100,000 due in 15 years assuming the fund will earn 4%?

Chapter 11 Sequences, Series, and Functions

11.1. Functions and Subprograms

The concept of function in mathematics is extremely important; yet it is very simple. It is so simple, in fact, that mathematicians tend to use set theory, which is basic to their field, to define in a logical way the notion of function. We will, however, avoid the temptation to provide a rigorous definition and state that a function is a rule which corresponds to each element of a set X a unique element of a set Y. The set X is called the *domain* of the function. For example, the formula $y = x^2$ defines a correspondence between the real numbers X (domain) and the nonnegative numbers Y. If $x = 2$, then $y = 4$; if $x = 3.1$, then $y = 9.61$; etc. We pick an x in the domain and use the formula to find the corresponding y. Since the x is any element in X, it is called a *variable* in computing circles, although this term is no longer too common in mathematics.

Functions can be defined for more than one variable; it merely requires a special form for the domain X (cartesian product if you know what that means). For example, $I = R \cdot P$ is a function defined for each choice of R and P. The value I corresponding to a particular P is calculated by the formula $R \cdot P$. (This can be interpreted as interest I on a principal P at a rate R for one period.)

The notation for a function of one variable is $Y = F(X)$. For example, if $F(X) = X^2$, then we can write $F(1) = 1$, $F(2) = 4$, $F(-3) = 9$, etc. This means the variable X has been replaced by one of the numbers in the domain and the corresponding Y value calculated. For functions of more than one variable, the notion is $F(R, P)$, $F(P, R, K, N)$, etc. For example, we can write $Y = F(P, R, K, N)$ where $F(P, R, K, N) = P(1 + R/K)^{K \cdot N}$.

To introduce a function in FORTRAN, this functional notation is used. We write the function simply as some name (no more than six letters) with parentheses enclosing a *dummy* variable. The last term refers to the fact that the variable is *not assigned a value prior* to the introduction of the function.

EXAMPLE.

a.

```
10         F(X)=X**2
20         DO 31 I=1,10
30         FLI=I
40         Y=F(FLI)
50         WRITE(6,*)I,Y
60      31 CONTINUE
70         END
```

b.

```
10         IDOIT(N)=N**2
20         DO 31 I=1,10
30         J=IDOIT(I)
40         WRITE(6,*)I,J
50      31 CONTINUE
60         END
```

In the first program, the dummy variable X, a real variable, has been used in the definition of the real valued function F. Since the variable X was real, we need to convert I to a real number (add the decimal) as in line 30 before substitution into the formula (line 40). The output by line 50 is 1, 1.; 2, 4.; 3, 9.; . . .; 10,100.; where the pairs appear in a column. For the second program, the dummy variable N is an integer variable. Hence, the function has as its domain the integers. Moreover, the function is IDOIT, which since it starts with I, has values that also are integers. The output is 1, 1; 2, 4; 3, 9; . . .; 10, 100 in a column of pairs.

It is possible within FORTRAN to define any function that can be explicitly written by using the operations +, -, *, /, and ** between constants or variables. Such a function is called an *arithmetic statement function*. For example,

$$F(X) = X^2 + 1/X^2, F(X) = (2X^2 + 7X + 9)/(X + 3),$$
$$F(X) = X^2 + 9, F(X) = X^{1/2} - X^{-3/2}$$

are arithmetic statement functions.

A program for the evaluation of the first of these functions at $X = 5.$ and $X = 10.$ does not require a subprogram. Indeed, one such program is simply the following:

```
10   F(X) = X**2+1./X**2
20   Y = F(5.)
```

```
30    Z = F(10.)
40    WRITE(6,*)Y,Z
50    END
```

The computer prints the values Y=25.04 and Z=100.01 when this program is RUN. It is necessary to evaluate and store (in position Y and Z) before the WRITE command. You cannot replace lines 20, 30, and 40 by a line 40 WRITE(6,*) F(5.), F(10.) since the computer does not perform a computation, namely the evaluation of the function F at 5. and at 10., in a WRITE command.

The arithmetic statement function used in a program should be defined at the beginning of the program. (There are only a few FORTRAN statements, to be introduced later, that can precede a function statement.) Any number of such functions can be defined for a program with one function defined per line.

There are many built-in functions that the computer remembers. A partial list of those functions that occur frequently in applications is as follows:

Function	FORTRAN Name
a. $\sqrt{x}$	SQRT(X)
b. $\lvert x \rvert$	ABS(X)
c. e^x	EXP(X)
d. $\log_e x$	ALOG(X)
e. $\log_{10} x$	ALOG10(X)
f. $\sin x$	SIN(X)
g. $\cos x$	COS(X)
h. $\tan x$	TAN(X)
i. $\arcsin x$	ARSIN(X)

The first two of these functions, $\sqrt{x}$ and $|x|$, are respectively the square root function and the absolute value function. In order to illustrate their use, consider the following programs:

```
10    X=2.                10    X=-3.712
20    Y=SQRT(X)           20    Y=ABS(X)
30    WRITE(6,*)Y         30    WRITE(6,*)Y
40    END                 40    END
```

The output for the first program is 1.14159 since the computer recognizes the square root function of line 20. The second program prints 3.712 the absolute (numerical) value of X.

Another simple function which is known to the computer is FLOAT(I). This function converts integers into their corresponding real numbers, for example, FLOAT(5)=5., FLOAT(13)=13., and it can be used instead of the implied conversion that we have been using in our programs. (Thus, we can use this function instead of a line like FLI=I which does the same thing.) For example, lines 30 and

40 in program a. (page 240) can be replaced by

```
30    Y=F(FLOAT(I))
```

a single line.

Built-in functions do *not* require the computer to remember a table of values, since this would be far too costly in terms of memory space. Instead, the computer calculates the desired function by some standard method for each choice of the variable you select from the domain. This calculation is done in a "function subprogram."

A function subprogram is a "baby" program which is placed at the end of the main program. Whenever in the main program you refer to a function, F(X) say, you can build a special program, the subprogram, to calculate the function F for various values of the variable X. We shall illustrate this with some functions that actually are built-in functions in the computer. The method, however, applies to other functions as well.

Consider first the ABS(X) function, which is the absolute value of X. It is defined by ABS(X)=X if $x \geqslant 0$ and ABS(X)=-X if $x < 0$. Thus, ABS(7.)=7. and ABS(-7.) = -(-7.) =7. (When $x < 0$, the sign of $-x$ is opposite to that of x and, hence, is positive.) A subprogram for the calculation of ABS(X) could be as follows:

```
1000      FUNCTION ABS(X)
1100      IF(X.LT.0)GO TO 5
1200      ABS=X
1300      GO TO 6
1400    5 ABS=-X
1500    6 RETURN
1600      END
```

The only things really new are lines 1000 and 1500. The first of these tells the computer that you have a function subprogram and the function's name is ABS with dummy variable X. When, in the main program, a calculation is encountered that involves the ABS function, control is transferred to the subprogram to compute this function. The line 1500 is necessary to return control back to the main program. The END statement is needed to tell the compiler that the subprogram work is done for the function being defined.

Note that the other lines in the subprogram constitute a program itself. If you read in a value for X and used lines 1100, 1200, 1300, 1400, 1600 as well as a new line "1500 6 WRITE(6,*)ABS" you would have a complete program for computing the ABS(X) function and writing the answer.

Another example of a subprogram is the computation of the SQRT(A) where $A > 0$. This is simply the function $\sqrt{A}$, the square root of A. This is a built-in function and the computer uses a subprogram somewhat like the following to calculate the square root. Assume $A > 0$ and we seek $\sqrt{A}$. Do it in steps: Pick a starting value X_0, say, which is positive. Compute

$$X_1 = \tfrac{1}{2}(X_0 + A/X_0)$$

Use this X_1 to compute a new approximate answer, namely,

$$X_2 = \tfrac{1}{2}(X_1 + A/X_1)$$

Use this X_2 to compute another approximation, that is,

$$X_3 = \tfrac{1}{2}(X_2 + A/X_2)$$

Continue this process. At the $(n + 1)$st stage (n is a counting number) we compute X_{n+1} from the previous value X_n by means of the formula

$$X_{n+1} = \tfrac{1}{2}(X_n + A/X_n)$$

Assume that as n gets large, the numbers X_n you are calculating get close to some definite number X. We write $X_n \to X$ as $n \to \infty$ to indicate this phenomenon. Then,

$$X_{n+1} \to X \text{ as } n \to \infty$$

and it follows that

$$\begin{array}{ll} \underset{\downarrow}{X}_{n+1} & = \tfrac{1}{2}\,(\underset{\downarrow}{X}_n + A/\underset{\downarrow}{X}_n) \\ X & = \tfrac{1}{2}\,(X + A/X) \end{array}$$

Multiply both sides of the last equality by $2X$ to obtain $2X^2 = X^2 + A$. Thus, transposing the X^2 with a sign change, we obtain $2X^2 - X^2 = X^2 = A$, that is, X is the positive number whose square is A. But this is the definition of the square root of A, that is, $X = \sqrt{A}$.

We have seen that if $X_n \to X$ as $n \to \infty$, then $X = \sqrt{A}$. The fact that the X_n's do get closer to a value X as n increases, regardless of the starting value $X_0 > 0$, is proved in more advanced courses in mathematics (calculus) and we will not concern ourselves with the problem here. Instead, we will be content to illustrate the method.

EXAMPLE. Let us compute $\sqrt{2}(\approx 1.414)$. Suppose our starting value is $X_0 = 1$. Then $X_1 = 1/2(X_0 + 2/X_0) = 1/2(1 + 2/1) = 3/2 = 1.5$. Next, $X_2 = 1/2(X_1 + 2/X_1) = 1/2(3/2 + 4/3) = 17/12 \approx 1.4166$. Next, $X_3 = 1/2(X_2 + 2/X_2) = 1/2(17/12 + 24/17) = 577/408 \approx 1.414$. Hence, after just three steps we have obtained the square root of 2 accurate to three places following the decimal.

A subprogram for computation of the square root function can be written as follows:

```
1000      FUNCTION SQRT(A)
1010      X=1.
```

```
1020        DO 99 I = 1,100
1030        X1=(X+A/X)/2.
1040        IF(ABS(X1-X).LT. .001) GO TO 33
1050        X=X1
1060    99  CONTINUE
1070    33  SQRT=X
1080        RETURN
1090        END
```

Line 1040 is a check to see if the value X1, just computed, and the previous value X are close together. If they are within .001 = 1/1000 of each other, we are stating that is good enough, so the computer would use this value for the square root of A. If they never get that close, the computer goes through the calculation of all 100 terms in the algorithm outlined on these pages and uses the last value of X for the SQRT(A).

For functions, like the square root function, that require an algorithm to evaluate, you should introduce a subprogram at the end of your program. For example, consider the function F(X) which is 1 when X >0; 0 when X $= 0$; and -1 when X <0. The following subprogram can be used to define it.

```
100         FUNCTION F(X)
110         IF(X) 98,99,100
120     98  F=-1.
130         GO TO 101
140     99  F=0.
150         GO TO 101
160    100  F=1.
170    101  RETURN
180         END
```

This subprogram is combined with a program that uses the function F. For example, it can be added after the END statement in the following program.

```
10   X1=-2.1
20   X2=3.2
30   Z=F(X1)-F(X2)
40   WRITE(6,*)Z
50   END
```

The output is 2.0. (The two END statements, lines 50 and 180 are necessary.)

Should you try to RUN just lines 10 through 50, the computer would transmit an error message. Indeed it would have no knowledge of the meaning of your function F. On the other hand, if you replace line 30 by

```
30   Z=ABS(X1)-ABS(X2)
```

the computer would print −1.1 since ABS is a built-in function.

Frequently, the function that is necessary in a problem is not complicated and it would be inconvenient to be required to build a subprogram for it. For example, $x^2 - 1$ is a simple function. Is a subprogram like the one below needed?

```
1000   FUNCTION F(X)
1010   F=X**2-1.
1020   RETURN
1030   END
```

Exercises

1. What is printed by each of the following programs?

a.

```
10        I(X)=X**2-X/4.
20        J=I(2.)
30        WRITE(6,*)J
40        END
```

b.

```
10        F(X)=X+1./X
20        DF(X)=1.-1./X**2
30        Y=DF(2.)/F(2.)
40        WRITE(6,*)Y
50        END
```

c.

```
10        I(N)=N**2
20        DO 22 J=1,5
30        ISQ=I(J)
40        WRITE(6,*)J,ISQ
50     22 CONTINUE
60        END
```

d.

```
10        F(I)=1.5**I
20        K=1
30        DO 23 J=1,3
40        G=F(K)
50        K=K+1
60     23 CONTINUE
70        WRITE(6,*)G
80        END
```

2. Write a subprogram for each of the following functions.

a. $f(x) = x^2 - 1$ if $x \geqslant 1$
   $f(x) = x^3$ if $x < 1$

b. $f(x, y) = x$ if $y > x$
   $f(x, y) = y$ if $y \leqslant x$

c. $f(x, n) = 1 + x + x^2 + \ldots + x^n$

3. What is wrong with each of the following programs?

a.

```
10   X=2
20   F=X**2+1.
30   Y=F(3.)
40   WRITE(6,*)Y
50   END
```

b.

```
10        X=2.
20        F(X)=X**2+1.
30        Y=F(3)
40        WRITE(6,*)Y,F(X)
50        END
```

c.

```
10      FUNCTION F(N)
20      F(N)=N**2
30      RETURN
40      DO J=1,10
50      K=F(J)
60      WRITE(6,*)K
70      CONTINUE
```

d.

```
10          X=1.2
20          DO 23 J=1,100
30          X=X+2./X
40          Y=F(X)
50          WRITE(6,*)X,Y
60       23 CONTINUE
70          FUNCTION F (X)
80          F(X)=X**2
90          RETURN
100         END
```

4. Given $y(x) = x + 1/\sqrt{x}$. What is $y(9)$ and $y(4)$? For what values of x is y defined; what is the suspected domain of this function?
5. Use three steps of the algorithm for the square root in this section to approximate $\sqrt{5}$.

11.2. The Search for Zeros

In many applications, you are asked to find a zero of a certain function, that is, a point x_0 in the domain of the function F such that $F(x_0) = 0$. We display, in this section, an algorithm which attempts to locate a zero of a continuous function, that is, a function with a connected graph.

If F is a continuous function on an interval and if $F(x_1) < 0$ while $F(x_2) > 0$ for two points x_1 and x_2 of this interval, then there must be a point c between x_1 and x_2 for which $F(c) = 0$. Indeed, it is necessary to cross the x-axis in con-

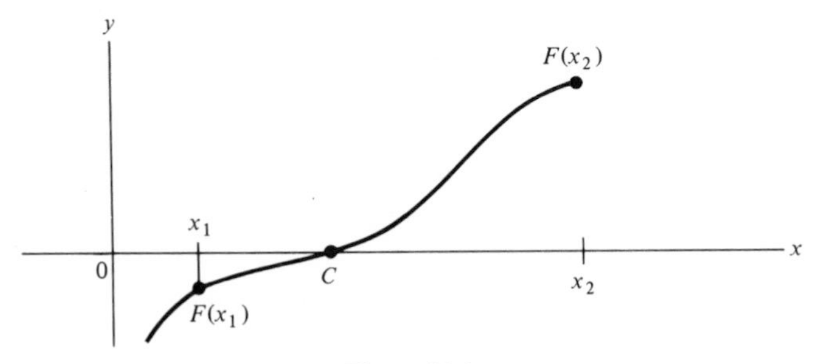

Figure 11.1

structing the graph from the negative $F(x_1)$ to the positive $F(x_2)$ axis since the graph is connected as in Figure 11.1. (This is sometimes called the location theorem and is proved in calculus.) Observe that if F is not continuous, this need not happen as illustrated in Figure 11.2.

To find a zero of the function F, consider the midpoint of the interval from

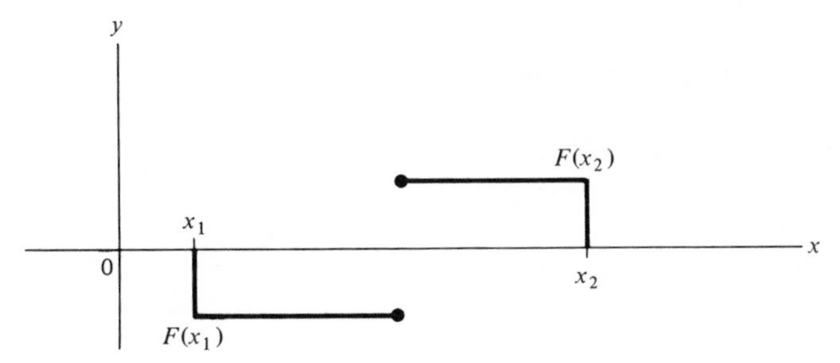

Figure 11.2

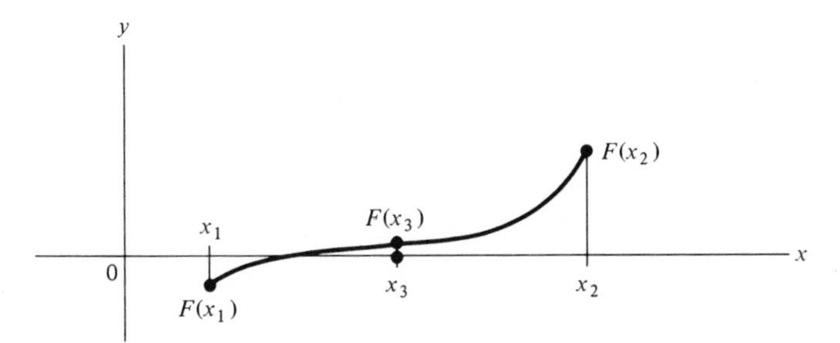

Figure 11.3

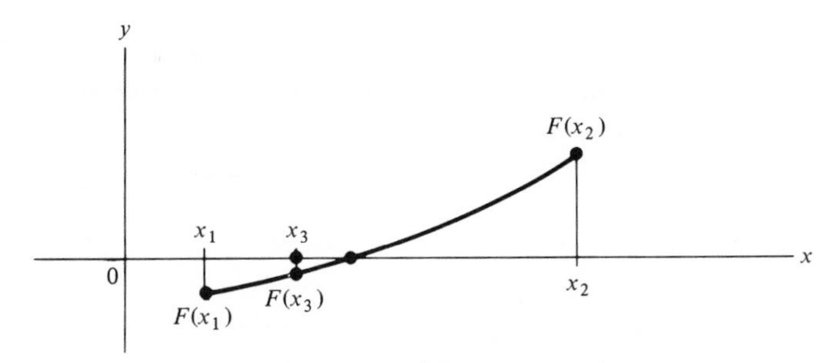

Figure 11.4

x_1 to x_2, that is, $x_3 = (x_1 + x_2)/2$. In the unlikely event that $F(x_3) = 0$, we are done. The zero of F has been found. See Figure 11.3.

Usually, however, $F(x_3)$ is a nonzero real number as in Figure 11.4. If $F(x_3) > 0$, and $F(x_1) < 0$, then, by the location theorem, there must be a zero c of F between x_1 and x_3. Hence, the interval containing the zero has been cut down by one half over the original interval from x_1 to x_2. On the other hand, if $F(x_3) < 0$, the zero of F must lie between x_3 and x_2 since $F(x_2) > 0$. In either case, therefore, we have cut down the size of the interval containing the zero.

By using the process again on the smaller interval containing the zero we can further cut down the size of the interval in which the zero of F lies. Repeated application of the process enables us to determine intervals $a \leqslant x \leqslant b$ where their length $b - a$ is as small as we wish. In particular if $b - a \leqslant 0.000001$, then either end point is within 0.000001 units of the actual zero, so for practical purposes, either end point can be listed as the zero. The approximation will be accurate to at least five places after the decimal. The process is easily programmed. For example, a program for finding the zero of $F(X) = X^2 - 3$ between 0 and 2 is as follows:

```
10         F(X)=X**2-3.
20         X1=0.
30         X2=2.
40     31  X3=(X1+X2)/2.
50         Y=F(X1)*F(X3)
60         IF(Y.GT.0) GO TO 21
70         X2=X3
80         GO TO 31
90     21  X1=X3
100        GO TO 31
110        END
```

In the program, line 40 computes the average X3 of the two points X1 and X2. The Y in line 50 is positive when F(X1) and F(X3) have the same sign. In this case the zero is between X3 and X2 so we redefine X1, in line 90, to be X3 and try the process again, computing a new average. If Y is negative, then F(X1) and F(X3) have opposite signs so the zero is between X1 and X3. Thus, we redefine X2 to be X3 in line 70 and compute a new average for these two points. (What happens when F(X3)=0.?)

The problem as now programmed has an infinite loop and will never print answers. Hence, we need a control to terminate computations when a certain accuracy has been reached and to print the result at this point.

In many cases the following statements suffice for these purposes:

```
65         IF (ABS(F(X3)) .LT. 0.0001)GO TO 41
95         IF (ABS(F(X3)) .LT. 0.0001)GO TO 41
105    41  WRITE (6,*)X3
```

Together with the other lines in the program, we now obtain a computation that will normally terminate and print the results.

Let us look at a few steps in the calculation. Since $X1 = 0$ and $X2 = 2$ we have $X3 = (0 + 2)/2 = 1$. $F(0) = -3, F(2) = 1$, and $F(1) = -2$. Thus $Y = F(0)*F(1) = 6 > 0$ in the first step and, hence, control is passed to 21 (line 90) where $X1$ is redefined as $X3$. Control then goes to 31 (line 40) and a new $X3 = (1 + 2)/2 =$

1.5 is computed. Since $F(1.5) = -0.75$, $Y = F(1)*F(1.5)$ is again positive. Hence X1 is redefined again to be 1.5 and a new $X3 = (1.5 + 2)/2 = 1.75$ is determined. This time $F(1.75) = 0.0625$ and $F(1.5) = -.75$ so $Y < 0$ and it is the X2 value that gets redefined. The process continues until $F(X3)$ gets within 0.0001 units of zero.

There are some extra controls that one might wish to place in a program of this type to avoid the unexpected. For example, suppose you made an error with your hand calculation and thought there was a zero between $X1 = 2.$ and $X2 = 3$. Since the graph is always above the x-axis in the interval $2 \leqslant X \leqslant 3$, the Y would always be positive in line 50. Hence, the calculation in the program would proceed indefinitely and you should, therefore, check your initial computations to be sure $F(X1)$ and $F(X2)$ have opposite signs. This can be done as follows:

```
35   IF (F(X1)*F(X2) .GT. 0) STOP
```

There are other ways, of course, that do not terminate computation; the program could be written so that you are asked to READ into the problem new values of $X1$ and $X2$ in case $F(X1)$ and $F(X2)$ have the same sign.

Exercises

1. Build a flow chart for the algorithm in this section.
2. Use four steps of the algorithm in this section to determine the zero of $F(X) = X**2 - 5$ between 2 and 3. Find the value of $\sqrt{5}$ using the algorithm on a computer.
3. What happens in the algorithm of this section when $X3$ is a zero of the function F? Illustrate with the FUNCTION F(X)=X**2-4. and with starting values X1=1. and X2=3.
4. If the program in this section is applied to F(X)=X**2 with X1=1. and X2=2. as initial values, what, if anything, will be printed? Suppose X1=0. and X2=1.
5. Consider the program:

```
10        X=1.
20        DO 99 I=1,100
30        X=2.+3./X
40     99 CONTINUE
50        WRITE(6,*)X
70        END
```

What number is being approximated by this program?

6. Replace, in the program of this section, lines 65 and 95 by:

```
65    IF (X2-X3.LT. .00001)GO TO 41
95    IF (X3-X1.LT. .00001)GO TO 41
```

   Discuss this control and compare it with the former lines 65 and 95. In particular, discuss which method is likely to use the more (costly) machine time.

7. Rework Exercise 4 with the controls introduced in Exercise 6.

8. Graph the function $F(x) = x^2$ and the function $G(x) = 2x + 1$. Then graph the function $H(x) = F(x) - G(x)$. Estimate from your graph the zeros of $H(x)$.

9. Suppose the supply function for a certain item is $S(P) = 7P$ and the demand function is $D(P) = 16 - P$. What will be the selling price (equilibrium point) if the price is in dollars? ($S(P) = D(P)$ at the equilibrium point.)

10. Galileo (1564–1642), who introduced the concept of function, also originated a mechanics of freely falling bodies. According to his theory, if $y(t)$ is the distance in feet from you to the ball at any time t (in seconds), measured from the time $t = 0$ when you first toss the ball, then

$$y(t) = -16t^2 + v_0 t$$

   where v_0 is the initial speed given in feet per second.

   The ball will return to you when $y(t) = 0$, for some $t > 0$. Toss a ball at an initial speed of 64 ft/sec straight upward. When will it return to you? Graph the distance from you as a function of time. What is the maximum height that the ball reaches and what is the time it takes to reach this height?

11.3. Maximum of a Function

Problems of finding a maximum or minimum of a function are discussed here with some reluctance. The fact is that there are improved methods using more advanced concepts of mathematics. However, the approach introduced is suitable for discrete problems and, at the same time, provides some additional experience in programming. The title of this section, perhaps should be "How to Find a Maximum before the Mathematician Comes."

Let $y = F(x)$ be a continuous function defined on an interval $a \leqslant x \leqslant b$. It is known (from calculus) that such a function has a maximum on this interval, that is, there is a point c, where $a \leqslant c \leqslant b$ and $F(x) \leqslant F(c)$ for all x in the interval. The value $F(c)$ is called the maximum value of the function F in the interval $a \leqslant x \leqslant b$. A minimum value for F in the interval $a \leqslant x \leqslant b$ is analogously defined.

Note that if $y = F(x)$ has a minimum value at $x = c$, then the function $y = -F(x)$ has a maximum at $x = c$. Hence, there is no special need to construct a program for minimums if we have one for maximums. Indeed, all that is necessary is to change the sign of the function (multiply $F(x)$ by -1).

EXAMPLE 1. An especially simple case is the problem of finding the maximum value of the function $y = 4 - x^2$ when $-100 \leqslant x \leqslant 100$. For, we know that $x^2 \geqslant 0$ for all real numbers x, with $x^2 = 0$ if and only if $x = 0$. Thus, $y = 4 - x^2 \leqslant 4$ with equality when $x = 0$, since we are subtracting the nonnegative number x^2 from 4. This proves that the value of x which yields a maximum y, is 0, that is, $F(0) = 4 \geqslant F(x)$ for *all* choices of x.

Although there are inherent limitations in the method, we can estimate the maximum value of a function by first evaluating the function at a finite set of points in the domain and then selecting the largest of these values of the function. This simple plan is easily programmed on a computer.

Suppose we seek the maximum of the function F on the interval $a \leqslant x \leqslant b$. The plan requires that we select certain points in $a \leqslant x \leqslant b$ and evaluate F at the chosen points. One way to make the selection is to subdivide the interval $a \leqslant x \leqslant b$ into small equally-sized subintervals and use the end points of these subintervals to evaluate the function F. If the subintervals are sufficiently small (and the function is well behaved) the largest value of the function F at these end points is close to the maximum value of F on the full interval.

For example, suppose the function F is defined on the interval $1 \leqslant x \leqslant 3$. We divide the interval into, say, 10 subintervals, each of which is 1/10th the total length of the interval. Thus, each subinterval is of length $h = (3 - 1)/10 = 0.2$. The end points of these subintervals are 1, 1.2, 1.4, 1.6, 1.8, 2.0, 2.2, 2.4, 2.6, 2.8, 3.0. The function F is evaluated at each of these points and the largest value is determined.

A computer program to carry out the plan for the function $F(x) = 4 - x^2$ on the interval $1 \leqslant x \leqslant 3$ is as follows:

```
10        F(X)=4.-X**2
20        A=1.
30        B=3.
40        BIG=F(A) <-----------  Initially the maximum
50        W=A      <-----------  value is taken at W=A.
60        H=(B-A)/10.
70        Z=A
80        DO 22 J=1,10
90        Z=Z+H    <-----------  Compute the end points of the
100       IF(F(Z) .LT.BIG)GO TO 22   subdivision of A ≤ X ≤ B.
110       BIG=F(Z) <-----------  This new value of BIG is assigned
120       W=Z                    only when F(Z) is not smaller than
130    22 CONTINUE               the older value of BIG.
140       WRITE(6,*)W, BIG
150       END
```

Let us evaluate a few steps by hand. We begin with BIG=F(1)=3., W=1 and compute, in line 60, H=(3–1)/10=.2. Since Z starts at 1 (line 70), the next value of Z is (line 90) 1 + .2 = 1.2. Now F(1.2)=2.56 SO F(1.2)<BIG=3. Hence, we go to 22 which continues the computation 90, using the next Z=1.2+.2=1.4. The process progresses in this way. The maximum value of F(X) is actually 3 at X=1 in the cited interval.

Suppose we replace lines 20 and 30 by A=–.2 and B=1.8. Then, again, H=(B–A)/10=.2 whereas the initial values of both W and Z are –.2. By line 90 the next value of Z is –.2+.2=0. Since BIG=F(A)=F(–.2)=3.96 is less than F(Z)=F(0)= 4. this time, line 100 is ignored and line 110 assigned a new value of F(Z)=4. to BIG. The next line (120) tells us the value of W that corresponds to the current value of BIG. Thereafter, we continue to the next Z=0.+.2=.2. Since F(.2)= 4–(.2)^2=3.96 is less than our current value of BIG, which is 4., we go to 22. This produces the next Z=.2+.2=.4. and a check to determine if F(.4) is less than the current value of BIG. The process progresses through the values of Z obtained by the DO LOOP of line 80, ten values in all. When completed, the computer passes to line 140 and prints the current (largest) values of BIG and the point at which this estimated maximum occurs.

In summary, the program begins with Z=A (line 70) and Z becomes, successively, A+H,A+2H,A+3H, . . . , A+9H,A+10H(=B) by the DO LOOP in 80 through 130. Line 100 sends the computation to the next value of Z whenever F(Z) is smaller than the current value of BIG.

If, however, F(Z) ≥ BIG at some point, we pass control to line 110 by the definition of an IF-statement. Then BIG is redefined to be this particular F(Z) and the W is taken to be this Z *before* we go on to the next value of Z.

We could improve the result by taking a "finer" subdivision of the interval AB, for example, we could take H=(B–A)/100. in line 60. The only other change would be line 80, which should now read "DO 22 J=1,100."

Another method for improvement of the approximation is to redefine the end points A and B at various stages of the program. For example, once the program of this section has been RUN you can take a smaller interval than $1 \leqslant x \leqslant 3$ which contains the approximate maximum. By redefining the end points A and B, therefore, you can check additional values of F that are closer to the suspected maximum value. The computer can be instructed to perform this task for you. Indeed, adjunct the following lines to the previous program in this section:

35	33 CONTINUE ←	This passes control to line 40. It can be used to add a number to the next FORTRAN statement. It saves retyping line 40.
132	A=X–H	
133	B=X+H	
135	GO TO 33 ←	Transfers control back to line 35.

Of course, there is now an infinite loop in the program. We leave the introduction of a control to avoid this problem as an exercise.

Exercises

1. Construct a flow chart for the initial algorithm of this section.
2. Find, with the initial algorithm of this section, the maximum of $F(x) = 3 + 7x - 2x^2$ on the interval $1 \leqslant x \leqslant 3$ by hand calculations (or with a mini-calculator).
3. Suppose the following lines are adjuncted to the initial program of this section: 35, 132, 133 in the text and

```
135    N=1
136    N=N+1
139    IF(N.LT.4) GO TO 33.
```

   Describe what is accomplished when the full program is RUN.
4. RUN the program in the text and the program in problem 3 to find the minimum of FUN(X)=X**4–4.*X**2–20.*X in the interval 1. ⩽ X ⩽ 3. The minimum value is approximately -40.1926 and the minimum point is approximately 2.0946.
5. It has been found useful in mathematics to supply the suitable last term of an expression $x^2 + bx$ or $x^2 - bx$ in order that the three terms make a perfect square. The process is called "completing the square." The task is quite simple to accomplish since we have the identities

$$(x + b/2)^2 = x^2 + bx + b^2/4, \qquad (x - b/2)^2 = x^2 - bx + b^2/4$$

   Thus, the square is completed by taking half the multiplier of x (coefficient of x) and adding its square. For example, $x^2 + 6x$ has $b = 6/2 = 3$ so we add $b^2 = 9$ to complete the square. Likewise, $x^2 - 4x$ has $b = 4/2 = 2$ so we add $b^2 = 4$; $x^2 + 3x$ has $a = 3/2$ so we add $b^2 = 9/4$. Use this method to fill in the blank:

   a. $x^2 + 8x +$ _____ $= (x +$ _____$)^2$ b. $x^2 + 10x +$ _____ $= (x +$ _____$)^2$
   c. $x^2 - 9x +$ _____ $= (x -$ _____$)^2$ d. $x^2 - 200x +$ _____ $= (x -$ _____$)^2$
6. We can find maximum and minimum values for quadratic functions by completing the square. For example, let $y = x^2 + 2x + 3$. Then $y = (x^2 + 2x + 1) + (3 - 1) = (x + 1)^2 + 2$. Hence, $y \geqslant 2$ since $(x + 1)^2 \geqslant 0$ and $y = 2$ only when $x + 1 = 0$, that is, $x = -1$. The value 2 is a minimum in this case. Apply this method to find the maximum or minimum of each of the following quadratic functions:

   a. $y = 9 - 4x - x^2$ b. $y = x^2 + 2x - 8$
   c. $y = 2x - x^2$ d. $y = x^2 + 3x + 5$
7. A farmer has 400 ft of fencing with which he plans to fence off a rec-

tangular plot for corn. What is the maximum area that can be enclosed by this fence?

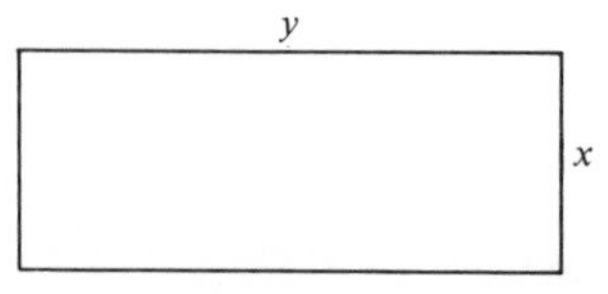

Figure 11.5

(Hint: The area of the plot is $A = xy = x(200 - x) = 200x - x^2$. Complete the square and apply an argument like that used in Example 1.)

8. A wire of length 24 in. is cut into two pieces. Both pieces are bent to form squares. How should the wire be cut if the sum of the two areas is to be (a) minimum and (b) maximum.

9. One side of an open field is bounded by a river as in Figure 11.6. How would you put 400 ft of fence around the other three sides of a rectangular plot in order to enclose as great an area as possible?

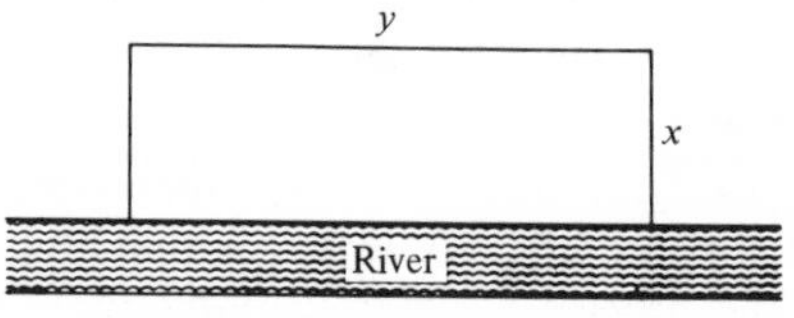

Figure 11.6

10. A box with a square base and an open top is to hold 32 cu in. Find the dimensions which require the least amount of material (neglect the thickness and the waste in construction). Solve this problem on a computer with the program in this section.

    (Hint: If x is the length of a side of the square base and if y is the height, then $x^2y = 32$ so $y = 32/x^2$. The sum of the areas of the four sides and the base is $A = 4xy + x^2 = 128/x + x^2$, which is the function to maximize.)

11.4. Sequences without Computers

A function with the natural numbers for its domain is of special interest to users of mathematics. Whenever there is a real number matched with each natural number (1,2,3,4, . . .), we call the correspondence a *sequence*. In each sequence there is a first member, a second, a third and so on. We denote this correspondence by subscripts and write $x_1, x_2, x_3, \ldots$ for the sequence. The nth member x_n in the sequence is called the general term. Whenever you are given the general term, it is a simple task to generate specific terms in the sequence. For example, if $x_n = 2^n$, then the first, second, third, and fourth terms ($n = 1,2,3,4$ respectively) are 2,4,8,16 in this order. If $x_n = 2n + 1$, then the early terms in the sequence are 3,5,7,9 corresponding to $n = 1,2,3$, and 4 respectively.

Given a few terms in a sequence, one cannot uniquely determine the general term. For example, if you are given 1,2,3 and 4 and asked to find the next term in this sequence it may be natural to answer 5 but this is not the only possibility. Indeed, another answer is 29 which arises if the general term is $n + (n - 1) \cdot (n - 2)(n - 3)(n - 4)$ instead of simply n. The first four terms are 1,2,3,4 in each case.

There are certain special types of sequences which have proved useful in mathematics and its applications. The first of these is the arithmetic progression in which each term is obtained from the previous term by addition of the same (fixed) number, the latter being called the common difference d. We begin the discussion with a tale, possibly a tall one, about a nineteenth century mathematical genius named Carl Friedrich Gauss. When Gauss was a young boy in his early years of school, his teacher decided to keep the class busy by asking the students to add all the counting numbers from 1 through 100 together. After a few moments, much to the surprise of the teacher, Master Gauss gave him the answer, namely, 5050. It was not that Gauss had completed the task previously, rather he undoubtedly reasoned as follows: "If I add 1 and the last number 100, the sum is 101; the sum of the second number 2 and the second from last 99 is also 101. In fact, if I add the third and the third from last I get 101 so this process continues, each time yielding 101. Since we have 50 such pairs from the 100 numbers, the answer must be $50 \cdot 101 = 5050$ because each pair totals 101."

The argument is quite general and can be used for any arithmetic progression (A.P.). For example, consider 100 odd numbers 1,3,5,7, ... The common difference $d = 2$ and the general term is $1 + 2(n - 1)$. Thus, the 100*th* term is $l = 1 + 2(99) = 199$. There are 100 terms, the sum of the first, 1, and the last, 199, is 200; the sum of the second term, 3, and the second from last, 197, is 200, etc. The answer is, therefore, $200 \cdot 50 = 10{,}000$ since there are 50 pairs which total 200 each.

Generally, if we have an arithmetic progression with first term a and common difference d, the last (general) or *nth* term will be $l = a + (n - 1)d$. For, the terms are (1) a, (2) $a + d$, (3) $a + 2d$, (4) $a + 3d$, (5) $a + 4d$, (6) $a + 5d$, etc. The 10*th* term is evidently $a + 9d$; the 50*th* term is $a + 49d$ so the *nth* term is $a + (n - 1)d$. Now let S be the sum of these n terms. Then

$$S = a + (a + d) + (a + 2d) + \cdots + (l - d) + l$$
$$S = l + (l - d) + (l - 2d) + \cdots + (a + d) + a$$

where the second sum is the same as the first except the terms are written in reverse order (commutative law of addition assures us the value is the same in each case). Next add the two sums in columns to obtain

$$2S = (a + l) + (a + l) + (a + l) + \cdots + (a + l)$$

where there is one $(a + l)$ term for each of the n terms in the original sum. Thus $2S = n(a + l)$ or $S = n(a + l)/2$. The formulas are

$$l = a + (n - 1)d, \qquad S = n(a + l)/2 \tag{11.1}$$

This proof differs from the argument of Gauss slightly since we added the series S to itself. The change was made so that we could cover, in an easy fashion, the case in which n is odd. You can apply Gauss' method to this case, but the middle term does not have a matcher to pair it with. Making a suitable adjustment, the same result as (11.1) is obtained. For example, try adding 99 terms of the sequence 1, 2, 3, . . . , 99 by the method of Gauss. The answer is 4950–you obtain 49 pairs that sum to 100 each (99 + 1, 98 + 2, 97 + 3, etc.) and one extra number, namely, 50. The formula (11.1) gives this answer also since $a = 1, d = 1, n = 99$ so $l = 1 + 98 \cdot 1 = 99$. Thus $S = 99(1 + 99)/2 = 99 \cdot 50 = 4950$.

EXAMPLE 1. A man obtains a position with a starting salary of \$6000 a year. If he gets a \$500 raise each year, what are his total earnings in 20 years?

Solution: $a = 6000$, $d = 500$, $n = 20$ so $l = 6000 + 500(19) = 15{,}500$. Thus, $S = n(a + l)/2 = 20 \cdot (6000 + 15{,}500)/2 = 10(21{,}500) = 215{,}000$.

EXAMPLE 2. In an arithmetic progression with first term 3 and common difference 4/3, what is the 49*th* term? What, also, is the sum of the 49 terms?

Solution: $a = 3, d = 4/3, n = 49$ so $l = a + (n - 1)d = 3 + 48(4/3) = 3 + 16 \cdot 4 = 3 + 64 = 67$. Also $S = n(a + l)/2 = 49(3 + 67)/2 = 49 \cdot 35 = 1715$.

EXAMPLE 3. Between 2 and 7 we wish to insert 7 numbers so that the resulting sequence is an arithmetic progression. (Thus, we are dividing the interval $2 \leqslant x \leqslant 7$ into *eight* equal sized subintervals.)

Solution: $a = 2$, $l = 7$, $n = 9$ so from $l = a + (n - 1)d$ we get $7 = 2 + 8d$, or $7 - 2 = 8d$, or $d = 5/8$. The end points in the subdivision are 16/8, 21/8, 26/8, 31/8, 36/8, 41/8, 46/8, 51/8, 56/8, each obtained from its predecessor by addition of 5/8. You can write these as mixed fractions if you wish.

Another type of sequence, frequently met in mathematics and applications, is the geometric progression. It is defined by the fact that each term is obtained from the previous term by multiplication by a common (fixed) number, called the common ratio r. For example, 2, 4, 8, 16, . . . is a geometric progression since each number is obtained from the previous number by multiplication by

$r = 2$. Thus, $4 = 2 \cdot 2$, $8 = 2 \cdot 4$, $16 = 2 \cdot 8$. The next term is $2 \cdot 16 = 32 = 64$ and so on. In particular, observe that if we divide any term by the previous one we obtain 2, that is, $8/4 = 2$, $4/2 = 2$, $16/8 = 2$, etc. It is for this reason that the name "common ratio" is used to describe the quantity of $r = 2$.

Notice the distinction between an arithmetic progression and a geometric progression. In the arithmetic progression, one *adds* a fixed number to get the next term, and in a geometric progression, one *multiplies* by a fixed number to get the next term. Which, therefore, of the following sequences are arithmetic progressions and which are geometric progressions? (a) $1, 3, 5, 7, \ldots$; (b) $1, 3, 9, 27, \ldots$; (c) $1, -1/2, 1/4, -1/8, \ldots$. The first is an arithmetic progression with common difference 2; $3 = 1 + 2$, $5 = 3 + 2$, $7 = 5 + 2$. The others are geometric progressions with $r = 3$ in (b) and $r = -1/2$ in (c). In these cases each term is obtained from the preceding one by *multiplication* by r.

In general, if a is the first entry of a geometric progression and r is common ratio, the terms of the geometric progression are $a, ar, ar^2, ar^3, \ldots$. Since the first is a, the second is ar, the third is ar^2, it is clear that the *nth* term is ar^{n-1}. Suppose we wish to add the first n terms of a geometric progression. Let

$$S = a + ar + ar^2 + \cdots + ar^{n-2} + ar^{n-1} \tag{11.2}$$

Multiply S by r and we obtain

$$rS = ar + ar^2 + ar^3 + \cdots + ar^{n-1} + ar^n \tag{11.3}$$

Subtract (11.3) from (11.2) and notice all the common terms in the sums like ar, ar^2, etc., cancel. Hence, we get

$$S - rS = a - ar^n, \quad \text{that is,} \quad (1 - r)S = a(1 - r^n)$$

This gives us the following formula for the sum, when $r \neq 1$,

$$S = a\,\frac{1 - r^n}{1 - r} = a\,\frac{r^n - 1}{r - 1} \tag{11.4}$$

which is obtained by division of both sides of $(1 - r)S = a(1 - r^n)$ by the multiplier of S. We have written (11.4) in two forms. The second, notice, is equal to the first since the sign change in the numerator is offset by the sign change in the denominator. The case when $r = 1$ requires special consideration since (11.2) and (11.3) have the same value. However, $S = a + a + \cdots + a$, where the summation includes n values a, when $r = 1$. Thus, $S = na$ in this case.

To illustrate what was just accomplished in general, consider the six terms of the progression $1, 1/2, 1/2^2, \ldots, 1/2^5$. We have the common ratio $r = 1/2$ and the sum is

$$S = 1 + 1/2 + 1/2^2 + 1/2^3 + 1/2^4 + 1/2^5$$

Now multiply by $r = 1/2$, that is, multiply each term by r to get

$$(1/2)S = 1/2 + 1/2^2 + 1/2^3 + 1/2^4 + 1/2^5 + 1/2^6$$

Subtraction of these last two results yields

$$S - (1/2)\,S = 1 - (1/2)^6 = 1 - 1/64 = 63/64$$

since all the intermediate terms cancel when subtracting. Thus, $S/2 = 63/64$ or $S = 63/32$ when we multiply through by 2. We have found this sum by a method which is exactly the same as the derivation for the general case. We could, however, use formula (11.4) directly. Indeed $a = 1$, $r = 1/2$ and $n = 6$. Hence,

$$S = (1 - (1/2)^6)/(1 - 1/2) = (1 - 1/64)/(1/2) = 63/32$$

EXAMPLE 1. Suppose you invest P dollars at 6% annual interest, compounded annually. How much would you have after 3 years?

Solution: Each year you add the principal to the interest to calculate the new investment. Hence, after the first year you have $P_1 = P + P(.06) = P(1 + .06) = P(1.06)$ as the new amount invested. After the second year, we get a new amount $P_2 = P_1 + P_1(.06) = P_1(1 + .06) = P_1(1.06)$. But $P_1 = P(1.06)$. Hence $P_2 = P_1(1.06) = [P(1.06)](1.06) = P(1.06)^2$. After the third year, you have $P_3 = P_2 + .06\ P_2 = P_2(1.06)$ or in view of the formula for P_2, you have $P_3 = P(1.06)^3$. The thing to notice is that the amounts form a geometric progression; we have $P(1.06)$, $P(1.06)^2$, $P(1.06)^3$ and the common ratio is 1.06. In four years the value of your investment would obviously be $P(1.06)^4$, in five years $P(1.06)^5$, etc.

EXAMPLE 2. Suppose that you invest P dollars *each year* into an account that pays 6% annually compounded interest. How much would you have in the account after 3 years?

Solution: The first P dollars remains in the account 3 full years so, by Example 1 above, it has value $P(1.06)^3$. The second P dollars is invested for only two years. Hence, its value is $P(1.06)^2$. Finally the last P dollars is in the account only for the last year so its value is $P(1.06)$. The total of these is $S = P(1.06) + P(1.06)^2 + P(1.06)^3$. This is the sum of a geometric progression so formula (11.4) applies with $r = 1.06$ and $a = P(1.06)$. The answer is $P(3.37)$ approximately. Of course, we could, in this problem, get the answer without formula (11.4). But, should you change from 3 to 20 years, using formula (11.4) would be much easier.

EXAMPLE 3. Suppose you are given a choice of jobs: Position 1. $100,000 a year but no raises. Position 2. 1¢ the first year, 2¢ the second year, 4¢ the third year, 8¢ the next year, 16¢ the next, etc. In 20 years which job would have paid you the most money?

Solution: Position 1 pays (100,000)(20) = \$2,000,000 in 20 years. Position 2 pays $S = 1 + 2 + 4 + 8 + \cdots 2^{19}$ cents in 20 years. This is a geometric progression with common ratio $r = 2$. According to (11.4), the sum is $S = (2^{20} - 1)/(2 - 1) = 2^{20} - 1$ since $a = 1, n = 20$, and $r = 2$. In dollars, this is $S =$ \$10,485.75. The first position is better.

In 30 years, however, the first position pays $S = 30(100{,}000) = 3{,}000{,}000$ whereas the second position pays $S = 1 + 2 + 4 + \ldots + 2^{29}$. According to (11.4) with $n = 30$, $r = 2$, $a = 1$ we have the last sum $S = (2^{30} - 1)/(2 - 1) = 2^{30} - 1 =$ 1073741823 cents or \$10,737,418.23. The second position is by far the better in a 30 year period.

EXAMPLE 4. Suppose that on an island there are initially two people, a male and a female. Suppose that the population of the island doubles every 20 years. In how many years will the population reach 1,000,000?

Solution: In 20 years the population is $2^2 = 4$, in 40 years it is $2^3 = 8$. Generally, in $(20)n$ years the population is 2^{n+1}. We want to find the value of n so that $2^{n+1} \approx 1{,}000{,}000$. Try values for n until you obtain the first choice which exceeds a million. For example, $2^{10} = 1024$ is too small, $2^{19} = 524{,}288$, again too small. But $2^{20} = 1{,}048{,}576$ which does exceed a million. Hence $n + 1 = 20$ or $n = 19$. This is 19 periods of 20 years each so it would take $(19)(20) = 380$ years to reach a million.

We next turn to infinite geometric progressions. Suppose that $0 < r < 1$ in a geometric progression. Then, clearly, $r^2 < r$ and $r^3 < r^2$ since higher powers of $r < 1$ produce smaller numbers. In fact the r^n gets closer to zero as n increases. For example if $r = 1/2$, then $r^6 = 1/64$ which is considerably closer to zero than $1/2$. We denote this phenomenon by $r^n \to 0$ as n increases.

Now consider formula (11.4). If $0 < r < 1$, then $r^n \to 0$ as n increases. Hence

$$S_n = a\,\frac{1 - r^n}{1 - r} \to \frac{a}{1 - r} \tag{11.5}$$

that is, the sum of the first n terms S_n gets closer to $a/(1 - r)$ as n increases since r^n gets closer to zero. In mathematics we assign this value $a/(1 - r)$ to the infinite sum

$$a + ar + ar^2 + \cdots + ar^{n-1} + \cdots \tag{11.6}$$

since the partial sums, $S_n = a + ar + \cdots + ar^{n-1}$, by formula (11.4) are getting closer to $a/(1 - r)$. The latter quantity is called the sum of the infinite series (11.6).

When $r > 1$ the sums (11.4) get larger and larger as n increases since powers of a number bigger than 1 grow as the power increases. In this case, the partial sums S_n do not get closer to any number; rather they grow to infinity. The series (11.6) is said to be divergent in this case.

EXAMPLE 1. Find the sum of the infinite series $1 + 1/2 + 1/2^2 + \cdots$.

Solution: $a = 1$, $r = 1/2$ and therefore (11.5) applies. We obtain from (11.5) that $S = a/(1 - r) = 1/(1 - 1/2) = 2$, which is the sum of the series.

EXAMPLE 2. Find the sum of $1 - 1/2 + 1/4 - 1/8 + \cdots$.

Solution: Here $a = 1$ and $r = -1/2$. We can use (11.5) to sum any series when $ABS(r) < 1$ as you can prove by an argument like the one used for $0 < r < 1$. Thus, we have $S = a/(1 - r) = 1/(1 + 1/2) = 2/3$ since $r = -1/2$ so $1 - r = 3/2$.

EXAMPLE 3. Sum the infinite series $.1 + .01 + .001 + \cdots = .111 \cdots$.

Solution: Here $a = .1$ and $r = .1$. Thus $S = a/(1 - r) = .1/(1 - .1) = .1/.9 = 1/9$. Notice that this series is usually written as a repeating decimal $.111 \cdots$.

EXAMPLE 4. Sum the infinite series $.3 + .03 + .003 + \cdots = .3333 \cdots$.

Solution: We have $a = .3$ and $r = .1$ so $S = .3/(1 - .1) = .3/.9 = 1/3$.

EXAMPLE 5. Sum the infinite series $.9 + .09 + .009 + \cdots = .9999 \cdots$.

Solution: We have $a = .9$ and $r = .1$. Hence, $S = .9/(1 - .1) = .9/.9 = 1$. At first it may seem strange that $1 = .9999 \cdots$ but it is less of a surprise if you multiply the $1/3 = .3333 \cdots$ by 3.

EXAMPLE 6. Suppose you drop a ball which bounces back $\frac{1}{2}$ the height it falls. If the initial height the ball falls is 2 ft, how far does it travel if you ignore friction?

Solution: The ball falls 2 ft, bounces up 1 ft and then falls 1 ft. On the next bounce it goes up $\frac{1}{2}$ a foot and falls the same distance. This process continues indefinitely. The total distance the ball falls is, therefore,

$$\begin{aligned} S &= 2 + (1 + 1) + (1/2 + 1/2) + (1/4 + 1/4) + \cdots = \\ &2 + 2 + 1 + 1/2 + 1/4 + 1/8 + \cdots = \\ &2 + 2/(1 - 1/2) = 2 + 2/(1/2) = 2 + 4 = 6 \end{aligned}$$

Formula (11.5) was used to sum all the terms, after the first, since they formed an infinite G.P. with $a = 2$ and $r = 1/2$. (In reality, the ball would not bounce forever. We could conclude that a real ball which bounces back approximately half the height it falls would not travel more than 6 ft if it were initially dropped 2 ft.)

The discussion of geometric and arithmetic sequences was primarily based on the fact that relatively simple techniques can be given to determine their sum. There are, however, many other sequences that arise from physical or social problems that require other, perhaps more complicated, methods for summation.

EXAMPLE. The intensity of light from a source decreases as the distance from the source increases. In fact, if I is the intensity of light at a point of distance d from the source, then $I = 1/d^2$ when I and d are measured in suitable units. If there is more than one source, the intensity at a point is the sum of the intensities due to each source.

Suppose that a man on the earth at night observes infinitely many stars, the first at distance 1 star-length, the second at distance 2 star-lengths, the third at distance 3 star-lengths, etc. If all the stars have the same light power, will the accumulated light received by the man blind him?

Solution: The light received from the first star has intensity $I = 1/1^2$ units of light, from the second star he receives $I = 1/2^2$ units of light, $1/3^2$ units are received from the third star, etc. Thus, the accumulated light from all the sources is

$$1/1^2 + 1/2^2 + 1/3^2 + 1/4^2 + \cdots$$

Using more advanced methods than we have here, it can be shown that this series has a finite sum of $\pi^2/6$. Thus the accumulated light is less than that received from two stars of distance one star-length from the man since $(\pi^2/6) < 2$. He will not be blinded.

Should, however, nature have a law that the light intensity at a point is $I = 1/d$ and accumulates additively from numerous sources, then the story would be different. For, in this science fiction example, the intensity of light received by the man is expressed by the series $1/1 + 1/2 + 1/3 + 1/4 + \cdots$. This series can be shown to be divergent, since the partial sums get larger as we add in more terms and never get closer to any number. Hence, our hypothetical man would receive an infinite amount of light from the sequence of stars and it would destroy him. (In fact, there would not be anything anywhere!) Hence, if the law of nature was $I = 1/d$ there could not be an infinite collection of stars if life is to exist.

Exercises

1. Find the sum of 100 terms of each of the following sequences.
   a. 3, 3.25, 3.50, . . . b. 9, 3, 1, . . . c. 5, 3.5, 2, . . .
   d. 5, 15/4, 45/16, . . . e. 2, 6, 18, . . . f. −1/2, +1/3, −2/9, . . .
2. Write each of the following in the form a/b where a and b are integers.
   a. .666 ⋯ b. .606060 ⋯ c. .712712712 ⋯
   d. 3.141514151415 e. $3 + 2 + 4/3 + \cdots$
3. Suppose you drop a ball which bounces $\frac{1}{3}$ the height it falls. If the initial height is 3 ft, how far does it travel if you ignore friction.
4. Suppose you invest $10 per month into an account which pays 6% compounded monthly. What is the value of your investment after 3 years?

5. Let
$$S_n = \frac{1}{1 \cdot 2} + \frac{1}{2 \cdot 3} + \frac{1}{3 \cdot 4} + \cdots + \frac{1}{n(n+1)}$$
What number does S_n approach as n increases?
6. How many ancestors do you have if you go back 30 generations? (Assume no ancestor is involved more than once in your family tree!)
7. Subdivide the interval $1/2 \leqslant x \leqslant 7/8$ into 7 equally sized intervals.
8. A man receives a job at an initial salary of $10,000. He receives a 5% raise each year. How much has he earned after eight years?
9. The man in problem 8 has an option. Either he can receive a 5% raise per year or he can obtain a raise of $1000 per year. Which option provides him with the greater total income for the initial eight years of his employment? What is the better choice for his initial 10 years?
10. Find the sum of each of the following:
   a. $1 - 2 + 3 - 4 + \cdots + 99 - 100$. b. $2 + 4 + 6 + \cdots + 2n$
   c. $1 + 3 + 5 + \cdots + (2n - 1)$

11.5. Sequences on the Computer

Since the computer recognizes functions and since sequences are functions defined on the positive integers, the computer should recognize sequences. For example, the general term of the arithmetical progression 7, 10, 13, . . . is $L(N) = 7 + 3 \cdot (N - 1)$. This function of N can be introduced in a program; thereafter, the computer can calculate $L(1), L(2), L(3)$, etc.

For many applications, however, the functional notation is inadequate to define a sequence. Indeed, many sequences are not obtainable from simple formulas. For example, we would not expect an explicit formula to compute the test score of each student in a class. It is for this reason that a special control statement is found in computer languages. In FORTRAN the statement is DIMENSION followed by the names of the entires in the sequences to be considered. Each named sequence is followed by the number of terms (expected) in this sequence enclosed in parentheses. For example,

```
DIMENSION X(50), J(100)
```

means there is a sequence X(1), X(2), X(3), . . . X(50) of real numbers and a sequence J(1), J(2), . . . , J(100) of integers in the program. *The* DIMENSION *command should appear on the first line*, except for comments, *in the program*. It reserves 50 memory locations for the sequence X(1), X(2), . . . , X(50) and 100 locations for the sequence J(1), J(2), . . . , J(100). You need not fill all locations with numbers but you must be certain that you have reserved *sufficient* space for your sequences. On the other hand, reserving a large number of spaces that you will not use is inefficient and limits the available space for some other users.

By way of illustration, consider the following program:

```
10        DIMENSION IFAC(10)
20        IFAC(1)=1
30        DO 35 J=2, 10
40        IFAC(J)=IFAC(J-1)*J
50     35 CONTINUE
60        END
```

Line 10 reserves 10 locations for the integer sequence IFAC. The DO LOOP computes successively IFAC(2)=2·1, IFAC(3)=3·2·1, IFAC(4)=4·3·2·1, etc. It ends when IFAC(10)=10! has been computed.

In naming sequences, it is wise to use more than one (up to six) letter names. If we call a sequence X(1), X(2), . . . , we cannot use the single letter X for any other variable in the entire program. If, instead, we called the sequence XIT(1), XIT(2), . . . , there is little chance we would want to use XIT for some other variable in the program. Note also the letter X represents a real number so the sequence would consist of real numbers rather than integers.

If a sequence has a large number of entries, it would be painful to be forced to enter the values with one entry per line. There is a way to READ in and WRITE out a sequence on a terminal so that a number of entries can appear on a single line. The simple READ command for a real sequence X(1), X(2), . . . , X(50) is

```
READ(5,*)(X(N),N=1,50)
```

When this command is reached during a RUN of a program on a terminal, a question mark is typed on the left. This is a request to enter values for the X-sequence. With this command, you are permitted to type as many numbers on a line as fit. A blank space, or a comma, can be used to separate one number from the next. When a line is filled and the return button is depressed, another question mark will appear on the left until all the locations X(1), X(2), . . . , X(50) are assigned values. Fifty numbers must be entered to complete the READ statement of this paragraph.

The simple WRITE statement for a sequence on the computer is illustrated by

```
WRITE(6,*)(J(N),N=1,100)
```

When this command is reached during a RUN on a terminal, the integers J(1) through J(100) are printed. As many integers appear on a line as fit and there are blank spaces between successive entries in the sequence. There is, however, one difficulty. The last number printed on a line can be split: for example, if 123456 is one of the entries in the sequence, it could be printed with 1234 on one line and, due to a lack of space, the digits 56 are printed on the next line of output. This minor inconvenience is overcome with the introduction of FORMAT, the subject of the next chapter.

With cards and a WATFOR or WATFIV compiler, the free form READ and WRITE commands have the form READ, (X(N), N=1,50) and PRINT, (X(N), N=1,50), respectively.

Exercises

1. Let $FX(1) = 1$, $FX(2) = 1$ and let $FX(N) = FX(N-1) + FX(N-2)$ for $N = 1, 2, 3, \ldots$. Write down the value of $FX(8)$.
2. Write a program for obtaining the first thirty entries in the sequence of Exercise 1.
3. Let $XIT(1) = \sqrt{2}$ and $XIT(N) = \sqrt{2+XIT(N-1)}$. Thus, $XIT(2) = \sqrt{2+\sqrt{2}}$, $XIT(3) = \sqrt{2+\sqrt{2+\sqrt{2}}}$. Build a program for obtaining a printout of the first forty entries in this sequence.
4. RUN the program of Exercise 3 on a computer. What does $XIT(N)$ approach as N increases?
5. Construct a program that computes the average $\overline{X}$ and the standard deviation S of a collection of twenty scores. (If SC(1), SC(2), . . . , SC(20) are the twenty scores, then $\overline{X}$=(SC(1) + SC(2) + . . . + SC(20))/20 while S^2 = (SC(1)**2 + SC(2)**2 + . . . + SC(20)**2)/20–$\overline{X}$**2.)

Chapter 12 How To Instruct a Computer in Communications

12.1. FORMAT

The simple WRITE (6,*) enables you to obtain answers to numerical problems you have programmed. You may wish, however, to space your answers in a special way like the columns in a table and you may desire to have the computer print instructions or column headings that are English words. To accomplish these objectives, another form of the WRITE statement is used. The word "FORMAT" is used to indicate a special arrangement of the input or output in a program. We begin the discussion with a simple way to have messages in the program.

Suppose you wish to inform the person using a program that he is to enter two integers. We then might begin the program with the following two statements:

```
10          WRITE  (6,15)
20    15    FORMAT (' TYPE TWO INTEGERS ')
```

The computer types the message TYPE TWO INTEGERS when it hits this instruction. It, of course, should be followed by a READ(5,*) I,J so the integers entered at this point can be used in the program. The first entered integer is then assigned to location I whereas the second integer typed by the user is assigned to location J in the computer's memory. The command is *not* changed when cards are used in place of teleprocessing. Notice the change in line 10 over the simple WRITE (6,*) statement. The number 15 (any positive integer less than 100,000 will do) refers to the statement numbered 15 in the program which, in this case, is the FORMAT in line 20.

An important observation is that there is exactly one space in line 20 between the apostrophe and the first letter 'T' of the message TYPE.... If you do not leave this space and type line 20 as 15 FORMAT('TYPE TWO INTEGERS'), the computer would print: YPE TWO INTEGERS, cutting off the first letter in the word TYPE. In each case, the printout starts at the extreme left margin set on your terminal.

The reason for the cutoff of the first character of the printed FORMAT statement is that this space is reserved for special printer controls. For example, you might want to insert some blank lines before printing or you may wish to have a blank page. We will not go into the special symbols used for this purpose since they really are not important for most beginning usage. The thing to remember is that the *first space is stolen by the computer in all* FORMAT WRITE *statements*.

Suppose you enter

```
20    15 FORMAT ('      TYPE')
               (six spaces)
```

in conjunction with line 10. In this case, the word TYPE is printed five spaces from the left margin. (The margin is set on the terminal just as it is on a typewriter.) Thus, you would have

_ _ _ _ _ TYPE

printed where _ is not written but is used here to indicate blank (empty) spaces from the left margin. The computer has stolen the first of the six blank spaces between the apostrophe and the first word in your message for its special printing control. All other spaces appear in the printout.

The next type of FORMAT we consider is the so-called I-field. It, as the use of the letter I indicates, is a FORMAT for integers and counts the number of spaces reserved for printing these integers, with the first space on the left used for control rather than printout. For example, consider the partial program

```
30        WRITE (6,33) J
40     33 FORMAT (I6)
```

This program prints whatever integer is stored in location J in the *five* spaces measured from the left, the first of the six reserved spaces on the line is taken for a printer control and is lost to the printout. The printout is "right-justified," which means that the number is written so that the last digit is in the fifth space, the next to last digit in the fourth space, etc. Thus, if J = 251 we get _ _ 251 while if J = 19 we obtain _ _ _ 19. If the number is negative, the sign of the integer uses a space itself. For example, should J = -351 we obtain from line 40 the printout _ -351.

Care must be taken to leave enough space for the number to be printed. For example, if J = 1234567 with lines 30 and 40, the number would overflow the five reserved spaces. This would be indicated by "*****" being printed. On the other hand, if J = -12345 the machine has enough space for the digits and would print them, but it has no space left for the minus sign and this would be lost in the printout. In other words, the machine prints asterisks rather than digits if J has more than 6 digits. Should J, however, have exactly 5 digits and be negative, the minus sign is *not* printed. If J has exactly 6 digits and is positive, something perhaps worse happens! Indeed, the computer prints only the last 5 digits and the first digit of the number (the hundred thousands place) is lost to the output.

You can write out more than a single integer in a row of the printout. For example, consider the following partial program:

```
50       WRITE   (6,29) J,K,L,M,N
60    29 FORMAT  (I5,I4,I4,I4,I3)
```

This reserves four spaces for J (remember one space of the five cited in the I5 is a *line* control). It also reserves the next four spaces of the line in the printout for K, the next four for L, the next four for M, and finally the next three for N. There is no stolen space in printing K,L,M, and N since only the *first* space of the line is used for line controls and these numbers all appear on the same line as J.

To illustrate this, suppose J = 23, K = 121, L = -315, M = 9 and N = -5. The printout would be, by lines 50 and 60,

_ _ 23 _ 121-315 _ _ _ 9 _ -5

Notice each number is right-justified in the spaces allotted. There are 19 spaces used in all; four spaces each for J (who had one stolen for line control from its five), K,L, and M while only three spaces are held for N. Since the numbers are physically too close together in this FORMAT, making the output hard to read, you could improve the printout for this example by using

```
60    29 FORMAT  (I5,I5,I6,I3,I4)
```

which produces

_ _ 23 _ _ 121 _ _ -315 _ _ 9 _ _ -5

The numbers are neatly separated in this case and the total number of spaces used (called the field width) is 22 = 4 + 5 + 6 + 3 + 4. Each integer printed is right-justified in the space allotted for it.

Although for simple problems it is not necessary, FORTRAN does permit a compact way to FORMAT a string of integers, such as the J,K,L,M,N in line 50. We could replace line 60 above by the following:

```
60    29 FORMAT (5I6)
```

This command states that there are five integers (J,K,L,M,N in our case) each of which is given *six* spaces. It prints

_ _ _ 23 _ _ _ 121 _ _ -315 _ _ _ _ _ 9 _ _ _ _ -5

for the above example. As usual we have lost that line control space in the first integer printed so $5 \cdot 6 - 1 = 29$ spaces are used to print the five integers.

You can mix commands and replace line 60 above by

```
29    FORMAT (I6,3I5,I3)
```

Then J is given five spaces ($6 - 1 = 5$), the next three integers K,L,M are allotted five spaces each, and the last integer N is given three spaces. The printout for the example is

_ _ _ 23 _ _ 121 _ -315 _ _ _ _ 9 _ -5

a total field width of $5 + 15 + 3 = 23$ spaces.

The maximum field width for an integer is 10 spaces. The computer without special programming will not work with integers containing more than 10 digits. If you know that an integer J has 10 or fewer digits, you can, however, write

```
70          WRITE   (6,32)J
80    32    FORMAT  (I20)
```

This merely right-justifies J to fit into the first 19 (line control, 20–1) spaces from the left margin. Thus, if J = -1234567, it prints

_ _ _ _ _ _ _ _ _ _ _ -1234567

As usual, the integer is right-justified in the 19 spaces allotted to it.

Exercises

1. a. Write a FORMAT command that would print the integer 7654321 with its first digit 20 spaces from the left margin.
   b. What FORMAT command was used to obtain the following printout?

      - 314 _ _ _ 5129 _ 31 _ _ _ 371 _ _ _ 75 _ -15

   c. What FORMAT command would you use to print 10 integers, the first four of which have 5 digits each and the last six have 2 digits each, if all are positive and you wish to have two blank spaces between consecutive integers?

2. Consider the following program:

```
10          WRITE (6, 51)
20     51   FORMAT ( ' ENTER THREE INTEGERS ')
30          READ  (5,*)  J,K,L
40          ISUM=J+K+L
50          WRITE (6, 52)
60     52   FORMAT (' WHAT IS THEIR SUM? ')
70     99   READ   (5,*) KSUM
80          IF      (ISUM=KSUM) GO TO 90
90          WRITE   (6, 53)
100    53   FORMAT       ( '              WRONG! TRY AGAIN. ')
110         GO TO 99
120    90   WRITE   (6, 54)
130    54   FORMAT ( ' GOOD!  NO MORE QUESTIONS. ')
140         END
```

a. What is printed when you RUN this program?
b. Suppose you enter 10,20,30. What happens next?
c. Suppose your sum for 10,20, and 30 is 40. What happens?
d. Suppose your sum is 60 in answer to the question about their sum. What happens?
e. Suppose you enter only one integer instead of three in answer to the first question in the program. What does the computer do?

3. Consider the following program.

```
10          READ(5,*)
20          ISUM=I+ J
30          WRITE (6,21)   I, J, ISUM
40     21   FORMAT('THE SUM OF', I6, 'AND', I6, 'IS', I7)
50          END
```

a. What is printed and how is it spaced when you enter 17 and 231?
b. Enter -1736 and 123. What is printed and how is it spaced?
c. Enter 50000 and 60000. What happens?
d. Enter 6131712 and -6131710. What happens?

4. Write a program that asks a question such as "Who did they bury in the Washington Monument?" on one line and then provides numbered multiple choices for the answer on the next line. The program should READ an answer from the numbered choices and comment about the correctness of the answer selected.

5. Write a program that asks the user to add two integers. Have the program replace the original problem by a harder problem if a correct answer is given and a simpler problem if the answer given is wrong.

12.2. FORMAT for Reals

One type of FORMAT, used for real numbers, is the F-field. An example is the following partial program:

```
10        WRITE   (6, 21) X
20     21 FORMAT  (F10.5)
```

This prints the real number stored in location X with five spaces after the decimal and right-justified. Ten spaces have been reserved for X, in total, but as before, the first space is lost to the line control. Thus, if X = -7.49286, the computer prints _ -7.49286. One space is lost to the line control so there are nine spaces left. The minus, the decimal point, and each digit use a space and, hence, eight spaces in all are used by the number. Should X = 3.1415925 it would print _ _ 3.14159 and should X = .0015999 it would print _ _ 0.00160, rounding the number appropriately and placing a "0" before the decimal point.

In an F-field an F20.3 reserves 20 spaces for the total length of the number and three spaces for the digits after the decimal point. An F5.6 does not make sense since only five spaces have been saved for the real number and yet you are requesting 6 digits after the decimal. The F-FORMAT, in fact, requires that the first integer n in F$n.m$ be at least three greater than the second integer m. This is due to the fact that we must have a space for the sign (minus), a space for the decimal point itself, and a space for the first digit, perhaps 0, before the decimal point. For example, in an F-field of F5.2 the real number X = -.123 becomes -0.12. However, if the number X is the first number of a line to be printed we should use F6.2 to give one space over to line control or else the minus sign would be lost.

The special controls of the previous section also apply here. For example, consider the following partial program:

```
30        WRITE   (6,22) X,Y,Z,W,U,V
40     22 FORMAT  (F10.3,F7.3,3F7.4,F9.5)
```

When X = -2.3512, Y = 371.1929, Z = .1, W = -.002, U = .00003, V = -8.235168, we have printed

_ _ _ -2.351371.193 _ 0.1000-0.0020 _ 0.0000 _ -8.23517

There are 9 + 7 + 3·7 + 9 = 46 spaces reserved for the numbers (count them!). The crowding of the numbers could be avoided by using a

FORMAT (F7.3, F9.3, 3F9.4, F10.5).

This yields

-2.351 _ _ 371.193 _ _ _ 0.1000 _ _ -0.0020 _ _ _ 0.0000 _ _ -8.23517 since the computer right-justifies each number so the decimal point will be placed at a position that leaves the commanded number of spaces following it.

Exercises

1. a. Let $A = -137.5291$. Write A in an F10.2 FORMAT and in an F8.2 FORMAT
   b. What is the above A in a F10.4 FORMAT?
   c. Write .00025 in an F10.3 and an F10.5 FORMAT. Do the same for -1234.0.
2. Consider the following program:

```
10         READ (5,*) X
20         Y=X**2
30         Z=X**3
40         W=X**4
50         WRITE (6,25) X,Y,Z,W
60      25 FORMAT (F11.4, 2F10.4, F11.5)
70         END
```

   For each of the following choices of X compute the values of Y, Z, and W and indicate the way in which the answer is printed. a. X = .2 b. X = 1. c. X = 1.1 d. X = 2. e. X = 3.001
3. Let I = 10, X = 3.9512, Y = −75.329. Then WRITE (6.22) X,I,Y followed by 22 FORMAT (F8.4,I3,F8.3) which spaces these numbers in what manner?
4. Let the FORTRAN statement WRITE (6,97) X be followed by 97 FORMAT (F8.4). What is printed in the first eight spaces from the left margin if (a) X = 35.1234 (b) X = 2.35912 (c) X = .00001 (d) X = 3459.5 (e) X = 192.5678 (f) X = .95919?

12.3. E for Exponent

With many problems, you are not sure in advance if the real number answer is large or small. Should you use an F10.5 FORMAT and the real number to be printed is .0000012345, you obtain a 0.00000 for your answer, losing all the significant digits in the number. The E-FORMAT solves this type of difficulty.

In order to discuss this FORMAT it is helpful to first observe how a computer operates with real numbers. Computations on computers are usually carried out in the base 2 or the base 16 (hexadecimal) because of the dependence on two-stage devices or sets of such devices in the computer's memory core. We will, however, assume that the computation is performed in the base 10, the standard base of numbers, since then the discussion is not cluttered with what might be foreign arithmetic of other bases.

Each positive number can be written as a mantissa times a suitable power of 10. The mantissa is a real number between .1 and 1. For example, 25 =

$.25 \times 10^2$, $3.75 = .375 \times 10$, $.0715 = .715 \times 10^{-1}$, $.0000012345 = .12345 \times 10^{-5}$. The negative numbers can be written in the same fashion except, of course, they are preceded by a minus sign.

In each case, the exponent on 10 counts the number of places one must move the decimal '.' to obtain from the original real number a mantissa between .1 and 1. Movement of the decimal to the left yields positive exponents whereas movement to the right produces a negative power of 10.

$$3075.2 = .3075 \times 10^4 \qquad .0003075 = .3075 \times 10^{-3}$$

four places to left   three places to right

The computer has a memory location for each number and, in real arithmetic (floating point), space is reserved for the mantissa and the exponent.

Suppose two real numbers are to be added in the floating point arithmetic of the computer. First of all, unless otherwise instructed, the computer retains only 7 digits in the mantissa of the number. When adding two such numbers it puts the numbers into a form in which both have the same exponent on 10 and then adds the mantissas. For example,

$$.3419127 \times 10^2 + .3511139 \times 10 = .3519127 \times 10^2 + .03511139 \times 10^2$$

The actual answer is $.38702409 \times 10^2$ but the computer rounds the answer to 7 digits. In fact, the computer keeps *one* extra decimal place, to make eight, in the mantissa of the smaller number before the addition is performed and then rounds the final answer to 7 digits. For example,

$$.3333333 \times 10^5 + .8888888 \times 10^2 = .3333333 \times 10^5 + .00088888 \times 10^5 = (.33422218 \times 10^5 =)\ .3342222 \times 10^5$$

In the conversion of the smaller number $.8888888 \times 10^2$ to a power of 10 an eight place mantissa .00088888 was used; this was added to the other mantissa and then the answer was rounded to seven places.

The effect of rounding is indeed dramatic. It is proved in mathematics that the probable error using rounding is the square root of the probable error without rounding. When a large number of additions occur in a problem, rounding helps slow down the accumulation of errors that result in a loss of significant digits.

Subtraction is performed like addition on the computer. Multiplication and division, however, are carried out by multiplying or dividing the mantissas and adding or subtracting the exponents. When the computer multiplies or divides two mantissas only 7 digits are retained and there is *no rounding*. For example, on the computer

$$(.66 \times 10^3) \times (.111111 \times 10^5) = .7333332 \times 10^8$$

where we obtain $.73333326 \times 10^8$ by hand computation.

The reason for not rounding in multiplication and division is that there are many more additions than these operations. It is not worth the extra computer time, therefore, to round multiplication and division. The addition accuracy does not have a significant effect on the final answer.

With this detour into floating point arithmetic, perhaps it is now apparent that it is worthwhile to have a FORMAT that prints the contents of a memory location without change except perhaps for the base of the numbers. The E-field is such a command. In the E-field, we write E$m \cdot n$ where m is the total number of spaces to be reserved and n is the number of digits after the decimal to be saved. Thus, E14.7 means 14 spaces are given for the printout of the number and 7 digits after the decimal point are printed. Since the machine saves only 7 digits without special instructions, this is the largest number you can use in the second position of the E-FORMAT.

Since a power of 10 requires half-spacing to type, FORTRAN uses a notation that is different from the usual mathematical symbolism. For $.35 \times 10^{20}$ and for $.27 \times 10^{-5}$ we write in FORTRAN .35E20 and .27E-5, respectively. In brief, the number 10 is replaced by E and the exponent is written on the same line as the mantissa. Owing to this notation one must reserve enough space in the E-FORMAT to permit the computer to print the E, the exponent, signs, and the mantissa. This requires that in an E$m \cdot n$ FORMAT the integer m should be at least *seven* greater than the n. Indeed, E requires one space; the mantissa printout requires three more than n, since the decimal, a "0" before the decimal point, and the sign of the number use one space each; and finally three spaces are necessary for the signed exponent itself, which may be any integer between roughly −78 and +75. For example, consider the partial program

```
10        WRITE (6,15) X,Y
20     15 FORMAT (E15.4, E14.6)
```

This, for X = −500.7 and Y = .00512311 gives

```
        1 2 3           4 5 6 7
        ↓ ↓ ↓           ↓ ↓ ↓ ↓
_ _ _ _ - 0 . 5 0 0 7 E _ 0 3 _ _ 0 . 5 1 2 3 1 1 E - 0 2
          ↑                       ↑                   ↑
  0 before decimal      ↑     0 before decimal    left for possible
                left for possible minus           two digit exponent
```

The printout again loses one space to the line control so only 28 total spaces are reserved for the numbers. The numbers are right-justified in the allotted space with the E four spaces from the end of this space; the remaining three spaces are kept for a possible signed 2 digit exponent. The mantissa appears immediately to the left of the E. We have also indicated in the printing of X why in E$m \cdot n$ the m must be at least seven greater than n.

Exercises

1. Write each of the following numbers in the E notation.

   a. 3.76×10^{15} b. -3.16×10^{23} c. 3.16×10^{-23}
   d. .00031716 e. 5,000,000 f. 157.321×10^{-5}

2. Write each of the following numbers as a decimal.

   a. 7.31E-01 b. .831E02 c. .0049E-02 d. -3.1271E-05
   e. .5E6 f. .001E-3

3. Consider the following partial program:

```
100        WRITE(6,11) X,Y,Z,U
110     11 FORMAT(E10.2,3E11.3)
```

   What is printed and how is it written when X = 3.71, Y = 2.999, Z = .01214, W = .0001?

4. Why is the letter E used for the E-FORMAT?
   Why is F used in the F-FORMAT? Why I in the I-FORMAT?

5. Consider the following partial program:

```
100        WRITE (6,22) X,Y,I,Z
110     22 FORMAT ( ' THE SUM OF', F5.2, 'AND', F5.2, 'TIMES  ',
                      I3, '=', E14.7)
```

   a. Let X = Y = 1. and I = 2. What and how is the printout?
   b. Let X = 3.198, Y = 4.239 and I = 100. What now is the printout?

12.4. READ FORMAT

READ statements can also be formatted but this is not wise if one is working on a terminal. Indeed, it requires that the numbers be entered in specified spaces which is difficult to accomplish without errors. With punch cards, however, a READ FORMAT is important.

The FORMAT for READ statements is similar to that for WRITE statements except there is no lost space for line control. For example, consider the following partial program:

```
10        READ(6,12) J,X,Y,Z
20     12 FORMAT (I8,2E14.7,F10.3)
```

This requires that the integer J be entered in the first eight positions, right-justified, of the punch card. X and Y are entered in E-form in the next 28 spaces, 14 for each with the exponents right-justified. Finally, Z is entered in the next ten spaces in usual decimal form. Right-justification need not be made for *real variables* in the F-FORMAT; all that matters is that they occupy the assigned spaces. No spillover of one number into another's allotted space is allowed.

Exercises

1. Consider the following partial program:

```
10          READ(6,19) I,J,K
20      19  FORMAT(I8,2I4)
```

Suppose the entries are typed in each of the following ways on the cards. Discover in each case the errors in entry if they exist.

```
a. 1 2 3 _ _ _ _ _ 1 2 _ _ 1 4 _ _
b. _ _ 1 2 3 _ _ _ _ _ 1 2 _ _ 1 4
c. _ _ _ _ _ 1 2 3 _ _ _ 1 2 _ 1 4
d. _ _ 1 2 3 4 5 6 1 2 3 4 _ _ 1 4
```

2. Are there any entry errors in each of the following cases? The partial program is

```
10          READ (5,39)X,Y,Z
20      39  FORMAT (E10.3,2F5.2)
```

```
a. _ 3 . E 0 2 _ _ _ _ _ 0 . 2 E - 0 5 _ _
b. 3 . 1 E - 0 2 _ _ _ _ 5 . 2 1 _ _ . 5 2
c. . 7 3 5 E - 1 2 _ _ _ 9 3 . 2 2 3 _ 5 .
d. - 0 . 3 1 E 3 _ _ _ 9 1 1 . 2 9 4 . 7 1
```

12.5. Sequence FORMAT

Since sequences involve considerable data, FORTRAN has ways of reading and printing the data in compact ways. Consider the partial program

```
10          DIMENSION X(50)
20          READ(5,17) (X(I),I=1,50)
30      17  FORMAT(5F14.7)
```

The second statement (line 20) is an implied DO LOOP and reads into the problem the 50 real numbers that comprise the sequence X(1), X(2), X(3), . . . , X(50). It is formatted with line 30 so that each real number to be read is in the F-form. The 5 before the F tells us to place five F-numbers on each card if the key punch is utilized. Furthermore, the numbers contain up to 7 digits after the decimal and must be spaced in intervals of 14 spaces each.

WRITE statements for sequences can be formatted in the same manner as the READ statement. Remember, however, the loss of the first space of each line to the control.

Exercises

1. Consider the following partial program.
   a. How would you arrange the data?

```
10        DIMENSION K(100)
20        READ (5,29) (K(I),I=1,100)
30     29 FORMAT (6I8)
```

   b. How many cards are needed for this entry?

2. a. Suppose X(1), X(2), . . . , X(75) is a real sequence. What and how is the sequence printed by the commands:

```
10        WRITE(6,37) (X(J),J=1,75)
20     37 FORMAT(5E14.7)
```

   b. Change line 10 to WRITE(6,37) (X(I), I=3,60). What now happens?

3. Consider the following program:

```
10        DIMENSION X(50)
20        READ(5,36) (X(I),I=1,50)
30     36 FORMAT(7F10.2)
40        Y=0.
50        Z=0.
60        DO 35 I=1,50
70        Y=Y+X(I)
80        Z=Z+X(I)**2
90     35 CONTINUE
100       AVE=Y/50.
110       STD=(Z/50.-AVE**2)**.5
120       WRITE(6,34)AVE,STD
130    34 FORMAT(2F10.2)
140       END
```

   a. What is being computed?
   b. How is the data entered?
   c. What is the spacing of the output?

4. Construct a program to find the greatest common divisor of two numbers. Do the same for three numbers.

5. Construct a program that lists the prime numbers that do not exceed 10,000.

12.6. Amortization Tables

As a final application and illustration, the methods of this chapter are quite suitable for building an amortization table of an installment contract with regular payments of a fixed amount. The table describes the proportion of the periodic payment which is interest due and the remainder that is used to reduce the principal. The monthly bank statement on a mortgage loan can be read from such a table.

Suppose, for illustration, that on April 1, 1984, you obtain a loan of $3600 at 1% per month for twelve months. In order to determine the monthly installment payment, the amount of the loan is divided by the present value of an annuity of $1 for 12 periods at 1% per period. For this case, this present value is

$$A = \left[1 - \frac{1}{(1+r)^n}\right] \Big/ r = \left[1 - \frac{1}{(1.01)^{12}}\right] \Big/ .01 = 11.2551 \qquad (12.1)$$

Hence, the monthly payments are $3600/A$ = $319.86. The first few lines of the amortization table are as follows:

DUE DATE	BEGINNING BALANCE	MONTHLY PAYMENT	INTEREST PAYMENT	PRINCIPAL PAYMENT	ENDING BALANCE
4/1/84	3600.00	---	---	---	---
5/1/84	3600.00	$319.86	36	283.86	3316.14
6/1/84	3316.14	$319.86	$33.16	$286.70	3029.44
7/1/84	3029.44	$319.86	$30.29	289.57	2739.87
..	...	...	...	...	

The monthly payment is fixed except for a minor adjustment in the last month to correct for rounding errors. The interest payment is one percent of the beginning balance each month. This is subtracted from the monthly payment to determine the amount paid toward the retirement of the principal, that is, the principal payment. The ending balance is obtained by subtracting the principal payment from the beginning balance of the month.

A program is easily written to build an amortization table. Except for the printout, which is left as an exercise, one such program is as follows:

```
10          READ(5,*) P,N,R          ← P = principal, N = no. of periods,
                                       R = rate
20          A=(1.-(1.+R)**(-N))/R
30          PMON=P/A                 ← Monthly payment computation
40      90  PINT=P*R                 ← Interest payment computation
50          RED = PMON - PINT        ← Principal payment computation
60          P = P - RED              ← New balance
70          IF(P.LT.PMON) GO TO 100 ← Control to terminate computation
80          GO TO 90
90     100  STOP
100         END
```

The formulas used in construction of an amortization table are also the bases of the *effective rate* of interest that is quoted when loans are made. Consider, for example, a loan of \$3600 for a one year period. The rate is stated as 13.5% so the simple interest each year is \$486. Now 3600 + 486 = 4086 and, hence, the monthly payments are quoted as 4086/12 = \$340.50. The *effective rate* for this problem is the *rate* that the borrower would have to pay if the loan were amortized like the loan at the beginning of this section. In other words, we have the monthly payments \$340.50 for a period of 12 months on a loan of \$3600. What is the rate of interest r that yields a zero balance in an amortization table after 12 payments? The computation of the monthly payment for the table is the key. Indeed, if P is the principal and M is the monthly installment, then $M = P/A$ where A is given by (12.1). By algebra, the $P = AM$ is equivalent to

$$rP = rMA = M\left[1 - \frac{1}{(1+r)^n}\right] = M\frac{(1+r)^n - 1}{(1+r)^n} \tag{12.2}$$

The unknown quantity in this formula is r, the rate per period, since $n = 12$, $M = 340.50$, and $P = 3600$. We can determine r by the method of finding zeros in the last chapter. For, equality holds in (12.2) if

$$F(r) = r(1+r)^n P/M - (1+r)^n + 1 = r(1+r)^{12}\,10.57 - (1+r)^{12} + 1 = 0 \tag{12.3}$$

where we have replaced P/M by its approximate value 10.57. The FINDZERO method indicates that the effective monthly rate is approximately $r = .02$ or the *yearly* effective rate is close to 24%. Recall, using the simple interest formula, the quoted rate was only 13.5%.

Exercises

1. Reprogram the amortization problem in this section so that the output is in columns with the following headings:

BEGINNING	INTEREST	PRINCIPAL
BALANCE	PAYMENT	PAYMENT

2. Introduce the additional columns:

NUMBER	MONTHLY
OF PERIOD	PAYMENT

   The last payment could be slightly different from the other monthly payments. Does your program take this fact into consideration in the final monthly payment printout?

3. How would you program a "DUE DATE" column if the loan can begin on the first of any month in 1976? Can the year also be READ into the program?

4. Find the monthly payments on a loan of \$3000 quoted at 6% (simple interest) for a two year period. What is the effective rate of this loan? (Use formula (12.3) and a computer for the last calculation. You must find a zero of the function (12.3) when $P = 3000$, $n = 24$, $M = 140$.)

5. Program the computer to build a table for the amount of an annuity and the present value of an annuity of one dollar per period for n periods, where $n = 1,2,\ldots,25$. The columns should be the percents: .5, .75, 1.0, 1.25, 1.5, 1.75, 2.0 and the rows should correspond to the periods.

Appendix X

If there is no free form READ or WRITE command in FORTRAN available at your computing center, there is an alternative. Whenever the text mentions a READ (5,*) or a WRITE (6,*) substitute READ (5,1) or WRITE (5,1), respectively. The number 1 that replaces the * in the code should not be used in numbering any FORTRAN statement except the following:

```
1    FORMAT(5G14.6)
```

Furthermore the above statement must be included before the END statement in every program or subprogram.

If this proposal is followed, the user will have no difficulty with the WRITE statements. However, there is a problem with the READ command. If an integer is read, the G-FORMAT behaves like the I-FORMAT. For the illustration, the G14.6, therefore, would require the entry of the integer right-justified in the field of 14 spaces. The .6 part of the format is ignored when considering integers. If a real number is read, the G-FORMAT behaves like the F-FORMAT for numbers in the proper range. A real number in the F-FORMAT anywhere in the field will be properly read. If the number is entered in the E-FORMAT, the exponent must be right-justified in the field. When using terminals, the G-FORMAT sometimes requires that each entered number be right-justified. You should check with your computing center to determine any special features that are present in its system.

Appendix Y

CUMULATIVE NORMAL DISTRIBUTION

$P(0 \leqslant t \leqslant x)$

x	.0	.01	.02	.03	.04	.05	.06	.07	.08	.09
0.0	.0	.0040	.0080	.0120	.0160	.0199	.0239	.0279	.0319	.0359
0.1	.0398	.0438	.0478	.0517	.0557	.0596	.0636	.0675	.0714	.0753
0.2	.0793	.0832	.0871	.0910	.0948	.0987	.1026	.1064	.1103	.1141
0.3	.1179	.1217	.1255	.1293	.1331	.1368	.1406	.1443	.1480	.1517
0.4	.1554	.1591	.1628	.1664	.1700	.1736	.1772	.1808	.1844	.1879
0.5	.1915	.1950	.1985	.2019	.2054	.2088	.2123	.2157	.2190	.2224
0.6	.2257	.2291	.2324	.2357	.2389	.2422	.2454	.2486	.2517	.2549
0.7	.2580	.2611	.2642	.2673	.2703	.2734	.2764	.2793	.2823	.2852
0.8	.2881	.2910	.2939	.2967	.2995	.3023	.3051	.3078	.3106	.3133
0.9	.3159	.3186	.3212	.3238	.3264	.3289	.3315	.3340	.3365	.3389
1.0	.3413	.3438	.3461	.3485	.3508	.3531	.3554	.3577	.3599	.3621
1.1	.3643	.3665	.3686	.3708	.3729	.3749	.3770	.3790	.3810	.3830
1.2	.3849	.3869	.3888	.3906	.3925	.3943	.3962	.3980	.3997	.4015
1.3	.4032	.4049	.4066	.4082	.4099	.4115	.4131	.4147	.4162	.4177
1.4	.4192	.4207	.4222	.4236	.4251	.4265	.4279	.4292	.4306	.4319
1.5	.4332	.4345	.4357	.4370	.4382	.4394	.4406	.4418	.4429	.4441
1.6	.4452	.4463	.4474	.4484	.4495	.4505	.4515	.4525	.4535	.4545
1.7	.4554	.4564	.4573	.4582	.4591	.4599	.4608	.4616	.4625	.4633
1.8	.4641	.4648	.4656	.4664	.4671	.4678	.4686	.4693	.4699	.4706
1.9	.4713	.4719	.4726	.4732	.4738	.4744	.4750	.4756	.4761	.4767
2.0	.4772	.4778	.4783	.4788	.4793	.4798	.4803	.4808	.4812	.4817
2.1	.4821	.4826	.4830	.4834	.4838	.4842	.4846	.4850	.4854	.4857
2.2	.4861	.4864	.4868	.4871	.4875	.4878	.4881	.4884	.4887	.4890
2.3	.4893	.4896	.4898	.4901	.4904	.4906	.4909	.4911	.4913	.4916
2.4	.4918	.4920	.4922	.4924	.4927	.4929	.4931	.4932	.4934	.4936
2.5	.4938	.4940	.4941	.4943	.4945	.4946	.4948	.4949	.4951	.4952
2.6	.4953	.4955	.4956	.4957	.4959	.4960	.4961	.4962	.4963	.4964
2.7	.4965	.4966	.4967	.4968	.4969	.4970	.4971	.4972	.4973	.4974
2.8	.4974	.4975	.4976	.4977	.4977	.4978	.4979	.4979	.4980	.4981
2.9	.4981	.4982	.4982	.4983	.4984	.4984	.4985	.4985	.4986	.4986
3.0	.4986	.4987	.4987	.4988	.4988	.4989	.4989	.4989	.4990	.4990

Directions: To determine $P(0 \leqslant t \leqslant x)$, where x is a three digit real number $a.bc$ (a,b,c digits), locate the row corresponding to the number $a.b$ and the column corresponding to 0.0c. The probability is given in the space to the intersection of this row and column. For example, $P(0 \leqslant t \leqslant 0.67) = 0.2486$ since $x = 0.6 + 0.07$.

Hints and Selected Answers

Chapter 1

SECTION 1.5

1. Three members, two proposals:

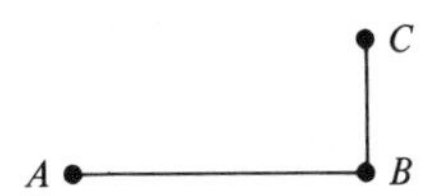

2. Two points and one line passing through, at most, one member. (There are other models.)
3. The axioms allow two members of the organization to favor many different proposals. Hence, there can be only a finite number of members (three at least) while there can be infinitely many proposals (lines, not straight lines, thru pairs of points).
4. Part, breadthless length, lies evenly, points on a line.
5. Seven members; Mr. 1, 2, 3, 4, 5, 6, and 7, say. There are three on each line so the lines are:
   123 145 167 246 257 347 356

SECTION 1.7

2. $8 \times 8 = 64$ squares whereas 3×1 cars cover three squares at a time. Since 3 does not divide 64 evenly, something is left uncovered. The $9 \times 9 = 81$ can be covered. Try it!
3. $9 \times 9 = 81$ so there must be at least one uncovered square by $2 \times 1 = 2$ square unit cars. When the squares are colored like the checkerboard, the corners all have the same color. Hence, removal of the diagonally opposite corners reduced the number of cars that can be parked to 39! When one corner is removed there is room for 40 cars. Try it!
4. If any car is parked diagonally near an edge, there is at least a triangle of area in the lot that is uncovered. Since the lot was fully covered by parallel parking, something is lost and fewer cars can be parked.

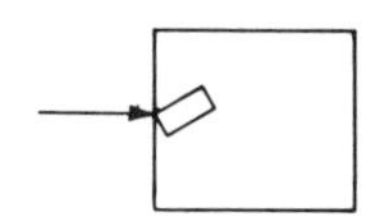

5. Thirty-six cars can be parked initially and 35 when one or opposite corners are removed.
6. It cannot be covered. 7. 24 cars. 8. 30 cars.
9. The word "entered" should be interpreted literally. Then it is impossible to carry out the instruction (b) while (a) is possible. (*Enter* the initial corner via a diagonal move in (a).)
10. It can be done. (Start with a 4×4 square.)

SECTION 1.8

1. At B. 2. At B. 3. Anywhere on the main street gives the same cost factor. 4. Use geometry. 5. The center of the square.
6. This problem is solved by first considering the minimum cost when you treat North–South travel only and then treat East–West directions. It depends on how you arrange the stores.
7. At the intersection of the diagonals.
8. 20 different ways.

SECTION 1.9

1. Try 13.
2. Try forty sticks!
3. Start the game with the removal of one stick. Thereafter, make certain five sticks (total) are removed with each *pair* of turns.
4. Remove one less than the remainder on the first turn and then always have $n + 1$ removed with each *pair* of subsequent turns.

5. Start by removing four sticks.
6. This is the same as a 100 stick game with no more than nine removed by any one player.
7. The troublemaker always picks a number that is one unit closer to five than the initial student. How and why?
8. You must select the numbers by some random process (like picking them from a hat) so no pattern is detected in your selection process.

Chapter 2

SECTION 2.1

3. a. What is meant by "short" or by "natural?" b. What is "cost?" Repeal all taxes and decrease costs! c. A ridicule argument. Is it inhuman not to recognize the type of ambition cited? d. "Many" could be used but it is likely to mislead. A percentage would be better if it were sufficiently accurate. e. Averages can mislead. One person with a very high income (who perhaps inherited a company) could distort the average.
7. The crook who did the counting in the problem.

SECTION 2.2

1. Only e is true if the sentences are considered collectively.
2. Only d. is a paradox. 3. Who knows (a paradox)? 4. Yes.
5. Of course. The way the problem is stated creates what seems to be a paradox. We never travel by "points."
6. This is a paradox. For, if the barber shaves himself the statement that he shaves "only those men who do not shave themselves" is violated. If he does not shave himself, the statement that he shaves "all men who do not shave themselves" is violated. Only women and boys in the village can have beards!
7. It is consistent but, psychologically, it indicates they are not sure the claim of the newspaper is false.
8. Sixty-two read the *Wall Street Journal*, 37 *more* (57 − 16) read *Time*, and 4 read neither. The total is 62 + 37 + 4 = 103% which is too large.

SECTION 2.3

1. a. {c}, b. {a, b, c, x, z}, c. all letters beyond c, d. {x, z}, e. all letters except c, f. all letters.
3. 65 + 30 = 95 were men or women. What were the other five?
4. 10. 5. 125. 6. a. 24 b. none c. 42 d. 20. 7. C is abnormal. 8. A paradox.

SECTION 2.4

1. FTFTTTTT. 3. Some, some, some, all, some, no, all, all. (Agree??)

SECTION 2.5

1. Only b., d., follow from the premises.
2. My doctor does not permit me to eat wedding cake; wedding cake is not suitable for supper, very rich food is not suitable for supper; wedding cake does not agree with me; very rich food does not agree with me; my doctor permits me to eat all types of food that agree with me.
3. You can also conclude that these problems give you a headache.
4. Mr. Dem. 5. Bill.
6. There are ways if you permit a man to be present with both ladies while they exchange the canoe. Otherwise they better return home.

SECTION 2.6

1. a. T b. T c. T d.T

2.

A	B	neither A nor B
T	T	F
T	F	F
F	T	F
F	F	T

3. $(8 \div 4) \div 2 = 1$, $8 \div (4 \div 2) = 4$

5.

A	B	C	(A or B)	(A or B) or C	B or C	A or (B or C)
T	T	T	T	T	T	T
T	F	T	T	T	T	T
T	T	F	T	T	T	T
T	F	F	T	T	F	T
F	T	T	T	T	T	T
F	F	T	F	T	T	T
F	T	F	T	T	T	T
F	F	F	F	F	F	F

6. The logical analog of the first is A or (B and C).

SECTION 2.7

1. Equivalents: a and b; d, f and h; c and g.
2. Some men are not against it. b. Some Playboy Clubs are not decadent. No politician is smart. d. All doctors are idiots. e. Some men are not equal. f. Nixon was not an excellent president. g. Some professor on this campus is human. h. Some administrator on this campus is not deaf. i. Some M.D. does favor medicare.

3.

A	B	$\sim A$	$\sim B$	$(\sim A)$ and $(\sim B)$
T	T	F	F	F
T	F	F	T	F
F	T	T	F	F
F	F	T	T	T

4. a. There will not be a Black vice-presidential candidate *and* there will not be a woman vice-presidential candidate. b. Either Nixon or Agnew is an ideal candidate for some office. c. We do not stop warfare and the world does not come to an abrupt end. d. Greed or government spending do not cause inflation.
6. ($\sim A$ or $\sim B$) and $\sim C$; $\sim A$ or ($\sim B$ and $\sim C$)
9. No; Yes; No

SECTION 2.8

1. a. T, b. F, c. T, d. T, e. T. f. T, g. F, h. F, i. F.
2. a. F, b. F, c. F, d. F, e. F, f. T, g. F, h. T, i. T. (Agree?)
3. d. *Converse:* If this campus is beautiful, then all our buildings are domes. *Contrapositive:* If this campus is not beautiful, then some of our buildings are not domes. h. *Converse:* If everyone is asleep, then life is a dream. *Contrapositive:* If someone is not asleep, then life is not a dream.
4. d. All our buildings are domes and this campus is not beautiful. h. Life is a dream and someone is not asleep.
5. a. All men are equal and some men are not more equal than others. b. You obtain an A and you do not know as much as the teacher. c. All students become students and grades should not be eliminated. d. Grades are the motivation for study and something positive comes from an education. e. You smoke Salems and you do not live or not love in a forest by a stream. f. You try it and don't like it. g. You want to know a person and you do not share an inheritance with him.
6. We wish to prove: "If there is a blue circle on a card, then the opposite side contains a red triangle." Use the contrapositive. Three cards must be turned over.
7. Only e is a tautology. 8. a. B is always true. b. $A \equiv C$.
9. $A' \cup B$. 10. The operations on the sets lead to the universal set I, for example, $A \cup A' = I$. 11. $A' \cup (A \cup B) = (A' \cup A) \cup B = I \cup B = I$.
12. "Is exactly one of the following true? You are a liar. This road leads to the village." Now that you know, do you believe this is an answer?
13. No. Ervin's second statement simply negates the antecedant of the first implication. Let A be the statement "these tapes establish that Dean is a liar" and let B be "the committee ought to have them" Senator Ervin's statements are, in symbols, "If A, then B" and "If $\sim A$, then B." Connected by "and" these statements are equivalent to B, that is, the committee ought have them in each case. It is not a tautology.

SECTION 2.9

3. c. If $x \in I$, then $x \in A$ or $x \in A'$. Hence, $x \in (A \cup B')$ or $x \in A'$. This implies $x \in (A \cup B') \cup A'$. 4. $x \in A \Rightarrow x \in B \Rightarrow x \in C$.

SECTION 2.10

1, 2. Use Venn diagrams to justify the following:

a $(A \cap B \cap C') \cup (C' \cap A') = C' \cap (B \cup A')$.

b $(A \cap B) \cup (A' \cap B) = B$.

c $(A \cap B) \cup (A \cap C) \cup (B \cap C) = (A \cap (B \cup C)) \cup (B \cap C)$.

d $(A \cap B \cap A') \cup (B \cap A) \cup (B' \cap A') = (A \cap B) \cup (B' \cap A')$.

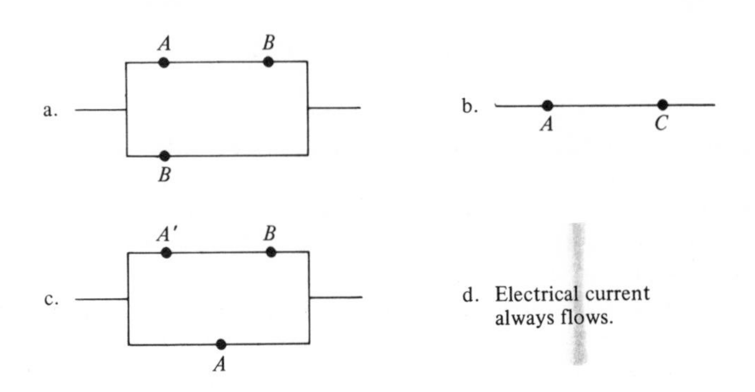

4. No.

Supplementary Problems

1. The butler did it. You can prove this by making a chart of the truth value for each of the three statements made by the individuals. Lady Shout is a good starting point since her third statement and her first statement disclaim guilt. Hence you find this statement must be true. The rest of the chart can then be filled in by the rule that exactly two statements of each character are true.

C.H.	_	_	_
B.	_	_	_
M.	_	_	F
L.S.	T	_	T
L.C.	_	_	_

2. List all cases.

	Mr. A.	Mr. B.	Mr. C. (Blind)
1.	W	W	W
x2.	W	W	R
3.	W	R	W
x4.	W	R	R
5.	R	W	W
6.	R	R	W
x7.	R	W	R

Mr. A would know he had on a white hat in case 4 because he would see the two red hats on the others. Since he did not answer, this case did not occur. Now Mr. B would also know he had on a white hat in case 7 for the same reason. Furthermore, he would know in case 2 that he had on white hat since he would see a red on the blind man. If he had on a red hat, Mr. A would have identified the color of his hat. Hence, case 2 is also out of consideration. The only remaining cases have the blind man in a white hat.

4. Keynes. 6. 64 8/13 mph. 8. There are infinitely many such places but not any place. 9. $1000.
10. 12 mph.

Chapter 3

SECTION 3.2

1. b and c are divisible by 4; b, c, and d are divisible by 6.
2. a. $x = 1, 4,$ or 7; b. $x = 0, 3, 6, 9$. Only $x = 7$ in case a and $x = 6$ in case b produce numbers divisible by nine.
3. Yes; the sum of the digits is unchanged by a rearrangement.
4. To be divisible by 5 the number must end with the digit 0 or 5. The rule for 6 is an even number divisible by 3. For 12 we apply the rule for 4 and the rule for 3.
5. Only a and e are divisible by 11. For large numbers, there is another similar rule. Essentially it is the rule of 11 in the exercise except the alternate "digits" are replaced by alternate blocks of three successive digits. For example, with 12,317,913,806 we add $806 + 317 = 1123$ and $913 + 12 = 925$. Now subtract these sums of alternate blocks of digits to obtain $1123 - 925 = 198$. The number is divisible by 11 if and only if the last number obtained, 198 in this illustration, is divisible by 11. Since $198 \div 11 = 18$, a is divisible by 11. We can also see immediately by the rule in the exercise that $11 | 198$. (The same construction as above also provides divisibility tests for 7 and 13. For example, the number 12,317,913,806 is divisible by 7 iff $7 | 198$. Since $7 \nmid 198$, the original number is not divided by 7. Similarly it is not divisible by 13 since $13 \nmid 198$.)
6. $2a + s = 2(a - s) + 3s$. Since $3 | (a - s)$ by the theorem in this section, we have $3 | (2a + s)$ by Law 1 and Law 2 of divisibility.
7. Take, for example, $a = 3, b = 6, c = 12$. $ab = 18 \nmid 12$ and $(a + b) = 9 \nmid 12$.
8. a. $\{3,9\}$ b. $\{11\}$ c. $\{2,3,5,6,9,10\}$ d. $\{2,3,4,6,11,12\}$ e. $\emptyset$ f. $\{3,5,9\}$.
9. The method works except when the circled digit is itself 9.
10. 37 days if the squirrel has two ears of his own.
12. a. $b|c$ means $c = bx$. $c|a$ means $a = cy$. Thus by substitution $a = cy = (bx)y = b(xy)$ so $b|a$. b. The converse is $b|a \Rightarrow c|a$ where $b|c$ is given. Take, for example, $b = 2, c = 8, 2 = 4$ to show the converse is not always true. c. $b|c$ means c is a multiple of b. The contrapositive of the two implications is "if $b \nmid a$, then $c \nmid a$." In words, if b does not divide a, then no multiple of b divides a.

SECTION 3.3

2. 1, 5, or 9.
4. Let x be a natural number. Then at least one of the numbers $x, x+1, x+2$ is even and one must be a multiple of three. Hence, their product is divisible by $3 \times 2 = 6$.
5. 1,000,000.

SECTION 3.4

1. a, b, c, e, f. 2. The longest string of composites is 90 thru 96. 3. Aside from 2 no prime can be even. A number that has 5 as a last digit is divisible by 5. Hence, only 1, 3, 7, or 9 are possible last digits for a prime. 139 and 149 are consecutive primes.
5. Take one of the three primes to be 2.
6. If p and q are consecutive primes, then their average $(p+q)/2$, which is between p and q, cannot be a prime. Hence, $(p+q)/2 = a$, a composite number, that is, $p + q = 2a$ where a is a composite.

SECTION 3.5

1. All even numbers are composites. 2. In each case only the factoral in the sum is divisible by 5, 6, 7, 8, and 9.
3. a. Divisible by 3. b. Divisible by 2, 3, and many more. c. Divisible by 3. d. Divisible by 7. e. Divisible by 9. f. Even number.
4. 3, 5, 7; 11, 13; 17, 19; 41, 43; 59, 61; 71, 73.
5. Suppose p is a prime larger than 3. Then three divides one of the numbers $p, p+1, p+2$. It cannot be the prime p. If $3|(p+1)$, then 3 divides $(p+1)+3 = p+4$ also. Thus 3 divides $p+2$ or $p+4$.
7. Let $n = ab$ where a and b are natural numbers larger than 1. Since $(n-1)!+1$ is divisible by n, it is divisible by a. But $1 < a < n$ so a divides $(n-1)!$ By Law 3 of Section 3.2, we conclude $a|1$ – absurd.
8. $1001!+2, 1001!+3, \ldots, 1001!+1001$ is a string of 1000 consecutive composite numbers. Take q to be the largest prime smaller than $1001!+2$ and p to be the smallest prime larger than $1001!+1001$.

SECTION 3.6

1. When $n = 17$ the expression has the value $17^2 + 17 + 17 = 17 \times 19$.
2. $n = 0, 1, 2, 3, 4, 6$. 3. $2 \times 3 \times 5 \times 7 \times 11 \times 13 + 1 = 30031$.

SECTION 3.7

1. a. $2^2 \times 7$, b. 2^6, c. $3 \times 5 \times 7$, d. $2^6 \times 5$, e. $2 \times 3 \times 71$, f. $2^3 \times 3^2 \times 5^2$, g. $2^4 \times 3^2 \times 5 \times 7 \times 11$.
2. a. 4; b. 1; c. 11; d. 360; e. 49; f. 72; g. 44.
3. The lcm. consists of the product of the primes that appear in the factoriza-

tion of *either* number and each prime is raised to a power equal to the *largest* power that this prime appears in either factorization.

4. $2^6 \cdot 7 = 448$; b. 140,154; c. 9999; d. 277200; e. 294; f. 432; g. 3872.
5. For example g, we have $3872 \cdot 44 = 170368$ and $352 \cdot 484 = 170368$.

SECTION 3.8

1. a. 32; b. 2; c. 43; d. 1; e. 19; f. 1. 2. 3, 5, 8, 13, 21, and 34.
3. Yes. For example, to obtain 3-qts, decanter the water twice from the 4-qt container into the 9-qt container. Then fill the 9-qt container from the 4-qt one, and 3 qts remain in the 4-qt container.
4. a. 3, 6, 9; b. 1 through 18; c. 1 through 11.
5. The gcd of a and b must be 1.
7. The algorithm for ca and cb is the algorithm for a and b where each equality of the latter has both sides multiplied by c.
8. $l(a, b) = 3$ implies $b = q_1 a + r_1$, $a = q_2 r_1 + r_2$, $r_1 = q_3 r_2$ where $0 < r_2 < r_1 < a$.
   The smallest possible value of r_2 is 1. Since $r_1 > r_2$, the smallest possible value of r_2 is 2 and since $a > r_2$ we must have $a > 2$ or $a \geqslant 3$. Finally, $b = q_1 a + r_1 \geqslant a + r_1 \geqslant 5$.
10. The ratio never exceeds 10 and is never less than 1.9. For example $80 \div 8 = 10$ and $19 \div (1 + 9) = 1.9$.

SECTION 3.9

1. 40 and 30 or 71 and 9.
2. 13 for the smaller size and 4 for the other.
3. 0, 2, 4, or 6 men and respectively 17, 12, 7, or 2 women.
4. $x = 44 - 4t$, $y = 3t$ $(t = 0, 1, 2, \ldots, 11)$.
5. If c/a and c/b, we have $c|(ax + by)$ by the laws of section 2. Hence, c must divide d if there is a solution.
6. a. $x = 5t, y = 5 + 3t$ $(t = 1, 2, 3, \ldots)$. b. No solution. c. (2, 3).
7. 41. 8. Three ways with nickels and dimes only. Twelve ways with pennies also.

SECTION 3.10

1. 100 miles. 2. It is impossible. A miniskirt of length 21 rather than 23 could be made from the cloth.
4. If $A^8 + B^8 = C^8$, then $(A^2)^4 + (B^2)^4 = (C^2)^4$ so $a = A^2$, $b = B^2$, $c = C^2$ would be a solution of Fermat's equation.
6. For $n = 2$ we have $2^a + 2^b = 2^c$ which can only be the case when $a = b$ and $c = a + 1$. In general, the equation is equivalent to $n^x + 1 = n^y$ (dividing by n^b). If $y \neq 0$, the right hand side is divisible by n whereas the left hand side is not when $n \neq 2$. Hence, there is no other n for which $n^a + n^b = n^c$ has a solution for a, b, c in integers.

7. See Table 3.1. 8. $n = 2PQ$. Take $Q = 1$ and $P = n/2$, which is an integer since n is even. Then the solution is n, $(n^2 - 4)/4$, $(n^2 + 4)/4$.
9. b. $64 = a^2 + b^2$ implies $62 = (a^2 - 1) + (b^2 - 1)$. But $8|(a^2 - 1)$ and $8|(b^2 - 1)$ whereas $8 \not| 62$. c. Set $x = c/2$. Then x is even. Thus $c - 2 = 2x - 2 = 2(x - 1)$ where $x - 1$ is odd.
10. p and q have no common factors.

SECTION 3.11

1. a. 10,001,000; b. 110,110; c. 10,000,111; d. 11,111.
2. a. 109 + 27; b. 109 − 55; c. 27 × 5; d. 93 ÷ 3.
3. a. 10,100,100; b. 10,000,111; c. 110110; d. 11,111; e. 111,110.
4. A number in base 2 is divisible by $(100)_2$ iff the last two digits are zeros.
5. Try a rule like the one corresponding to 11 in base 10. (See Exercise 5, Section 3.2.)
6. 20 digits, $2^{20} = 1048576 > 1{,}000{,}000$.

SECTION 3.12

2. Your friend should start. 3. Each person picks up a one stack pile so you lose. 4. a. The winner; b. The winner.
5. Select the person who starts. Leave your opponent with two piles containing the same number of sticks unless one pile has no sticks.

Chapter 4

SECTION 4.1

1.

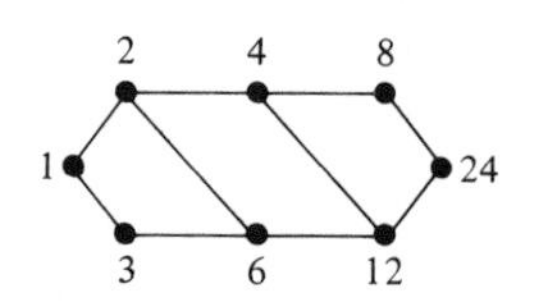

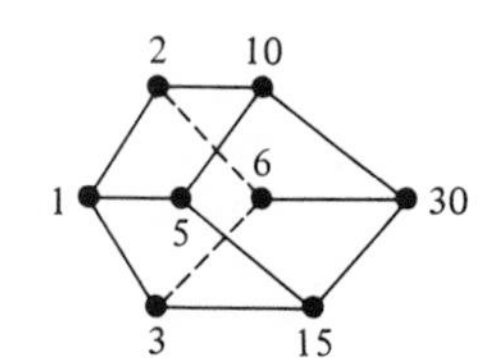

2. The second graph involves no conflict but each is getting only his second preference. Tom and Dick could exchange dinners in the second case for a better solution.

4.

5. The dates are Harry-Steak; Dick-Ham; Tom-Turkey.

SECTION 4.2

1. (a), (b), (c), (e).
2. The problem is equivalent to drawing a graph in which the land is replaced by vertices and the bridges by edges. The graph has four vertices, one of order 5 and three of order 3. A bridge can be placed anywhere that joins two vertices (since all are odd) and it will then be possible to travel all bridges once and only once.

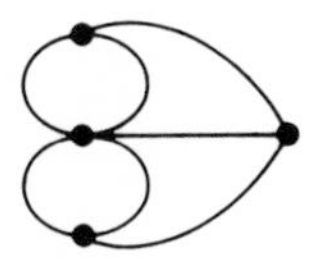

3. Each edge is counted twice since each edge has two vertices.
4. If there is an odd vertex, then its contribution to the sum in problem 3 is an odd number. Since the final sum is even (twice the number of edges), there must be at least one other odd number in this sum. Indeed, each odd number in any sum of integers can be paired with another odd number in the sum whenever the sum is even.
5. There are many such routes, for example, P-T-O-M. 6. (c), (d), (e) only.
7. See (e) in problem 6. 9. If A has at most two odd vertices, then the graph can be drawn as indicated in the text. If there are four odd vertices, you must lift your pencil once before you can complete the graph. Twice for six, three times for eight, . . . , $(n - 1)$ times for $2n$ odd vertices.
10. Try AB, BD, DC, CA; a total of 820 miles.
11. If you start a graph at an even vertex, you must end the graph at the vertex. There are an odd number of uncovered edges emanating from this starting even vertex after one edge has been used to begin the graph.

SECTION 4.3

1. There is a path for (b), (c) only. 3. (c), (d), (e). Illustration:
4.

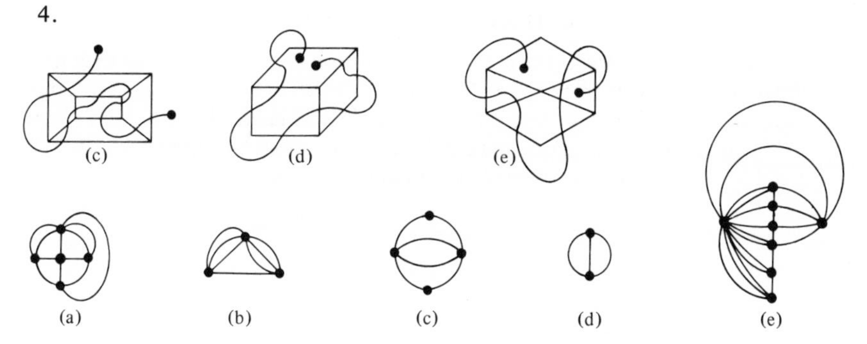

5. (a) Three doors must be traversed more than once. (b) No door need be walked through more than once. (c) One door.

SECTION 4.4

1. a. No effect. b. Yes, it is now 10 days. 2. 10 quarters.
4. Minimum time 23 days; critical path, Tasks: 1-3-4-6-7.

SECTION 4.5

1.

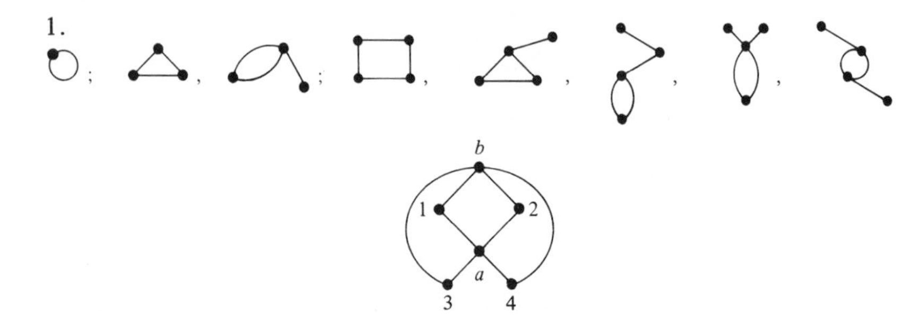

2. Yes. 3. It cannot be done.
4. The conditions (1) + (2) imply $3V = 2E$. Also, if n is the minimum number of edges bounding any face in the figure, then $nF \leqslant 2E$ by (3). Since $V - E + F = 2$, we have $F = 2 + E/3 \geqslant 2 + nF/6$ or $6 - 12/F \geqslant n$. The largest possible value of n is 5 since a positive number $12/F$ is being subtracted from 6 in this inequality.
5. four, five. For n disconnected graphs, the quantity $V - E + F = n + 1$.
6. First observe, at the final stage there is only one vertex of order two; the remaining are of order 3. Counting edges, we obtain $3(V - 1) + 2 = 2E$ or $3V - 2E = 1$ from these observations. This equation has the solutions (in integers) $E = 1 + 3n$, $V = 1 + 2n$ ($n = 0, 1, 2, \ldots$). The construction in the problem starts with $E = 2$, $V = 3$ and always adds one vertex and two edges. Thus after m such applications of the construction there are $E = 2 + 2m$ edges and $V = 3 + m$ vertices. Now $2 + 2m = 1 + 3n$ and $3 + m = 1 + 2n$ only when $m = 4$, $n = 3$. This gives $E = 10$, $V = 7$, $F = 5$.

   This analysis indicates that the starting position is important. For example, if we start with •——•——•——•, that is four vertices and three edges rather than three vertices and two edges, then the problem changes. This new problem requires five steps until it is resolved. The final figure has $V = 9$, $E = 13$, $F = 6$. However, sometimes it cannot be drawn! Example:

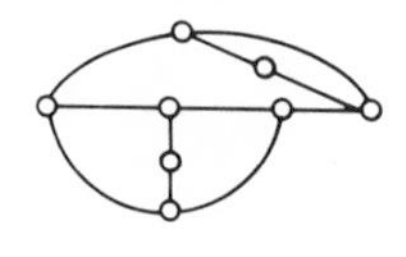

SECTION 4.6

1. (b), (c) are balanced. 2. (b) seems to be the better relationship. 3. Yes, let all the one-way streets be say North to South or East to West.
4. The existence of at least two triangles of the same color is independent of how you (or anyone) colors the edges in the figure.

SECTION 4.7

1. Let I, II, III denote the piles of 1, 2 and 3 sticks respectively. Then (1, I) denotes the removal of one stick from pile I, etc. If A starts, the first part of the tree diagram is as follows.

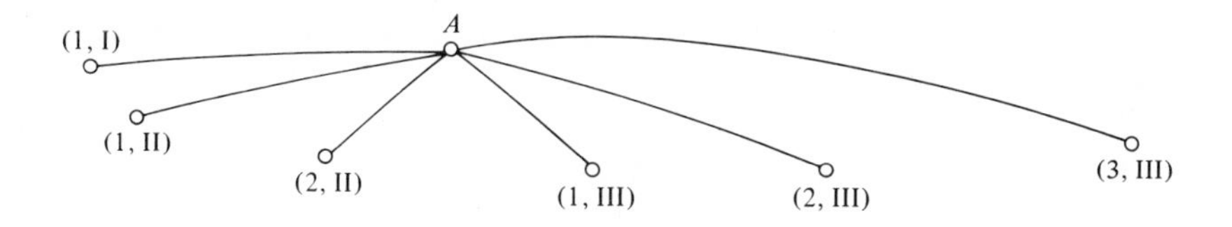

Next, the player B extends each branch according to the available moves. For example, if A selected (2, III), then

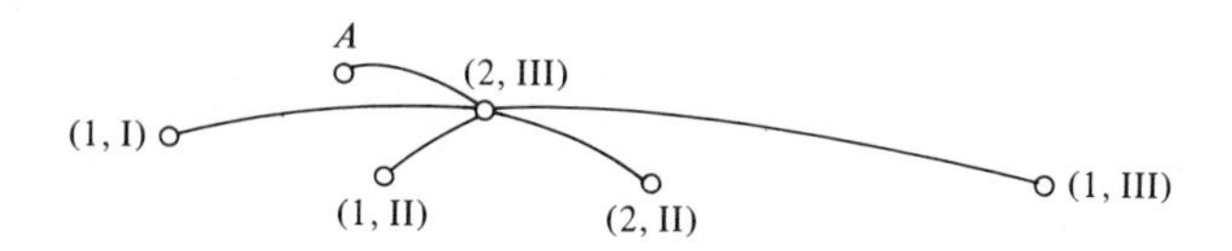

This process is continued until the entire tree is drawn.
3. AD, AB, BC is the economy tree.
4. AD, AB, AC and AD, AB, BC are economy trees.
8. This can be done if there is at most one cycle in the graph.

SECTION 4.8

1. Interchange the V and the F of each solid to form a dual.
2. Only an octahedron. 3. All. 4. See reference [3] of Chapter 4.

Chapter 5

SECTION 5.2

1. $5 \cdot 4 = 20$. 2. $8 \cdot 16 \cdot 10 \cdot 4 = 5120$.
3. $8 \cdot 3 \cdot 2 \cdot 2 = 96$; eight choices of basic pizza, three sizes, with or without extra cheese or extra sauce.

4. There are $2^5 = 32$ different ways to answer the questions. With more than 32 students, there must be at least two identical papers when each answers all questions. Should they not be required to answer all questions, there are three choices for each question, namely, true, false, or no answer. This means there $3^5 = 243$ different ways to turn in the test. We must have 244 or more students to *assure* us that at least two papers are identical.
5. a. $3 \cdot 3 \cdot 3 \cdot 3 = 81$ b. $4 \cdot 4 \cdot 4 = 64$.
6. a. $4 \cdot 3 \cdot 2 = 24$; b. 24; c. $4 \cdot 4 \cdot 4 = 64$ for a and $64 \cdot 4 = 256$ for b.
7. a. All except one, that is, $4 \cdot 3 \cdot 2 \cdot 1 - 1 = 23$ incorrect ways, 1 correct way. b. $3 \cdot 3 = 9$. Count them by enumeration or with the aid of a tree diagram. (There are 8 ways to have exactly one letter correctly sent and 6 ways to have two.)
8. $5 \cdot 4 \cdot 3 \cdot 2 \cdot 1 = 120$ weeks or $120/50 = 2.4$ years (50 work weeks per year with a two week vacation). With seven ties, the answer is $7 \cdot 6 \cdot 5 \cdot 4 \cdot 3/50 = 50.4$ years.
9. $2^{10} = 1024$ if the empty set is counted.
10. $24 \cdot 24 \cdot 10 \cdot 10 \cdot 10 \cdot 10 = 5{,}760{,}000$. (The system is at least good enough for Nevada next year.)

SECTION 5.3

1. a. $5 \cdot 4 \cdot 3 = 60$ b. $2 \cdot 4 \cdot 3 = 24$ c. $3 \cdot 4 \cdot 3 = 36$ d. $2 \cdot 5 \cdot 5 = 50$.
2. $6 + 5 + 4 + 3 = 18$. 3. $4 \cdot 3 \cdot 2 \cdot 1 = 24$. 4. 3 ways. Keys, unlike tables can be turned over. Try it.
5. $3{,}456 = 4 \cdot 3 \cdot 2 \cdot 1 \cdot 6 \cdot 6 \cdot 2 \cdot 2$
6. a. $50 \cdot 49 = 2450$; b. $15 \cdot 49 = 735$; c. $15 \cdot 35 + 35 \cdot 15 = 1050$; d. $35 \cdot 35 + 15 \cdot 14 = 1400$.
7. To draw the ace of spades there are $3 \cdot 51 \cdot 50 \cdot 1 = 7650$ ways (3 orders—first, second, or third, 51 other cards and one ace of spades). To draw three red cards, there are $26 \cdot 25 \cdot 24 = 15600$ ways. The second is more likely; in fact more than twice as likely.
8. It is easier to calculate the number of ways not to pick a club ($39 \cdot 38 = 1482$) and subtract this from the total number of ways to pick two cards ($52 \cdot 51 = 2652$). 1170 is the answer.
9. a. $2 \cdot 4 \cdot 3 \cdot 2 \cdot 1 \cdot 5 \cdot 4 \cdot 3 \cdot 2 \cdot 1 = 5760$ b. $1 \cdot 2 \cdot 3 \cdot 4 \cdot 1 \cdot 2 \cdot 3 \cdot 4 \cdot 5 = 2880$.
10. $2 \cdot (3 \cdot 2 \cdot 1) \cdot (3 \cdot 2 \cdot 1) \cdot (2 \cdot 1) = 144$ (There are two ways to arrange the group; the union people must be either to the *right* or the *left* of management.)

SECTION 5.4

1. a. $C_4^{12} = 495$; b. $P_4^{12} = 11{,}880 = (4!)(495)$.
2. $P_4^{26} = 358{,}800$ 3. $9 \cdot P_6^9 = 544{,}320$ have distinct digits. There are 9,000,000 possible numbers.
4. a. 999,000 b. 2520 c. 210 d. 720 e. 21 f. 35 g. 4950 h. 120 i. 70.

5. a. 2,598,960 b. $52 \cdot 12 \cdot 11 \cdot 10 = 68{,}640$ (The first card selected fixes the suit.)
6. $34{,}650 = 11!/(4!4!2!)$. 7. 34 at last count.
8. a. $C_3^6 = 20$ b. *Three!* Count them directly!
9. $2^8 - 1 = 255$. 10. Each selection of a subset of r objects from the original set of n *leaves* a subset of $n - r$ objects.

Chapter 6

SECTION 6.2

1. a. No place is safe.; b. Events are not equally likely.; c. See b.; d. The chance for a boy still is theoretically even.; e. The coin is as likely to be heads as tails. Hence, it does not matter what is called.; f. These events are not equally likely.
3. $2,12 \rightarrow 1/36$; $3,11 \rightarrow 2/36$; $4,10 \rightarrow 3/36$; $5,9 \rightarrow 4/36$; $6,8 \rightarrow 5/36$; $7 \rightarrow 6/36$.
4. None correct, one way; one correct, 4 ways; two correct, 6 ways; three correct, 4 ways; all correct, one way. For at least two correct, there are $6 + 4 + 1 = 11$ ways.
5. 1/8, 1/8, 7/8, 3/8, 1/2. 6. $525/1225 = 3/7$.
7. 37/40, 27/40, 28/40. 8. 7/52, 8/13, 5/13. 9. 2/11 (Enumerate!)
10. 32/221, 33/221, 80/221, 1/17, 1/221.

SECTION 6.3

1. Play numbers. There are only 25 odd numbers in the list of 51 numbers. 2. 2/7, 1/2, 3/8, 1/3, 1/10. The total is greater than 1 so something is wrong with the cited odds. 3. It is easier. 4. 3 to 25. 5. 79 to 21 or 8 out of 10. 6. I would hate to be his wife!

SECTION 6.4

1. 1/2. 2. (a) No male can have a defective X-chromosome passed on from his mother; a female can have, at most, one defective X-chromosome. (b) The males have a defective X-chromosome from the mother; no female can have two defective X-chromosomes since one comes from the father. (c) All the X-chromosomes are defective. (d) Yes. The female children carry a defective X-chromosome. Should they marry, this X-chromosome has an equal chance of being passed to their children.
4. 3/4. 5. For example, in a the probability is 1/2 that the offspring has brown hair.

SECTION 6.5

1. .26, .86. 2. (a) The probability of an A is 1/2 while that of a B is 3/4. Their total is larger than 1. b. The events are not mutually exclusive c. How can "at least one" be smaller than "one"?

3. a,b,c are mutually exclusive. 4. a. 28/40. b. 14/40. c. 22/40. d. 30/40. 5. 5/12, 5/12. 6. a. .7. b. .35. c. .95. d. 1. e. 0. f. .05.

SECTION 6.6

1. .33 = .21 + .27 − .15. 2. .25, .6. 3. a. .75. b. .73. c. .27. d. .85. e. .48. f. 52. 4. $A - C$ and C are mutually exclusive. Thus, $P(A) = P((A - C) \cup C) = P(A - C) + P(C)$.
5. $P(A \cup (B - A \cap B)) = P(A) + P(B - A \cap B) = P(A) + P(B) - P(A \cap B)$.

SECTION 6.7

1. .87. 2. a. .325. b. .15. c. .175. d. .575. e. .825.
3. a. 1/2. b. 1/3. c. 2/5. d. 1/2. 4. a. .009. b. .54 c. .639. 5. a. .698 b. .844. c. .742. d. .626

SECTION 6.8

1. c, f are independent. 2. a. 1/169. b. 1/221. 3. 125/216, four rolls. 4. 1/2. 5. 64% are red, 32% pink and 4% white.
6. The proportion of genes of each type in the cross-fertilization remains unchanged throughout each generation (80% red and 20% white).

SECTION 6.9

1. 7. 2. 5. 3. 2/9, 12/25. 4. $671/1296 \approx 518$. 5. $(39 \cdot 26 \cdot 13)/(51 \cdot 50 \cdot 49) \approx .1054$.

SECTION 6.10

1. a. .0145. b. .34482. 2. a. 009671. b. .0703.
3. a. .340. b. .7059. c. .1324. 4. a. .975. b. .97436.
5. a. .62. b. 28/31. 6. 55. 7. 1/6, 1/4. 8. Three boxes, with one red in each of the two and the remaining balls in the third box. 9. .548, .4744.

SECTION 6.11

2. 1/4. 3. 5/8. 4. 2/9. 5. $6/36 + 2/36 + (2/36)(1/4) + 2(3/36)(1/3) + 2(4/36)(4/10) \approx .5068$. 6. .4929. 7. .49689.

Chapter 7

SECTION 7.1

1. a. Not all people who read the *Wall Street Journal* have "average" incomes. b. Was the survey correctly conducted? Is the number using sunglasses who

had accidents greater or less than the number using sunglasses who do not have accidents? Was the accident rate higher than on a cloudy day?

c. Was the sample of doctors surveyed biased? What was the size of the sample? (Perhaps only four doctors were contacted.)

d. The lake could be very deep in certain places.

2. a. Median. b. Median or Mode. c. The union could have used the median while the management used the mean. d. Mean. e. Mean.
3. There are two "middle" items. The median is a value half way between these middle items.
4. a. 2. b. 2. c. 5. d. 32. e. 2.

SECTION 7.2

1. \$21 dollars to pay six games. 2. \$2. 3. \$4.11 (rounded).
4. \$.25. 5. \$.25 6. \$70,000. 7. a. −10/37 dollars. b. −16/37.
8. a. \$1.73. b. \$2.56 for two, \$2.19 for three, \$1.22 for four. c. Two is maximal.

SECTION 7.3

1. The first game. 2. a. Prepare 3000. b. Prepare 2000. c. $p = 4/9$.
3. Bare. 4. a. The second hotel. b. The second hotel. c. 5/6.

SECTION 7.4

1. a. (1, I). b. (1, I). c. (3, II). d. (2, II). Each strategy is the best possible. 2. For example, a and c has the following payoff matrix for B respectively:

A \ B		
	−2	10
	−3	−5

A \ B			
	−2	1	−4
	1	−2	−2
	2	−3	−2

3. No difference. 4.

B \ A	I	II
1	−1	0
2	0	1

The minimax strategy is (I, 2)

5.

A	I	II
1	−1	1
2	1	−1

A should vary his choice so that half the time he plays heads.

SECTION 7.5

1. Each should play a strategy half the time.
2. We write $A(p,q)$ for the fact A plays the first strategy with probability p and the second with probability q. a. $A(0, 1), B(1, 0), E = 3$ (saddle point); b. $A(5/11, 6/11), B(4/11, 7/11), E = 13/11$; c. $A(3/4, 1/4), B(1/6, 5/6), E = 5/2$; d. $A(1/2, 1/2), B(1/2, 1/2), E = 0$.
3. b. Use a bag with 5 balls of one color and 6 of another color. c. Toss a pair of coins. d. Toss a coin.
4. $E\text{I} = 4p + 6(1 - p), E\text{II} = -p + 8(1 - p)$. $E\text{I} = E\text{II}$ implies $p = 2/7$. The minimax strategy is best because it maximizes B's gain and minimizes A's loss.
5. $p = 1/2, q = 1/3, r = 1/6, E = 0$ so the game is fair.
6.

5	-8
-6	9

1*st* (17/28, 11/28), 2*nd* (15/28, 13/28), *E*: - 3/28.

7. a. J(3/2, 1/4), b. \$1.25, c. S(1/4, 3/4).
8.

-3	4
1	-3

Gray should defend the supply position and Blue should attack the other position.

SECTION 7.6

3. 18, 5. 4. \$5.70. 5. Country Gents get $\frac{1}{2}$ hour and the crabgrass does not get invited. The comedian performs for $2\frac{1}{2}$ hours. The university pays only \$105 this time and is likely not to dare run such a program in the future. 6. \$15 profit, $x = 1/2, y = 0$. 7. \$139 profit, $x = 1/2, y = 2$. 8. If strictly inside the region, you can change (x, y) to increase $ax + by + c$. The same is true when you are on a line segment of the boundary unless you are at a corner.

Chapter 8

SECTION 8.1

1. 53/70; 62/70. 2. 56/1024. 3. .033 (Bayes formula has been used.) 4. 1/4; 1/6; 2/30. 5. The probability of exactly x heads in 10 tosses of the coin is $p = C_x^{10}/1024$. The chart gives this p for various x.

x	0	1	2	3	4	5	6	7	8	9	10
p	.0009	.0097	.0439	.1171	.2050	.2460	.2050	.1171	.0439	.0097	.0009

6. Reject the hypothesis at the .01 level if the number of heads x is larger than 9 or less than 1. For a .20 level $x > 7$ and $x < 3$ causes rejection. We dis-

tribute the .20 so that there is .1 on each end of the chart in the answer to problem 5. The type I error in the first case is the probability that $x = 10$ or $x = 1$, that is, .0018. In the second case, it is 2(.0439 + .0097 + .0009) = .109. Additional information is needed to check for type II errors. For example, if we knew that $p = .8$, where p is the probability of a head, then some computation of type II error could be made.

SECTION 8.2

1. .7734, .2266, .5468, .9104, .1915, .1915, .4332, .4332. Each represents the area under the standard normal curve in the interval specified.
3. $P(t \leqslant -x) = P(t \geqslant x)$ by symmetry. Thus, $P(t \leqslant x) + P(t \leqslant -x) = P(t \leqslant x) + P(t \geqslant x) = 1$.
4. $x = .85$, $x = 1.96$, $x = .85$.

SECTION 8.3

1. a. .1587. b. .3085. c. .0250. 2. .7745. 3. a. .0228. b. 257 (approximately). c. .4435.
4. a. 1.4142. b. .7071. c. .6324. 5. a. A, 6.68%; B, 24.17%, C, 38.30%. F, D are the same percent as A, B respectively. b. A $\geqslant .975 >$ B $\geqslant .825 >$ C $\geqslant .675 >$ D $\geqslant .525 >$ F.
7. We hope you are not disappointed. If you are, change S to .5.

SECTION 8.4

1. (0, 1/32), (1, 5/32), (2, 10/32), (3, 10/32), (4, 5/32), (5, 1/32).
2. 12/125, 13/125, 61/125, 64/125.
3. a. $C_{40}^{100}/2^{100} \approx .011$. The estimate uses the theorem in this section. We find the area under the standard normal curve $t = (r - 50)/5$ between $-9.5/5 (r = 39.5)$ and $-10.5/5$ $(r = 40.5)$. b. 02. c. nearly zero. Remark: It is best to compute in b the value $t = (r - 50)/5$ for $r \geqslant 59.5$ rather than $r \geqslant 60$ in the approximation. In this example, however, there is an insignificant difference in the answers.
4. At the 5% significance level the coin is biased in each case.
6. Yes. ($t = (r - 100)/(5\sqrt{3}) \approx 2.3$, a 5% significance level was used. At a 1% level, the answer is no.)

SECTION 8.5

1. $t = 10.8$ (Definite trend). 2. $t = 7.7$ (Definitely, beer drinkers are fat.) 3. $t = -5.9$ (Good students find college dull.) 4. $t = .10$ (No difference). 5. $t = -.84$ (Freshman are like other college students!). 6. $t = 1.4$ (No significant difference. The Dean was wrong.)

Chapter 9

SECTION 9.5

2.

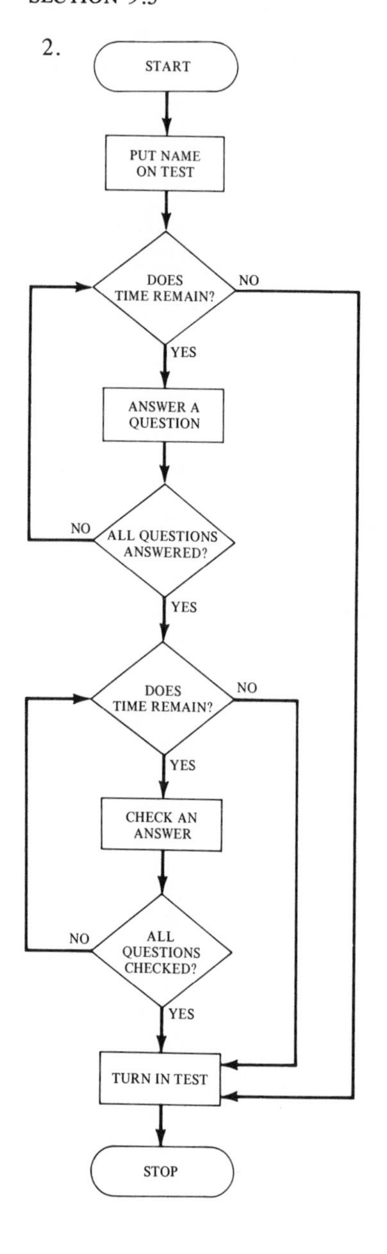

3.

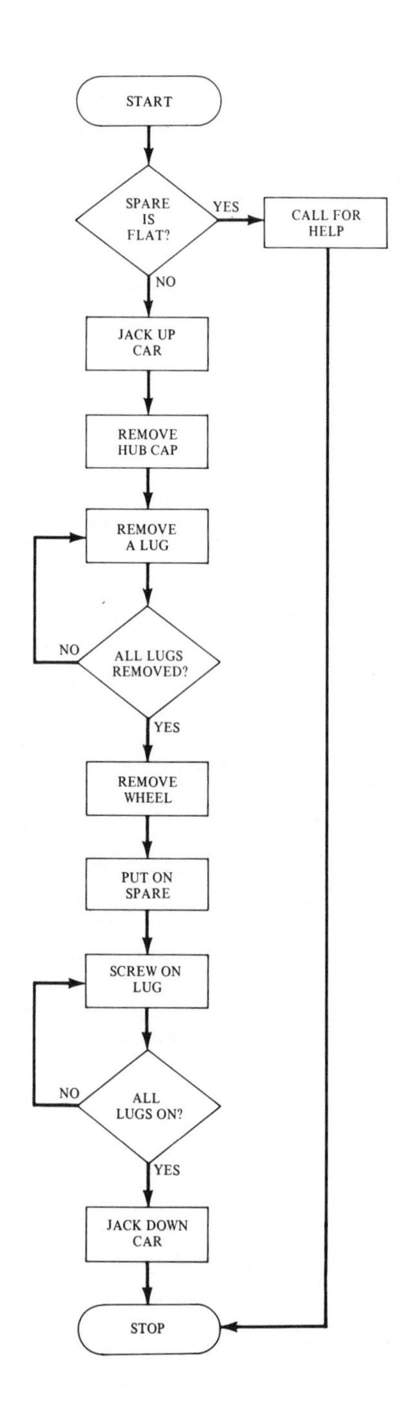

6. Yes

SECTION 9.7

1. a. 64, b. 36, c. 18, d. 81, e. 648, f. 4, g. 26.
2. a. Y=2.*X**2+3.*X-35.;
   b. Y=3.*X+(2.*X**2+1.)/(3.*X-1);
   c. Y=5.*X**X;
   d. Y=1.+X/(1.+(X/(1.+X))
3. 256. 4. a. X=3.2; Y=3.2; b. X=5.1, Y=5.1.
5. a. 2*(-3), b. 4+3**2, c. 3**(-2), d. 4*(3+2-(7+1)), e. 6/(-3), f. 2*(3+2)*2.

SECTION 9.8

1. a., b., c., g., h.
2. a. 4.4 2.2 b. 5.
3. a. 0. b. -2. c. -20.
4. a.
   ```
   10 READ(4,*)X
   20 Y=X**2
   30 WRITE(6,*)Y
   ```
   b.
   ```
   10 READ(5,*)B
   40 END
   ```
   c. line 20 should be before line 10

SECTION 9.9

1. a. 3 b. 0 c. -3 d. 3 e. 9 f. 12 g. 0 h. 0 i. 1 j. 0
2. a. X=3. The computer would first do 7/2 to get 3 in integer arithmetic and then add the decimal and store in the real location X. b. 6 since KISS is an integer variable.
3. NUMBER3 has more than 6 characters.
4. a. 3 b. 3.0 c. 3.5 d. 3
5. a. 8 b. 0 c. 0 d. 5 e. 19 f. 3
6. a. 1 b. 1.6 c. 0 d. 1.6 e. 1 f. 3
7. Line 10 does not use the proper code for READ; it should be READ(5,*) NATURE. Line 20 assigns X to the NATURE location but X has not been defined. It likely should read X=NATURE which would add a decimal to the integer variable NATURE. (X is a real variable.) The next line does not have the required line number. Line 40 has an undefined variable ZNATURE that exceeds the 6 character limit. Line 40 also has this problem and an improper WRITE statement. We should have a WRITE(6,*). In the second program, I is undefined. Furthermore line 20 removes what was in Y and rereplaces it by whatever is in I (which is likely to be 0 on a terminal). Line 40 should read WRITE(6,*)J. The END statement does not carry the required line number. Line 30 is missing an operation between Z and I.

SECTION 9.11

1. a. 1.1 Line 30 is skipped and control is transferred to statement 30 which is line 40 by the GO TO 30 statement.
   b. 2. Line 30 does nothing since X is *not* greater than 10. When we first meet this command. The IF statement is ignored.
   c. 11. We start with $X = 1.$ and then add 1. to obtain $X = 2.$ The IF statement transfers control back to line 20 (statement 35) which adds another

1. to X to get $X = 3$. The process repeats until X becomes larger than 10., which first occurs when $X = 11$.

d. This problem would never print anything. Indeed, adding of 1.'s to an initial X of 1.1 will never produce (exactly) 10. The problem has an infinite loop.

2. a. There is no statement number 20. Admittedly there is a *line* 20, but to transfer control in a program there must be a statement with 20 in front of it. Control is never transferred to the line. The program should always make sense when line numbers are removed!

b. There are two statements numbered 27 in the program. The computer would not know which you desire for the transfer of control. It, thus, would reject the program.

c. This program has an infinite loop. Line 30 transfers control back to line 20 and the computation continues indefinitely.

d. The IF statement is unnecessary. J, an integer, is never going to be 1.21. Also there is no printout so the whole program is a waste of computer time.

3. a. 4.0. b. -1.0. c. 1.0. f. 11.75.

5. a. 35 IF(I.GT.25)GO TO 30; b. 7;2,4,8;3,7,21.

SECTION 9.12

1. This prints the number stored in J for each step of the iteration. Thus, the printout is 1,3,6,10,15,21,28,36,45,55,66,78,91,105,120,136,153,171,190, 210,210, printed in a column. The last sum 210 is printed twice, once in the loop and once by line 50.

2. a. 231. The J has been computed with $I = 21$ before line 60 is executed.

b. 1.283333. The line 30 was added to convert the integer K into a real number. The problem computes the sum 1./2.+1./3.+1./4.+1./5. and prints the answer.

c. The program begins with 2, squares it to get 4, squares 4 to obtain 16, and finally squares 16 to print 256. The variable I here was strictly a counter and does not itself appear in the computation.

d. Control is transferred from within the loop to the WRITE statement *before the* loop is completed. Note the successive values of JSUM after the initial value of 0 are 1, 1 + 2 = 3, 3 + 3 = 6, 6 + 4 = 10, 10 + 5 = 15, and, hence the printout is "15 5." Exit from the loop occurs when $M = 5$ and the value of JSUM is 15 at that point.

3. a. Line 20. There is no control number and the counting must begin with a positive integer, not with zero. Line 40, the end of the loop should have the control number. A corrected program is to replace lines 20 and 40 by "20 DO 15 M=1,5" and "40 15 CONTINUE."

b. Line 20. You must count by integers, not real numbers A.

c. There is no 40 in the *program*. Control is not transferred to line numbers. Also the counting must progress from smaller to larger positive integers—counting cannot progress backwards from 10 to -5.

d. The loop variable N should not be redefined with the loop as done in line 20.

SUPPLEMENTARY PROBLEMS

1. a. IF(A.GT.B) GO TO 152; b. A.EQ.B; c. End the parentheses before the word GO, the names are longer than six characters; d. No comma permitted in 2300; e. IF(A.NE.B) GO TO 12 (≠ may or may not be available.); f. PRINT, A,B,C in WHATFOR or WHATFIV; g. The "DO" should contain the letter "O" not the number "0"; h. Here the letter "O" should be replaced by the number "0." (Some terminals print the letter "O" as "Ø" to avoid the type of problem illustrated by g and h.
2. a. Remove line numbers and replace the WRITE by PRINT,J. Also the IF statement must be changed to IF(I.LT.21) GO TO 15. (b), (c), (d) have similar changes.
3. 27 times since once a card is used, in part, the card reader goes to the next card.
4. a. 3., b. 5., 7.
5. a. 4, b. 4.33333, c. 11., d. 5., e. 2., f. 0, g. 256.
6. a. The B should be given an initial value before line 20. Use of equals here leads to an infinite loop. B will never equal 10. (Equals should be avoided in an IF statement using real numbers. Round-off errors can cause two numbers, judged to be equal in your conception, to be ever so slightly unequal after a calculation.)
   b. Missing END statement.
   c. A should be given an initial value before line 10. Mixed-mode since I is an integer and A is real in line 20.
   d. The value of 1/I is zero. If there was a WRITE statement after 40, the value of J would be 0.
7. a. I=2*J. b. X=2.*(3.**2) (unnecessary parenthesis). c. No control: "DO 15 I=2,212" is correct if 15 CONTINUE appears later in the program. d. CONTINUE spells the word. e. Maybe PRINT,A,B,C (on cards). f. Enclose I.NE.J in parenthesis. g. I is an integer while 3. is real. h. No computation permitted in a WRITE statement. i. No computation permitted in the range of the loop (2*N would have to be previously computed and stored as, say, N2=2*N.) Then "DO 29I=1,N2" is correct. j. No location named for the READ. k. You must GO TO an actual number and not a variable like N. l. Range of a DO LOOP must begin with an integer greater than or equal to 1. m. 99. is a real number and, as such, is not a legal number to end the counting of the loop. n. No comma in the 3400.
8. a. You should not *enter* a DO LOOP from outside the Loop as in line 60. b. Do you redefine the loop variable I within the loop (line 30): c. You should not use the loop variable L outside the loop (line 50) unless it is redefined. d. Loop variable L has been previously defined (lines 10 and 30). The same storage (L) is used in the loop so the result in 20 may never be satisfied. Old numbers are dumped for the L in the loop.

Chapter 10

SECTION 10.1

1.
```
10      READ(5,*)N,R,P
20      FLN=N
```

2.
```
10      READ(5,*)N,R,P
20      L=12*N
```

```
30        XINT=P*R*FLN
40        WRITE(6,*) XINT
50        END
```

```
30        DO  31 I=1,L
40        P=P+R*P/12
50     31 CONTINUE
60        WRITE(6,*)P
70        END
```

3.
```
40        J=1
50     23 P=P+R*P
60        WRITE(6,*)J,P
70        J=J+1
80        IF (J≤N)GO TO 23
90        END
```

4. a. 1.6,2.56,4.096,6.5536;
   b. 1.6,2.6,4.2,6.7;
   c. 1.6,2.5,4.0,6.4
5. $1000. Too many digits for the computer to obtain more than $1000 in each step.

SECTION 10.2

1. a.
```
10        P=1000.00
20        DO 22 J=1,360
30        P=P+.05*P/360.
40     22 CONTINUE
50        WRITE(6,*)P
60        END
```
   b.
```
10     P=1000.*(1.+.05/360.)**360
20     WRITE(6,*)P
30     END
```
2. The 5% annual rate always has the higher yield.
3. *P* dollars at 5% annual interest, compounded annually, has a value after *N* years of *PN* = *P* * (1.05)***N* according to formula (10.6). On the other hand, 4.8%, compounded quarterly, has the value *PL* = *P* * (1.012)**(4**N*) since $R/K = .048/4 = .012$. Now

$$(1.012)^{4N} = [(1.012)^4]^N = (1.048870932736)^N$$

which is always smaller than $(1.05)^N$! Hence $PN > PL$ for all positive integers N.

4.
```
10     READ(5,*)N,R,P
20     XN=N
30     Y=P*EXP(XN*R)
40     WRITE(6,*)Y
50     END
```
5. 5% compounded annually is the better investment regardless of the length of the investment.
6. $54.60, $2980.96
7. Around $6001

SECTION 10.3

1. $306.04, $410.10, $2724.32
2.
```
10        A=0
20        DO 17 I=1,24
30        A=200.*(1.+.045/360.)**(30*I)+A
40     17 CONTINUE
50        WRITE(6,*)A
60        END
```
3. We obtain $2514.29. In the DO LOOP the term 100.*EXP(FLI*.05/12.), where FLI=I, should appear.

4.
```
10   READ(5,*)B,R,N
20   RMONTH = R/12.
30   L=12*N
40   PAYMO=B*RMONTH/(1.-(1.+RMONTH)**(-L))
50   WRITE(6,*)PAYMO
60   END
```
5. 5 years. $(32 = 31/(1-2^{-L}))$
6. DELETE 60 in the program in answer 4 above. Then add the following steps.
```
 60        P=B
 70        DO 22 I = 1,L
 80        XINT = P*RMONTH
 90        P=P-(PAYMO-XINT)
100        WRITE(6,*)XINT,P
110     22 CONTINUE
120        END
```

SECTION 10.4

1.
```
10   READ(5,*)R,N
20   S=100.*((1.+R)**N-1.)/R
30   WRITE(6,*)S
40   END
```
2. In 49 years, the account is $119,055.60. It is $125,600.89 in 50 years.
3. Try 5%

SECTION 10.5

1.
```
10   READ(5,*)R,N,P
20   A=P*(1.-1./(1.+R)**N)/R
30   WRITE(6,*)A
40   END
```
2. $194,481. 3. $115,824. 4. $55,527.00.

Chapter 11

SECTION 11.1

1. a. 3; b. .3; c. 1,1;2,4;3,9;4,16;5,25 in a column of pairs; d. 3.375
2. b.
```
      FUNCTION F(X,Y)
      IF(X.LT.Y) GO TO 12
      F = Y
      GO TO 13
   12 F=X
   13 RETURN
      END
```
c.
```
      FUNCTION F(X,N)
      A = 1.
      DO 14 I=1,N
      A = A+X**I
   14 CONTINUE
      F=A
      RETURN
      END
```

3. a. The function should be on the first line of the program.
   b. Same as a. Also F(3) should read F(3.) and the WRITE statement should not contain F(X).
   c. The function subroutine should be at the end of the program.
   d. Line 80 should be (F=X**2).
4. $y(9) = 9\frac{1}{3}$, $y(4) = 4\frac{1}{2}$. Here $X > 0$.
5. $X_0 = 1$; $X_1 = 3$; $X_2 = 2.3333$; $X_3 = 2.2381$

SECTION 11.2

2. The zero is between 2,3; 2,2.5; 2,2.25; 2.125,2.25; 2.1875,2.25.
3. The $X3$ is a zero, the process continues with $X2 = X3$. It does not notice the zero as programmed.
4. With line 35 included, the calculation would stop. Otherwise we have an infinite loop. When $X1 = 0$ and $X2 = 1$ it would approximate the zero at 0 after some time.
5. 3. 6. Which is more costly depends on the function. The new control will terminate computation after a certain number of steps even if a zero has not been found.
7. Each problem stops after 17 calculations ($1/2^{17} \approx .000008$).
8. The zero is $\sqrt{2} + 1 \approx 2.414$. 9. 2. 10. It returns in 4 sec. The maximum height is 64 ft after 2 sec.

SECTION 11.3

2. Maximum of 9.12 at $X = 1.8$ (Actually it is 9.125 at $X = 1.75$).
3. The program now adjusts the interval in which the maximum appears to be assumed three times. Each time the calculation is performed on a smaller interval.
5. a. 16, 4. b. 25, 5. c. 81/4, 9/2; 10000, 100.
6. a. Maximum 13 at -2. b. Minimum -9 at -1. c. Maximum 1 at 1. b. Minimum 11/4 at $-3/2$.
7. 4000 when $x = 200$.
8. a. 3 in. from an end. b. There is no maximum although the sum of the areas cannot exceed 36.
9. $x = 100$, $y = 200$
10. $x = 4$, $y = 2$ (A computer or calculus is necessary to solve this problem.)

SECTION 11.4

1. a. 1537.50. b. $27(1 - 1/3^{101})/2$. c. -6925. d. $20(1 - (3/4)^{101})$. e. $3^{101} - 1$. f. $(-3/10)(1 - (2/3)^{101})$.
2. a. 2/3. b. 20/33. c. 712/999. d. 31412/9999. e. 9.
3. 6 ft. 4. \$392. 5. Write $\frac{1}{2\cdot 3} = \frac{1}{2} - \frac{1}{3}$, $\frac{1}{3\cdot 4} = \frac{1}{3} - \frac{1}{4}$, etc. From this you can deduce $S_n \to 1$ as n increases.
6. 2,147,483,648 7. Let $a = 1/2$ and $d = 3/56$ in an arithmetic progression.

8. 110,265.58 (total); he has a $14,774.55 salary in the eighth year.
9. The first choice is better from the fourth year onward.
10. a. −50; b. $n(n+1)$; c. n^2.

SECTION 11.5

1. 21 2.

```
10         DIMENSION FX(30)
20         FX(1)=1
30         FX(2)=2
40         DO 21 J=3,30
50         FX(J)=FX(J-1)+FX(J-2)
60         Y=FX(J)
70         WRITE(6,*)Y
80      21 CONTINUE
90         END
```

3.

```
10    DIMENSION XIT(40)          60       WRITE(6,*)Y
20    XIT(1)=SQRT(2.)            70    19 CONTINUE
30    DO 19 J=2,40               80       END
40    XIT(J)=SQRT(2.+XIT(J-1))   25       Y=XIT(1)
50    Y=XIT(J)                   26       WRITE(6,*)Y
```

4. The sequence approaches 2.
5.

```
10         DIMENSION SC(20)
20         READ(5,*) (SC(J),J=1,20)
30         PAVE=0
40         PSTD=0
50         DO 97 I=1,20
60         PAVE=PAVE+SC(I)
70         PSTD=PSTD+SC(I)**2
30      97 CONTINUE
90         AVE=PAVE/20.
100        STD=SQRT(PSTD-AVE**2)
110        WRITE(6,*)AVE,STD
120        END
```

Note the READ statement in line 20. This command picks up twenty real numbers entered in any fashion, for example, 4 rows of five each, and it assigns them to the locations SC(J) in a natural reading order. A similar command, namely, WRITE(6,*) (SC(J),J=1,20) will write out the full sequence.

Chapter 12

SECTION 12.1

1. a. FORMAT(I28). b. FORMAT(I5,I7,I3,I6,I5,I4). c. FORMAT(I6,3I7,6I4).
2. a. ENTER THREE INTEGERS and then a '?' in the left margin of the next line. b. WHAT IS THEIR SUM is printed and another ? on the left of the

next line. c. WRONG! TRY AGAIN followed by a ? on the next line. d. GOOD! NO MORE QUESTIONS. and the machine stops. e. It puts another ? on the next line and repeats this process until it has all the data requested. (Three numbers in this example.)

3. a. HE SUM OF _ _ _ _ 17AND _ _ _ 213IS _ _ _ _ 248
   b. HE SUM OF _ -1736AND _ _ _ 123IS _ _ -1613
   c. HE SUM OF _ 50000AND _ 60000IS _ 110000
   d. HE SUM OF******AND******IS _ _ _ _ _ _ 2

   In d the computer does the problem but cannot enter the original number in the allotted field width 2I6.

4.
```
 10       WRITE(6,17)
 20    17 FORMAT(' WHO DID THEY BURY IN THE WASHINGTON
          MONUMENT?')
 30       WRITE(6,19)
 40    19 FORMAT(' 1.NIXON, 2.GRANT, 3.THE ARCHITECT')
 50       CORECT = 3.
 60       READ(5,*)ANS
 70       IF(CORECT.NE.ANS) GO TO 21
 80       WRITE (6,23)
 90    23 FORMAT(' REALLY!)
100       GO TO 25
110    21 WRITE(6,27)
120    27 FORMAT(' NO, BUT THEY SHOULD HAVE!')
130    25 STOP
140       END
```

SECTION 12.2

1. a. _ _ _ -137.53, _ -137.53; b. -137.52910;
   c. _ _ _ _ _ 0.000, _ _ _ 0.00025; _ -1234.000, **********.

2.
a.	_ _ _ _ 0.2000	_ 0.0400	_ _ 0.0080	_ 0.00016
b.	1.0000	1.0000	1.0000	1.00000
c.	1.1000	1.2100	1.3310	1.46410
d.	2.0000	4.0000	8.0000	16.00000
e.	3.0010	9.0060	27.0270	81.10805

3. _ 3.9512 _ 10 _ -75.329

4. a. 35.1234
   b. 2.3591
   c. 0.0000
   d. *******
   e. 92.5678
   f. 0.9592

SECTION 12.3

1. a. 3.76E15; b. -3.16E23; c. 3.16E-23, d. .31716E-3; e. .5E7; f. .157321E-2

2. a. 731; b. 83.1; c. .000049; d. −0.000031271; e. 500,000; f. .000001
3. _0.37E_01 _ 0.300E_01 _ 0.121E-01 _ -0.100E-03
4. Exponent. Floating point. Integer.
5. Using the FORTRAN hierarchical rank for the arithmetic operations suggested in the message, we have the following printout:
   a. THE SUM OF _ 1.00AND _ 1.00TIMES _ _ 2 = _ 0.3000000E _ 01
   b. THE SUM OF _ 3.20AND _ 4.24TIMES100 = _ 0.4272000E _ 03

SECTION 12.4

1. a. I=12300000 (The computer puts in the zeros since the number was not right-justified.) J=1200, K=1400 for the same reason.
   b. Here only I(=123000) is not right adjusted.
   c. J has spilled over into the space for K.
   d. No error if I=123456,J=1234,K=14.
2. a. The second number is an E-FORMAT rather than an F. There also is no third number. b. No error. c. There is spillover from Y into the space for Z. d. No error if $X = -0.31E3$, $Y = 911.20$, $Z = 94.71$.

SECTION 12.5

1. a. Six integers per line right-justified in the spaces allotted for each integer. The final line contains only four integers—addition integers would be ignored.
   b. 17 cards are necessary, 16 with 6 integers each and the last with 4 integers.
2. Five real numbers in the E-FORMAT on each line, 15 lines. Each number is printed in 14 spaces allotted except the first number of each line is restricted to 13 spaces.
3. a. The average and standard deviation of the sequence of real numbers.
   b. Seven numbers on each of the first seven lines with one number on the eighth line. The numbers are in the 10 spaces allotted for each and are in the F-FORMAT.
   c. The AVE is printed in the F-FORMAT with two decimal places in the first 9 spaces followed by the STD in the same FORMAT in the next 10 spaces.
4.

```
10       READ(5,*)I,J              100       IF(KREM.EQ.0) GO TO 31
20       IF(J.GT.I) GO TO 11       110       II=JJ
30       II=I                      120       JJ=KREM
40       JJ=J                      140       GO TO 11
50       GO TO 21                  150    31 WRITE(6,*)JJ
60    11 II=J                      160       END
70       JJ=I
80    21 KQNT=II/JJ
90       KREM=II-KQNT*JJ
```

   For three numbers you can find the gcd for two and the gcd of the answer with the third numbers.

5.
```
 10          DIMENSION IPRIME(60)
 20          IPRIME(1)=2
 30          I=1
 40          DO 21 J=3,9999,2
 50          IEND = SQRT(FLOAT(J))+1
 60          DO 22 K=3,IEND,2
 70          ICHEK=J/K
 80          IF(K*ICHEK.NE.J)GO TO 22
 90          GO TO 21
100      22  CONTINUE
110          I=I+1
120          IPRIME(I)=J
130          IF(I.EQ.60)WRITE(6,23)
             (IPRIME(N),N=1,60)
140      23  FORMAT(6I7)
150          IF(I.EQ.60)I=0
160      21  CONTINUE
170          IF(I.NE.0)WRITE(6,91)
             (IPRIME(N),N=1,I)
180      91  FORMAT(6I7)
190          END
```

← (line 40) This line does the augmentation of J in steps of size 2, that is, J=3,5,7,9, ... We thus eliminate the even numbers which, aside from 2, cannot be prime.

← (line 50) This is to check to determine if J has a divisor $\leqslant \sqrt{J}$. The 1 is put in to adjust for truncation of the number in case some square root has a sequence of 9's after the decimal point.

← (line 140) Six primes are printed on a line.

SECTION 12.6

1. Add the following lines

```
12          WRITE(6,11)
13    11       FORMAT(' BEGINNING    INTEREST    PRINCIPAL')
14          WRITE(6,12)
15             FORMAT(' BALANCE    PAYMENT    PAYMENT')
51          WRITE(6,13) P,PINT,RED
52    13       FORMAT(F9.2,F13.2,F12.2)
```

2. We must introduce a counter for the number of periods as well as adjust the FORMAT in lines 13 and 15 of the previous answer. Let N=1 on line 16 and then N=N+1 in line 61. The monthly payment PMON can be introduced and suitably formatted in lines 51 and 52.

   In order to adjust for the final payment, we should replace line 90 so that it prints out the last month's payment which equals $P + P * R$ where P is the beginning balance of that month.

3. This requires two counters, NMON for the month and N YEAR for the year. Once NMON reaches 12 NMON+1 must return to 1. This can be programmed as follows:

```
IF(NMON=12)NMON=1
IF(NMON=12)NYEAR=NYEAR+1
```

   The due date can be formulated by something similar to WRITE(6,101)NMON, NYEAR followed by 101 FORMAT(13,'/2/',I4) if the payment is due on the first of each month and if the year is printed with four digits (like 1984). Of course, to have the date printed on a line with other information, you must replace lines 51 and 52 by the longer formulated WRITE statement that includes NMON,NYEAR, and the other information desired.

4. $140 per month (2*3000*.06/12). The effective rate is approximately 12.13 percent per year.
5. You should be able to do it without an answer. Try!

Index